Implementing 802.11 with Microcontrollers:

Wireless Networking for Embedded Systems Designers

By

Fred Eady

ELSEVIER

AMSTERDAM • BOSTON • HEIDELBERG • LONDON
NEW YORK • OXFORD • PARIS • SAN DIEGO
SAN FRANCISCO • SINGAPORE • SYDNEY • TOKYO

Newnes is an imprint of Elsevier

Newnes

Newnes is an imprint of Elsevier
30 Corporate Drive, Suite 400, Burlington, MA 01803, USA
Linacre House, Jordan Hill, Oxford OX2 8DP, UK

 Recognizing the importance of preserving what has been written,
Elsevier prints its books on acid-free paper whenever possible.

Library of Congress Cataloging-in-Publication Data
Eady, Fred.
    Implementing 802.11 with microcontrollers : wireless networking for
      embedded systems designers / by Fred Eady.
      p. cm.
    ISBN 0-7506-7865-8 (pbk. : alk. paper)
    EAN 978-0-7506-7865-0 (pbk. : alk. paper) 1. Wireless communication systems.
      2. Embedded computer systems--Design and construction. I. Title.

TK5103.2.E33 2005
004.6'8--dc22

2005014188

British Library Cataloguing-in-Publication Data
A catalogue record for this book is available from the British Library.

For information on all Newnes publications
visit our website at www.books.elsevier.com

05 06 07 08 09   10 9 8 7 6 5 4 3 2 1

Contents

Preface .... *ix*

What's on the CD-ROM? .... *x*

Chapter 1: Why Are We Doing This? *1*
Selecting a Suitable Microcontroller 2
Selecting a Suitable 802.11b Communications Device 3
802.11b Hardware Overview 3
AirDrop Basics 4

Chapter 2: The AirDrop-P *9*
The AirDrop-P Hardware 9
Learn to Play Guitar and Become Famous 15

Chapter 3: The AirDrop-A *17*
The AirDrop-A Hardware 17
Bowing Out 23

Chapter 4: 802.11b CompactFlash Network Interface Cards *25*
They Were Not Designed To Do This 25
The TEW-222CF 25
Never Ignore an Inquisitive Author with Hand Tools 26
Unwrapping the TEW-222CF 30
An Undercover Look at the Zonet ZCF1100 32
What's Behind Door Number 4 34
RF, Witchcraft, Pointy Hats, Ghouls, Goblins...Same Thing 35

Chapter 5: Talking With 802.11bCompactFlash NICs *37*
Physically Connecting a Microcontroller to a CompactFlash Card 38
Musical Overtones 43

Chapter 6: Touring the Card Information Structure *45*
Talking in Tuples 46
First Steps with the AirDrop-P 48
Walking the Tuple Chain 50
CIS Reconnaissance 59

Dumping Linksys WCF12 Tuples 65
Dumping Netgear MA701 Tuples 68
Dumping Zonet ZCF100 Tuples 70
Enabling the 802.11b CompactFlash NIC 74
The Value of Parsing the CIS 77
Full Throttle 77

Chapter 7: Learning to Talk to 802.11b CompactFlash NICs 79

What the 802.11b NIC Does for Us 79
The 802.11b CompactFlash NIC I/O Drivers 83

Chapter 8: Setting Up An AirDrop Wireless Network 93

Setting Up the AP 94
Something's in the Air 95
Guitars and Hollywood 104

Chapter 9: AirDrop Driver Basics 105

BAP 105
FID 106
RID 107
Reading a RID 118
Stringing Up the SSID 125
Good RIDdance 128
Retrieving the MAC Address 130
Status Check 134

Chapter 10: Putting an AirDrop on a Wireless LAN 137

Bogie Number 1 – Allocating Transmit Buffers 137
Bogie Number 2 – Enabling the MAC 147
Authenticating the AirDrop Wireless LAN Station 158
Associating with the AIRDROP_NETWORK AP 160

Chapter 11: Processing 802.11b Frames with the AirDrop 167

AirDrop Frame Structure 168
AirDrop-P Frame Reception 184

Chapter 12: PINGING the AirDrop 209

Examining the IP Header 231

Chapter 13: Flying Cargo with UDP and the AirDrop 243

Running a UDP Application on the AirDrop-P 243
The EDTP Internet Test Panel and the Code Behind It 245
Exercising the AirDrop-P with the EDTP Internet Test Panel 249
Notes 274

Chapter 14: Flying Cargo with TCP/IP and the AirDrop 275

TCP and the AirDrop-P 275
The TCP/IP Stack's Physical Layer 284
The TCP/IP Stack's Data Link Layer 284

The TCP/IP Stack's Network Layer..284
The TCP/IP Stack's Transport Layer 284
The TCP/IP Stack's Application Layer 285
TCP/IP – The Big Ugly 285
You've Done It! 334

Chapter 15: WEP and the AirDrop 335

Incorporating WEP into the AirDrop 802.11b Driver................................335
The New Experimental AirDrop Hardware 345
An Experimental AVR AirDrop Variant..345
The Experimental AirDrop Firmware...348
Coding a Simple 802.11b Web Server 355
The AirDrop SRAM 358

Chapter 16: A New Kid in Town Who Calls Himself ZigBee........... 361

Zig What???..361
Making ZigBee Talk 363
The Microchip ZigBee Stack 366

Chapter 17: Parting Frames 371

Numeric Notation..373
Source Code Presentation..373

Conventions 373

Sub Snippets 374
Netasyst Sniffer Capture Text Presentation 375
Mini Sniffs 375

Index 377

Acknowledgments

Microchip's Eric Lawson and Lucio Di Jasio were essential elements in the production of this book. On the software side, the folks at HI-TECH Software, Katie Cameron, Megan Cairney, George Combis and Clyde Stubbs, did their part to make sure all of the AirDrop source code bits compiled successfully.

Preface

This book is intended to provide you with everything you need to know to create and deploy a microcontroller-based 802.11b wireless network. You read it correctly, I did indeed say *everything* you need to know. I've spent the last year being rejected, ignored and hung up on. When I wasn't being subjected to any of the aforementioned disrespectful acts, I was being lied to, promised to and conveniently forgotten. Some of the folks holding the 802.11b Holy Grail had no scruples and performed all of the despicable acts I've mentioned against my person. All of that angst was directed at me (or rather not directed at me) because I wanted to learn how to implement 802.11b in the world of microcontrollers. Well, I have seen the 802.11b light and I am here to spread the word to all in microcontrollerdom. 802.11b communication with inexpensive off-the-shelf microcontrollers is possible for I have done it and you will do it too.

I have successfully designed, built (from scratch) and programmed two variants of microcontroller-based 802.11b devices. These devices are called AirDrops and are available in PIC or AVR configurations. The AirDrop 802.11b drivers for both variants of AirDrop are available to you uncensored on the CDROM that accompanies this book.

I have written the base AirDrop 802.11b drivers in C. You will find that the AirDrop 802.11b driver source is very easy to follow and can be easily ported to many other C compiler platforms. I have opted to present all of the AirDrop 802.11b driver source code using source lines written to be compiled by the HI-TECH PICC-18 C compiler. A companion version of the AirDrop 802.11b driver for the AVR was ported from HI-TECH PICC-18 C compiler to ImageCraft ICCAVR for those of you that wish to use the AVR variant of the AirDrop module. A complete set of working AirDrop 802.11b driver source code for the AVR is also included on the enclosed CDROM.

An AirDrop User's Group has already been established and a grass roots AirDrop community has been built. Thanks to citizens and founding fathers Julian Porter, William Welch, Paul Curtis and Dave Comer, the AirDrop forum is informative as well as entertaining.

The knowledge you will need to understand and deploy 802.11b with a microcontroller is contained within the pages of this book you're about to read. I sincerely hope you have as much fun reading this book as I have had writing it.

What's on the CD-ROM?

I'm old. I remember having to copy source code manually from the pages of a book or magazine into a text editor for use in my projects. I've always been a fair typist but the transcription of the source code from text format to electronic format still consumed a great amount of time. Authors in the old days knew that some of their "one-finger" typist customers would never get the source code copied into a text editor for compilation. So, to solve the problem for the "one-finger" types, the author would sell the source code on a diskette. Well, I'm going to do that one better. At no additional charge, the CD-ROM that accompanies this book contains the complete set of C source for the AirDrop-P and AirDrop-A modules described within the pages of this book.

To prevent you from having to lug the book around on your workbench, I've also included a full set of printable AirDrop-A and AirDrop-P schematic diagrams in PDF format. For those of you that want to dig a bit deeper into what makes 802.11b NICs work, there's a group of data sheets I collected from here and there on the CD-ROM too.

I figure since you have laid out your hard-earned money for this book, you deserve access to the EDTP AirDrop FTP site. So, I've included the AirDrop FTP site login and password on the CD-ROM. I've also provided the AirDrop-A User Group information for those of you that wish to join the AirDrop forum. I strongly recommend that you join the AirDrop forum and carefully read the earlier posts. There is a wealth of highly valuable information there to help you get up to speed with your AirDrop module. There are also ports of the original AirDrop code to other mainstream C compilers in the AirDrop User Group files area. The guys on the forum are also working on porting the AirDrop code to 802.11b NICs with special features. The best part of being a member of the AirDrop User Group is the excellent support the members provide for those that post requests for help.

If you couple the contents of the CD-ROM with the book text and supplement all of that with the AirDrop User Group posts, you'll find that you have EVERYTHING you need to be successful with your AirDrop-A or AirDrop-P. Wireless is wonderful.

Why Are We Doing This?

Most every communications gadget that you see in science fiction flicks is wireless and capable of radiating a combined video, audio, and complex sensor-based data signal across endless parsecs of distant galaxies by means of some futuristic high-tech communications method designed by a drunkard scientist marooned on a mining planet in the 24th century. In reality, we're not too far removed from what in the past was once deemed as science fiction. We don't yet have access to inebriated scientists on remote mining planets. However, in to-day's world you talk on the phone, control your television and stereo, start your car, and send email without the need for a physical wired connection. Your laptop computer most likely incorporates a wireless 802.11a/b/g LAN setup and Wi-Fi "hotspots" are popping up all over the place. If you're "important," you most likely use a wireless PDA on a daily basis. Speaking of wireless PDAs, I know of a local high-class restaurant where the servers use wireless 802.11b-enabled handheld computers to enter and retrieve their customer's dinner and drink orders.

So, with all of the wireless technology in our everyday lives, why am I writing this book and why are you sitting (or standing) there reading it? Simple...I'm here to explain how to make embedded microcontroller-based 802.11b wireless hardware and firmware work in plain terms and you're a student, an engineer, a sober earthling scientist or just a gadget-building geek that is interested in taking embedded wireless technology to the microcontroller level for the benefit of all mankind (and yourself).

If you're really interested in doing 802.11b stuff with itty-bitty off-the-shelf microcon-trollers, I'm going to show you how to be successful in implementing inexpensive and simple embedded 802.11b wireless hardware and firmware through the code examples, pictures and theoretical explanations contained within this book.

Plain talk technical information behind this wireless stuff is really hard to find. In fact, it is nonexistent. For the past year or so, I've been hacking at getting enough pertinent and use-ful information put together to realize and build an inexpensive 802.11b embedded platform based on simple and cheap off-the-shelf microcontrollers executing rudimentary 802.11b driver firmware.

After months of reading microcontroller data sheets, wireless communications white papers and pouring over microcontroller and wireless device application notes and source code, some of the concepts of 802.11b connectivity started to creep out of the mist. I then turned to some of the ready-to-roll 802.11b solutions on the market only to find out in the end that I couldn't get any useful technical information about the gadgets from the manufacturers. In some cases, I couldn't even get the manufacturer to sell me the items. After months of scratching my head (and other parts of my body), I got desperate and turned to the Linux open-source websites looking for clues to the 802.11b driver mystery. My hopes of conquering the 802.11b demon using Linux source as a foundation were quickly dashed as the Linux code proved to be strung out to the point that I couldn't get a handle on even the most basic of 802.11b concepts. As it turns out, most of the Linux 802.11b stuff is aimed at getting various manufacturers' PCMCIA 802.11b network cards to work with Linux desktop machines. The 802.11b basic routines I needed to implement 802.11b with microcontrollers are indeed buried in the Linux source code. However, my goal was to implement a microcontroller-based set of clean and simple 802.11b drivers, not to become a Linux 802.11b guru.

The good news is that, unlike Bono of U2 fame, I did finally find what I was looking for. In fact, my discovery of useful 802.11b driver documentation that I could apply to a microcontroller prompted the beginning of this book. The next step involved gathering enough good information about 802.11b hardware to make an intelligent decision as to how to proceed with my embedded 802.11b projects.

Selecting a Suitable Microcontroller

The first hurdle I had to clear was which microcontroller would be the best for the 802.11b job. There are a multitude of microcontrollers that I could have used, each of which has more than sufficient resources to handle doling out the simple logic that I would need to drive an 802.11b radio card. I figured it would be a useless effort to try to use some expensive, high-powered exotic microcontroller when there are so many good low-cost microcontrollers capable of doing the job that can be had from trustworthy mail order and Internet distributors. And, if I went with a wild-horse microcontroller, I would most likely have to procure a new and probably expensive compiler to go with it. I always try to step into the reader/user's shoes when I make these kinds of decisions and I figured you wouldn't like that. The logical thing to do here would be to determine which microcontroller is most popular with folks that would be reading this book. Fortunately, I can draw on the experiences from my prior book, *Networking and Internetworking with Microcontrollers,* in which I featured Microchip PIC® and Atmel AVR® microcontrollers participating in a wired Ethernet network environment. I found that readers of my *Networking and Internetworking with Microcontrollers* book that wanted to use a microcontroller other than a PIC or AVR had little trouble porting the original PIC and AVR source code provided in the book to their preferred target microcontroller. I also discovered that most of my readership already owned AVR and/or PIC toolsets. I major in writing simple uncomplicated code for microcontrollers and I have a comprehensive library of microcontroller-specific assemblers and compilers to assist me. However, I realize that lots of you don't. So, the selection of a microcontroller came around the circle to the PIC and the AVR once again. I figure if I can make the 802.11b driver work on a simple microcontroller using a standard programming language such as C, the idea behind my microcontroller-based 802.11b driver code can be easily ported across various C compilers or to a more complex microcontroller of the reader's choice.

Selecting a Suitable 802.11b Communications Device

The entire 802.11b design is dependent on the 802.11b radio. There are multiple 802.11b radio designs to choose from out there and all of them can be easily obtained commercially via local office supply warehouses or through online vendors. In addition to the CompactFlash® and PCMCIA 802.11b radio cards that can plug into your laptop or PDA, there are various vendors that offer stand-alone solutions. Most of them are very nice. The only drawback to a vendor-controlled device is that you're locked into their system, their way of doing things and their I/O interfaces. So, I turned away from the "all-in-one" 802.11b solutions in favor of my open 802.11b microcontroller-based design idea. Nothing came easy. After a couple of dozen or so unreturned phone calls, rejections and unanswered emails, a choice had to be made about whose 802.11b radio technology I would use. A couple of weeks more of agonizing indecisiveness followed. I finally threw in my hand and settled on incorporating the PRISM architecture into my embedded 802.11b design. I was drawn to the PRISM decision because I could positively identify and easily purchase CompactFlash 802.11b products that contained the PRISM chipset.

During my 802.11b research period, I found that there are a number of internet websites that contain a list of the various manufacturers' 802.11b cards and the chipset they contain. Some of the sites even went as far as to tell you how the cards behaved and what worked and what didn't. I had already been exposed to the term PRISM while trying to glean 802.11b information from the Linux wireless internet sites. I have a heavy respect for Linus Torvalds and his Linux following. The Linux coders turned over millions of rocks and leaves in search of 802.11b knowledge. I guess I didn't have the time or patience to go that route with them. As I became more familiar with the 802.11b architecture, the Linux 802.11b information that used to confuse me actually became helpful.

In my sojourn (that's a Moody Blues word for travels) through various internet sites, I noticed that most of the 802.11b cards I encountered that I deemed compatible with my 802.11b design idea were also highly recommended for use in the Linux environment. To me that meant I had a chance to get this 802.11b microcontroller thing working because the Linux guys and gals most likely had to figure it out mostly on their own, but I also know that in the beginning the Linux community had the backing of the PRISM chipset founders (Intersil) for an open source Linux 802.11b project. In addition to positive identification of PRISM-based radio cards, some of the Linux 802.11b sites provide a one-sentence translation of some of the PRISM terms found in the open source Linux PRISM driver source code comments and Linux-generated 802.11b error messages, which equates to yet another plus for using the PRISM architecture. I also managed to come across some 802.11b-related things that I had previously deemed "TOP SECRET" laid out in the open for all to see by the Linux 802.11b open source coders. That prompted me to don my Ray-Bans.

802.11b Hardware Overview

I will present embedded 802.11b to you using off-the-shelf embedded hardware from EDTP Electronics, TRENDnet™, Zonet®, Netgear® and Linksys®. The EDTP Electronics-specific

802.11b hardware I will use to develop a working embedded 802.11b microcontroller-based device is called *AirDrop*. The EDTP Electronics AirDrop products use the 802.11b radio services of the aforementioned wireless 802.11b CompactFlash cards, which are all built around the PRISM chipset.

There are many advantages to this particular hardware arrangement. If you recall, in my previous book, *Networking and Internetworking with Microcontrollers*, I just made up a MAC (Media Access Control) address of "00EDTP" that identified the Easy Ethernet hardware to the LAN. The reason for this was that I didn't want to have to purchase a chunk of MAC addresses. The hardware (MAC) address is normally a purchased item that is regulated by the IEEE. I also knew that if I had to use the Internet as a communications conduit, I could hide the Easy Ethernet embedded Ethernet interface behind something that had a valid MAC address such as a router or personal computer NIC. Using the 11 Mbps CompactFlash Network Adapters solves the buy-a-MAC-address problem I encountered in my previous book as each 11 Mbps CompactFlash Network Adapter has its own certified and unique IEEE-issued MAC address, which is imprinted on the back side of the card and safely concealed within the 802.11b CompactFlash NIC's ROM.

Another advantage to using 11 Mbps CompactFlash Network Adapters is the form factor of the CompactFlash Network Adapter package. The 802.11b CompactFlash NIC radio cards are small and lightweight. This makes life easier if you decide that you want to port the base AirDrop hardware and firmware model to a unique portable device of your own design.

The 802.11b CompactFlash NIC's are based on standards that allow the CompactFlash 802.11b radio cards to participate in standard Ethernet LAN environments. That means that although the 802.11b CompactFlash NIC is wireless, it can still be accessed from standard wired LAN architectures via a wireless router or access point (AP). And, TCP/IP is still the same old TCP/IP even in a wireless network. Only the way we access the network has changed, as the way we manipulate all of the well-known Internet and Ethernet protocols remain relatively unchanged.

In my mind, the biggest advantage to using 11 Mbps CompactFlash Network Adapters is that they all adhere to the CompactFlash standard. That means we can call on the publicly available CompactFlash documentation to get answers about interfacing an 11 Mbps CompactFlash Network Adapter to the AirDrop. Most of the 802.11b CompactFlash NIC manufacturers also provide some information about their card on their website.

AirDrop Basics

In this book we will be working with two AirDrop variants, the AirDrop-P and the AirDrop-A. The AirDrop-P is based on the Microchip PIC18LF8621, while the AirDrop-A is driven by Atmel's ATmega128L. Except for the microcontroller engine, the AirDrop-A and the AirDrop-P are logically identical. To optimize power consumption and to guarantee compatibility with a number of PRISM-based CompactFlash NICs, both AirDrops operate with a +3.3 VDC regulated power supply. Both the AirDrop-A and AirDrop-P microcontrollers are clocked to a safe maximum rate at the +3.3 VDC power level. For those that wish to push the envelope, the AirDrop-P microcontroller can be clocked up to a bit over 25 MHz. The Air-Drop-A runs at 7.37 MHz, and the AirDrop-A's ATmega128L can be clocked at a maximum rate of 8 MHz. Note also that both the AirDrop-A and the AirDrop-P are equipped with a regulation RS-232 port built around a +3.3 VDC SP3232 RS-232 converter.

The TRENDnet TEW-222CF data sheet states that the TEW-222CF can operate at +3.3 VDC or +5 VDC. I have not tested the TEW-222CF or any other CompactFlash NIC at +5 VDC and a +5 VDC version of the AirDrop will not be explored in this text. The reason I mention the TEW-222CF's dual voltage support is that if you want to experiment with the TEW-222CF at +5 VDC, you can run the PIC18LF8621 or a substituted PIC18F8621 at a maximum clock rate of 40 MHz. If you decide to soup up your AirDrop-A, you will also have to ditch the ATmega128L, which maxes out at a clock rate of 8 MHz, and substitute an ATmega128 before attempting to run the replacement ATmega128 in an 802.11b application at its maximum clock speed of 16 MHz. Although the higher voltage operation may work, I don't recommend you try it unless you have some bucks to burn on smoked-out Compact-Flash cards and AirDrop hardware.

The AirDrop-P can be programmed and debugged using any Microchip-compatible development hardware. A standard 6-pin RJ-12 programming/debugging interface is included on the AirDrop-P board. The Microchip MPLAB ICD2 along with MPLAB IDE was used as the programming/debugging platform when developing the AirDrop-P. You will not need to purchase an expensive in-circuit emulator, as the MPLAB ICD 2 is all you need for both programming and debugging. I highly recommend getting an MPLAB ICD 2 if you plan to get serious about doing some custom AirDrop 802.11b coding. My MPLAB ICD 2/AirDrop-P configuration is shown in Photo 1.1.

Photo 1.1: The MPLAB ICD 2 is easy to setup and use. When used with Microchip's MPLAB IDE, the MPLAB ICD 2 becomes either a debugger engine or a PIC programmer with the click of a mouse button. The MPLAB IDE contains downloadable MPLAB ICD 2 drivers that instantly adapt the MPLAB ICD 2 to service many of the medium to high-end PIC microcontrollers. Here the MPLAB ICD 2 is shown attached to an EDTP AirDrop-P.

The original development C compiler for the AirDrop-P is HI-TECH PICC-18. I found the HI-TECH PICC-18 C compiler to be full-functioned and free of bugs. And, since the HI-TECH PICC-18 C compiler isn't based on proprietary PIC-targeted macros, I found it very easy to port the AirDrop 802.11b driver source code between the Microchip and Atmel platforms. In fact, you'll see that the Microchip PIC and Atmel AVR source code images are very similar. HI-TECH PICC-18 C compiler is so easily portable that all of the AirDrop 802.11b driver code in this text will consist of HI-TECH PICC-18 C source.

The Atmel JTAGICE mkII and AVR Studio were used to develop the AirDrop-A. The original development AVR C compiler for the AirDrop-A is ImageCraft's ICCAVR. ICCAVR is a low-cost, high quality C compiler for the Atmel AVR devices. As I mentioned previously, I had little trouble with porting ICCAVR source code changes to the Microchip PIC platform and the HI-TECH PICC-18 C compiler. A standard pair of AVR 10-pin ICSP programming and JTAGICE mkII interfaces are included on the AirDrop-A board. The AirDrop-A is also compatible with the legacy JTAGICE, which does not support a USB interface. The JTAGICE mkII that I developed the AirDrop-A with is shown in Photo 1.2.

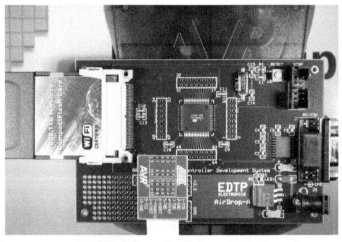

Photo 1.2: The JTAGICE mkII differs from the legacy JTAGICE at the personal computer interface. The original JTAGICE only supports an RS-232 personal computer interface while the JTAGICE mkII supports both RS-232 and USB personal computer interfaces. The JTAGICE mkII also supports the new debugWIRE interface for smaller AVR devices.

The PIC18LF8621 interface to the 802.11b CompactFlash NICs is a standard port I/O interface. This scheme allows the simplest of configurations to be employed and reduces the complexity of the AirDrop 802.11b driver code. The 802.11b CompactFlash NICs that can be used by the AirDrop series of 802.11b devices are shown in Photo 1.3.

Photo 1.3: Everything 802.11b that we will need to participate in a wireless Ethernet LAN is crammed into these tiny CompactFlash modules. Although transmitting and receiving packets are a priority, the CompactFlash cards you see in this shot are more than just an 802.11b RF devices. There are also microcontroller-like "smarts" inside each 802.11b CompactFlash NIC that allow the CompactFlash card to automatically perform tasks specified by the 802.11b standard and process commands from the AirDrop's on-board microcontroller.

Pilot training involves more than just learning to fly the aircraft and the same is true for putting an AirDrop on the air. So, in the next chapter we'll begin our AirDrop "flying lessons" with some "ground training" and an examination of the AirDrop "aircraft" variants.

The AirDrop-P

As you continue reading further into the pages of this book you will find that I have attempted to make everything hardware and everything firmware as simple and straightforward as possible. You will not find any complex hard-to-follow C code or fancy constructs, as I am not assuming that you are a seasoned C coder. The idea here is to use HI-TECH PICC-18 C source code to convey the basic building blocks and procedures that make the PRISM-based CompactFlash radio cards work. Likewise, the EDTP-based 802.11b hardware I present will be ample for the 802.11b task but as simple as I can make it. With that, let's examine the AirDrop-P hardware in detail.

The AirDrop-P Hardware

The AirDrop series of 802.11b modules started out as a universal microcontroller platform incorporating 802.11b capability that could be deployed using the 802.11b designer's microcontroller of choice. However, I couldn't see myself writing 802.11b drivers for every microcontroller that happened along. I also realized that handing you a piece of universal 802.11b development hardware with no supporting 802.11b driver code examples would serve no purpose. So, instead of coding to a "virtual" microcontroller, I drove a stake into the ground and produced the AirDrop-P, which is based on the very popular Microchip PIC microcontroller. The very first AirDrop was an AirDrop-P. The original AirDrop prototype is shown in Photo 2.1.

The Microchip PIC variant of the AirDrop modules is identified with a "-P" suffix. This chapter will focus on the production AirDrop-P hardware that you see in Photo 2.1.

Don't be fooled by the AirDrop-P's seeming lack of intricate electronic parts. The AirDrop-P is a powerful yet simple device that is based on one of Microchip Technology's family of high-pin-count PIC microcontrollers, the PIC18LF8621. With a total of nine pinned-out I/O ports, the PIC18LF8621 has more than enough available I/O pins to fabricate a microcontroller-based 802.11b solution. The PIC18LF8621's standard on-chip memory subsystem includes 3.8K of SRAM and 64K of high-endurance program Flash. The high-endurance program flash allows you to reprogram the PIC18LF8621 up to 100,000 times. To be precise, the PIC18LF8621 is a microcontroller incorporating 69 I/O lines, a 10-bit analog-to-digital converter, two enhanced USARTs, 5 timers, a SPI module and an I²C module all native to the PIC18LF8621 silicon. The PIC18LF8621's resource arsenal is housed in a compact 80-pin TQFP package. That all adds up to enough microcontroller resources to support a CompactFlash wireless NIC (Network Interface Card) with I/O left over for things

Photo 2.1: Everything on this prototype was hand soldered by yours truly. The only major change from prototype to production model is the replacement of the 20 MHz powered oscillator by a 20 MHz standard crystal.

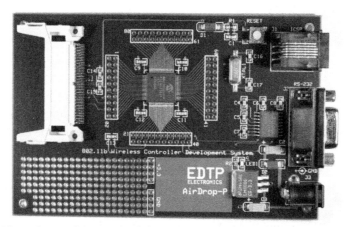

Photo 2.2: Did you know that some of the most powerful rock music chords are made up of only two notes? Do you know what these two-note chords are called? While you're thinking about that, I have only two words for the production AirDrop-P model shown in this photo: Simple, Powerful.

such as analog-to-digital conversion and digital control. That means the AirDrop-P can do the normal things you do with a microcontroller (read switch states, generate waveforms, monitor voltages, control relays, drive displays, scan keyboards, digitally communicate with other devices, etc.) and cooperatively communicate with other wireless or wired devices on an Ethernet LAN. A schematic depiction of the AirDrop-P hardware is rendered in Schematic 2.1.

Reference Schematic 2.1 and let's begin our tour of the AirDrop-P hardware with a look at the AirDrop-P power supply subsystem. For those of you that build electronic gadgets from scratch, there's nothing here you haven't seen before. For those of you that don't know which

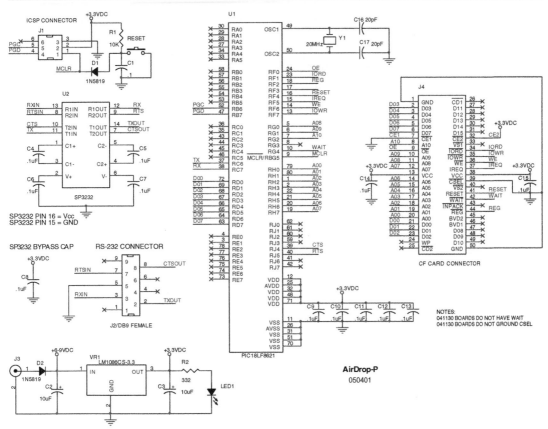

Schematic 2.1: What you see here is a typical PIC microcontroller system complete with a regulated +3.3 VDC power supply, a 20 MHz clock and a regulation RS-232 serial port. The bulk of the 802.11b work is being done within the 802.11b CompactFlash NIC under the direction of the PIC18LF8621.

end of a soldering iron to hold, the AirDrop-P regulated power supply is an industry standard power supply circuit that is taken almost directly from the LM1086CS-3.3 voltage regulator data sheet. A National Semiconductor LM1086CS-3.3 linear regulator supplies the +3.3 VDC power that energizes the 802.11b CompactFlash NIC, the Sipex SP3232 RS-232 converter IC and the PIC18LF8621 microcontroller. Any +9 VDC center positive power brick that can deliver a minimum of 400 mA is suitable for powering the basic AirDrop-P setup, which entails the AirDrop-P electronics and the 802.11b CompactFlash NIC. A high-current switching diode (D2) is inserted inline with the incoming bulk DC voltage to protect the AirDrop-P circuitry from the accidental application of a reverse polarity voltage. Otherwise, a negative-center power brick plugged into the AirDrop-P's power receptacle would severely damage or destroy the AirDrop-P circuitry. In that all of the AirDrop's active electronic components are surface mount packages and the AirDrop printed circuit board is 4-layers deep, putting a crater in the AirDrop printed circuit board would be fatal for the AirDrop module. Filtering the AirDrop-P's +3.3 VDC power supply is "by the book" and is done with a couple of 10 uF tantalum capacitors at both ends of the LM1086CS-3.3. LED1 provides a visual indication as to the health of the AirDrop-P's regulated power supply. During testing of one of the Air-Drop-P prototypes I accidentally installed the SP3232 backwards. On powerup I noticed that LED1 was quite dim and of course nothing worked. It didn't take long for me to find the goof but it was a bit too late for the SP3232's roasted innards.

A standard RS-232 serial port is realized on the AirDrop-P with the inclusion of a 3.3-volt Sipex SP3232 RS-232 converter IC. Capacitors C4-C7 are mandatory ingredients of the AirDrop-P serial port as they enable the SP3232's on-chip charge pump. A close-up view of the SP3232 and it supporting charge pump capacitors is shown in Photo 2.3.

Photo 2.3: The RS-232 female connector, the SP3232 and the five capacitors that surround it are all the hardware you need to deploy a regulation RS-232 serial port on the AirDrop-P.

Utilizing charge pump technology, the SP3232 can generate RS-232 voltage levels that are well beyond the RS-232 high-noise voltage band. Good design practice demands that all active digital component power supply rails be properly bypassed to reduce electrical noise. Capacitor C8 provides the power supply bypass function for the SP3232. The SP3232 is capable of driving two pairs of RS-232 transmit and receive lines. While testing the AirDrop-P I never encountered a situation where I was forced to use data traffic management. However, that doesn't mean that someone reading this and applying an AirDrop-P may not need to pace the serial data. So, just in case handshaking is required by an application, the second pair of the SP3232's transmit and receive lines are dedicated to RTS/CTS handshaking duty. The AirDrop-P's RS-232 9-pin D-shell connector is wired as DCE (Data Communications Equipment) to eliminate the need for a crossover serial cable or null modem when connecting to DTE (Data Terminal Equipment) devices like your personal computer's serial ports.

The inclusion of a true RS-232 serial port on the AirDrop-P module has many advantages. If you need to wirelessly pipe data between a device that can only speak RS-232 and an Ethernet LAN, the AirDrop-P's serial port is essential. A working serial port is also good to have during the 802.11b firmware development process. Using the AirDrop-P's serial port in conjunction with a personal computer terminal emulator makes easy work of writing little snippets of debugging code that can literally speak to you via the AirDrop-P's serial port. The AirDrop-P's serial port can also earn its keep while an application is running. You can use the AirDrop-P's serial port to send status messages to the personal computer terminal emulator or spit out 802.11b signal strength reports from the 802.11b CompactFlash NIC. That pretty much covers the AirDrop power supply and RS-232 port. Let's spend some time talking about the microcontroller.

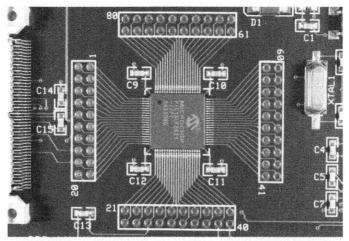

Photo 2.4: Just in case you want to do your own thing with the PIC18LF8621, all of the PIC18LF8621's I/O pins are pulled out to standard .1-inch-spaced header pads. The original idea was to not have a microcontroller here at all and allow the user/programmer to mount their own microcontroller using the header pins.

The PIC18LF8621 that you see in Photo 2.4 is the low-voltage, lower-power cousin of the PIC18F8621. Both the PIC18LF8621 and the PIC18F8621 are capable of operating with a maximum clock frequency of 40 MHz with a Vdd of +5.5 VDC. However, the AirDrop-P is powered with a 3.3 VDC power supply and the PIC18LF8621's maximum clock frequency is reduced to a bit over 25 MHz. Running the PIC18LF8621 at its maximum clock speed was tempting but I decided not to include any logic level shifting hardware in the AirDrop designs to make that happen as it would add unnecessary complexity and cost. Here's the PIC18LF8621 maximum clock frequency versus maximum applied Vdd equation:

$$\text{Fmax} = (16.36 \text{ MHz/V}) (\text{Vddappmin} - 2.0 \text{ V}) + 4 \text{ MHz}$$

Where:

Fmax = Maximum operating clock frequency

Vddappmin = Minimum Vdd power supply voltage

Fmax = (16.36 MHz/V) (3.3 V – 2.0 V) + 4 MHz = 25.268 MHz

To guarantee stability, I decided to run the AirDrop-P clock at 20 MHz. If you can handle changing the CompactFlash card's mandatory I/O delays, you can run your AirDrop-P at the clock rate ceiling of 25 MHz if you desire.

The AirDrop-P's PIC18LF8621 ICSP programming interface is standard Microchip issue. The AirDrop-P is designed to be programmed and debugged easily using the MPLAB ICD2 in combination with the latest version of Microchip's MPLAB IDE. Any Microchip-compatible programming/debugging system can be used to enable the AirDrop-P.

The business end of the AirDrop-P is the 50-pin CompactFlash connector that carries the 802.11b wireless LAN network interface card and interfaces the CompactFlash card's electronics to the PIC18LF8621. Since the PIC18LF8621 is an 8-bit microcontroller, the 16-bit amenities of the CompactFlash connector are unused and thus not connected. To save PIC18LF8621 I/O, only the CompactFlash connector I/O lines that are absolutely necessary for 802.11b operation are connected and used. For instance, the CD1 and CD2 pins are used to determine if a card is mounted correctly. Those pins are left disconnected as we most likely won't be hot swapping 802.11b CompactFlash cards and writing code to sense these pins would be a waste in our particular application of the CompactFlash card. A memory-mapped approach to driving the AirDrop-P was considered. However, to keep the circuitry and driver code as simple as possible a simple I/O interface was adopted. Note that there is absolutely no "glue" logic on the AirDrop-P board. One of the AirDrop-P design points was to leave as many of the PIC18LF8621's subsystems open for use and to leave as many open microcontroller I/O lines as possible without resorting to additional special purpose components. Also, using the I/O interface allowed the use of simple I/O code routines to move data back and forth between the CompactFlash card and the PIC18LF8621.

All of the firmware to drive our 802.11b application and the CompactFlash NIC is contained within the PIC18LF8621's program flash. The CompactFlash NIC is responsible for taking data we give to it and broadcasting it wirelessly to another wireless station or access point. The original wireless LAN card of choice for the AirDrop series of 802.11b development systems was manufactured by TRENDnet and is shown along with the other AirDrop-P-compatible CompactFlash NICs you saw in Chapter 1 (Photo 1.3). The TRENDnet TEW-222CF is an 11 Mbps wireless CompactFlash network adapter card that can transmit and receive up to 60 meters indoors and up to 250 meters outdoors. I particularly like the TRENDnet card because it has an activity LED that let's you know if something is being sent or received by the TRENDnet CompactFlash NIC. Also, the TRENDnet TEW-222CF is certified for use anywhere in the world. Unfortunately, the TEW-222CF was discontinued while I was in the middle of writing this book. However, I managed to find an exact replacement for the TEW-222CF that you can obtain from many internet vendors. The good news is that there are plenty of other Type 1 (thin format) 802.11b CompactFlash NICs to choose from that will also work well with the AirDrop modules.

Learn to Play Guitar and Become Famous

At least that's what John Mellencamp keeps telling us. This chapter has revealed the hardware that makes up the AirDrop-P. Along the way you've also been introduced to some theory behind what it takes to play rock guitar. The following chapter will explore a companion 802.11b device based on a microcontroller from Atmel, the AirDrop-A.

Here's an easy one. What was John Mellencamp's stage name?

The AirDrop-A

The AirDrop-A module is basic AVR ATmega128L microcontroller hardware combined with specialized EDTP Electronics 802.11b hardware and firmware. Logically, the AirDrop-A is a hardware/firmware clone of the AirDrop-P with only the microcontroller and its programming and debugging points differing.

The AirDrop-A Hardware

The AirDrop-P came to physical realization first and soon after, the AirDrop-A hardware was fashioned after that of the AirDrop-P. The original AirDrop-A prototype is shown in Photo 3.1.

Photo 3.1: A JTAG port was added to the AirDrop-A variant to support debugging with the JTAGICE mkII.

Like the AirDrop-P, the AirDrop-A is designed to allow the 802.11b designer to use his or her microcontroller of choice. This is done by not populating or depopulating the native ATmega128L and mounting the desired microcontroller using a daughterboard that pins to the AirDrop-A's microcontroller I/O header pads. Of course, the AirDrop-A is more suited for AVR microcontrollers because of the inclusion of the AVR JTAG interface. In this chapter we will focus on the production AirDrop-A hardware that you see in Photo 3.2.

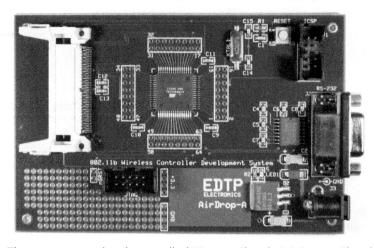

Photo 3.2: Those two-note chords are called "Power Chords." A Power Chord consists of the root note of the chord and the 5th interval note. The 3rd major or minor interval note is left out of the Power Chord structure. The octave of the root note is usually played as well. Hard rock and metal rock are Power Chord havens. As for the AirDrop-A, it is based on the most powerful 8-bit microcontroller currently in the AVR 8-bit arsenal. It doesn't have as much available I/O as the comparable PIC on the AirDrop-P but it rocks just as hard.

The AirDrop-A is based on the Atmel ATmega128L, the low-power, low-voltage relative of the venerable ATmega128 shown at close range in Photo 3.3.

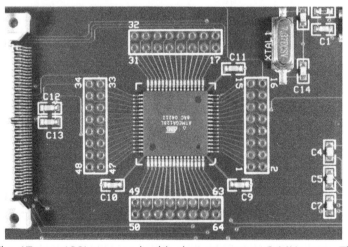

Photo 3.3: The ATmega128L you see in this shot can run at 8 MHz tops. The AirDrop-A clocks the ATmega128L at 7.37 MHz.

With seven pinned-out I/O ports, the ATmega128L offers up plenty of available I/O pins to drive its version of the AirDrop 802.11b solution. The ATmega128L comes with just a bit more standard memory than the PIC18LF8621. The memory complement of the ATmega128L includes 4K of SRAM and 128K of program Flash. However, the ATmega128L microelectronics are housed in a higher pitch 64-lead TQFP package, which slightly reduces the ATmega128L's physical I/O capacity. The ATmega128L I/O subsystem is comprised of 53 I/O lines, a 10-bit analog-to-digital converter, an analog comparator, dual USARTs, 5 timers, an SPI module and a 2-wire serial interface (I^2C) module. With a maximum clock speed of 8 MHz (half that of the standard ATmega128), the ATmega128L still carries enough firepower to support an 802.11b CompactFlash card NIC (Network Interface Card) with I/O left over

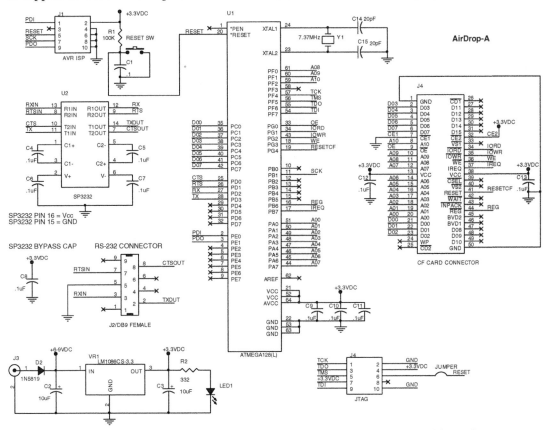

Schematic 3.1: I built my first electronic circuit, a single-transistor audio amplifier, when I was 10 years old and wrote my very first commercial magazine article at the age of 29. Do you know how old Mozart was when he wrote his first piece of music and when his first performance occurred? While you're humming tunes from Wolfgang's "Marriage of Figaro," note that again in the AirDrop-A schematic you see a simple microcontroller system, but this time it is AVR-based. The AirDrop-P's PIC18LF8621 tops out at about 5 MIPS (Millions of Instructions per Second) with a 20 MHz clock. The AirDrop-A's ATmega128L can deliver up to 7 MIPS running with its 7.37 MHz clock. Regardless, the bulk of the 802.11b work on the AirDrop-A is being done within the 802.11b CompactFlash NIC under the direction of the ATmega128L.

to allow the deployment of analog-to-digital conversion, pulse width modulation (PWM) and digital control. That means like the AirDrop-P, the AirDrop-A can do the normal things you do with an AVR microcontroller (read switch states, generate waveforms, monitor voltages, control relays, drive displays, scan keyboards, digitally communicate with other devices, etc.) and communicate cooperatively with other wireless or wired devices on an Ethernet LAN. A schematic depiction of the AirDrop-P hardware is rendered as Schematic 3.1.

The AirDrop-A power supply subsystem is an exact replica of the AirDrop-P power supply circuitry. The "slower" ATmega128L was incorporated into the AirDrop-A because the standard ATmega128 cannot operate reliably below +4.5 VDC. The song remains the same with the AirDrop-A as sung by the AirDrop-P. A National Semiconductor LM1086CS-3.3 linear regulator supplies the +3.3 VDC power that feeds the 802.11b CompactFlash card NIC, the Sipex SP3232 RS-232 converter IC and the ATmega128L microcontroller. Any +9 VDC center positive power brick that can power the AirDrop-P (400 mA minimum) is suitable for powering the AirDrop-A as well. The magic smoke that makes a microcontroller work can be released from the ATmega128L if the incorrect power conditions warrant, and a high-current switching diode (D2) is inserted inline with the incoming bulk DC voltage to protect the AirDrop-A circuitry from the accidental application of a reverse polarity DC voltage.

A standard RS-232 serial port is realized on the AirDrop-A exactly like is on the AirDrop-P. If you skipped the AirDrop-P serial port hardware description, you can read about the AirDrop-A's RS-232 serial port details there as there is no difference in the two circuits. A nose-to-nose view of the AirDrop-A's serial port electronics can be seen in Photo 3.4.

Photo 3.4: The AirDrop-A serial port is an exact copy of the AirDrop-P serial port circuitry. The identical hardware designs found on the AirDrop-P and AirDrop-A carry over to the 802.11b driver firmware as it is possible to write almost identical 802.11b driver code for each of the AirDrop module variants.

While testing the AirDrop-A (and the AirDrop-P) I never encountered a situation where I was forced to use the modem control lines (RTS and CTS) to initiate data traffic management. Without a doubt, some of you will push the RS-232 speed envelope to that point. So, just in case your pet 802.11b project requires RTS/CTS handshaking, the second pair of the SP3232's transmit and receive lines are dedicated to RTS/CTS handshaking duty. If handshaking isn't in your future, there's nothing to stop you from cutting the AirDrop-A's serial port RTS/CTS traces and using them to interface to the ATmega128L's second serial port. Like the AirDrop-P, the AirDrop-A's RS-232 9-pin D-shell connector is wired as DCE (Data Communications Equipment) to eliminate the need for a crossover serial cable or null modem when connecting to DTE (Data Terminal Equipment) devices like your personal computer's serial ports.

The inclusion of a true RS-232 serial port on the AirDrop-A module provides the same advantages that are gained with the AirDrop-P serial port. You can wirelessly pipe data between a device that can only speak RS-232 and an Ethernet LAN, and write snippets of debugging code that use the AirDrop-A's serial port as a "speaker," announcing when a certain area of code has been entered or exited.

The decision to run the AirDrop-A at a somewhat slower clock speed of 7.37 MHz instead of the maximum clock speed of 8 MHz is justified in the computations rendered by Photo 3.5 and Photo 3.6. 57600 bps is a standard baud rate for the EDTP products that support RS-232 serial interfaces. Although the 8 MHz clocks a perfect 1mS timer interval, the 7.37 MHz clock works perfectly with the 56K baud rate used by the AirDrop-A.

The ATmega128L can be programmed using an inexpensive AVR ISP dongle. The AVR ISP interface to the ATmega128L is a 10-pin male header that is positioned just above the serial port and to the immediate right of the reset switch in Photo 3.2. I've drawn the ISP electrical interface into Schematic 3.1 as well. In addition to the standard AVR ISP programming interface, the ATmega128L can also be programmed via its JTAG interface, which is yet another 10-pin male header found on the AirDrop-A printed circuit board just below and to the right of the CompactFlash card connector in Photo 3.2. When used with Atmel's AVR Studio (the synonym of Microchip's MPLAB IDE), the ATmega128L's JTAG interface provides both a programming and debugging interface similar to what is provided by the Microchip MPLAB ICD 2.

The essential 50-pin CompactFlash card connector connects the 802.11b wireless LAN network interface card and interfaces the CompactFlash card's electronics to the ATmega128L. Like the PIC18LF8621, the ATmega128L is an 8-bit microcontroller. That means the 16-bit functionality of the CompactFlash connector will again go unused and thus will not be connected. There are fewer ATmega128L I/O pins and only the CompactFlash connector I/O lines that are absolutely necessary for 802.11b operation are connected and used. In a conscious effort to keep the AirDrop-A circuitry and driver code as simple as possible, a simple I/O interface identical to the AirDrop-P was adopted. Again, note that there is absolutely no "glue" logic on the AirDrop-A board and no special purpose ICs other than the RS-232 converter IC. By implementing this simple I/O interface on both the AirDrop-P and the AirDrop-A, I was able to port most of the AirDrop-P 802.11b driver code directly to the AirDrop-A with very little modification.

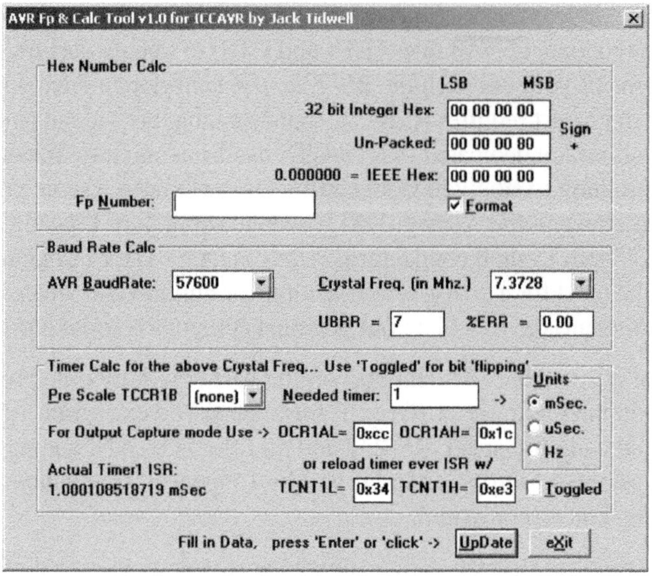

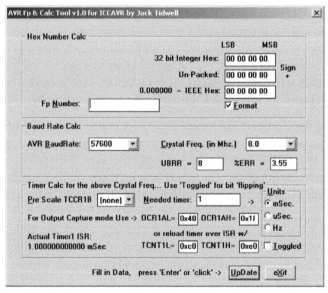

Photo 3.5: The one ten-thousandth of a millisecond error is well within the limits of the delays that will be used in the AirDrop-A 802.11b driver code. It is more important to assure a good RS-232 serial port data stream as the AirDrop-A could be asked to become an RS-232-to-802.11b converter. Mozart was a child prodigy and wrote his first piece of music at age five. He then went on to begin performing at the tender age of six. My first musical "debut" was a trumpet solo (the Beatles' Yesterday) in a concert at my elementary school at age twelve.

Photo 3.6: If the delay timing were critical, the 8 MHz clock speed would be ideal. Even with an error of almost 4%, the AirDrop-A serial port would still function satisfactorily.

The ATmega128L is a very "normal" microcontroller and its behavior is predictable. Thus, all of the firmware to drive our AirDrop-A 802.11b application and the CompactFlash NIC is contained easily within the ATmega128L's 128K of program flash. The CompactFlash NIC is responsible for taking data we give to it and broadcasting it wirelessly to another wireless station or access point. The original wireless LAN card of choice for the AirDrop-A series of 802.11b development systems is the same radio card used by the AirDrop-P, the TEW-222CF, which you now know is out of production. The clone of the TEW-222CF that is now shipping with the AirDrop-A works perfectly. Although the TEW-222CF is for all intents and purposes not available, the basic technology behind the replacement CompactFlash card and other compatible CompactFlash NICs is the same.

Bowing Out

This chapter has presented a bit of Fred Eady trivia along with some historical notes on Wolfgang Amadeus Mozart. You've also been introduced to the hardware design behind the AirDrop-A. As you can see, I was no child prodigy like Wolfie, but by the time you've finished reading this book you will be an 802.11b virtuoso.

If you're not chapter hopping, you've gotten all you really need to know at this point about the AirDrop-A and AirDrop-P hardware. So, let's move on and explore the details of the TEW-222CF CompactFlash NIC and other 802.11b CompactFlash NICs that the AirDrop modules support.

By the way, John Mellencamp was originally introduced to us as Johnny Cougar. He hated that name but the show had to go on.

802.11b CompactFlash Network Interface Cards

It would be next to impossible to get a microcontroller on the air with 802.11b without the PRISM radio hardware contained within an 802.11b NIC like the TEW-222CF. I'm sure none of the TEW-222CF designers (or any other 802.11b CompactFlash NIC designers for that matter) were thinking that their CompactFlash 802.11b products would be used by an 8-bit microcontroller as a tool to enter the 802.11b wireless LAN sanctum that has previously been reserved for Linux and Windows® operating system. With the sudden demise of the TREND-net TEW-222CF, I was forced to test and certify other manufacturer's 802.11b CompactFlash NICs for AirDrop compatibility.

They Were Not Designed To Do This

If you read the marketing stuff that describes the TEW-222CF, it was designed to be used in PDAs, Palm PCs and Pocket PCs. With an available PCMCIA-to-CompactFlash converter, the TEW-222CF was also intended to replace an 802.11b PCMCIA NIC in your laptop personal computer. All of the downloadable and CDROM TEW-222CF drivers are geared towards Windows and I didn't see anything in the marketing data sheet that mentioned it was compatible with many of the popular off-the-shelf 8-bit microcontrollers from Microchip and Atmel. Come to think of it, I didn't see any references to Linux drivers either. Well, things are about to make a radical change. TRENDnet, Netgear, Linksys, Zonet, Xterasys. Welcome to Smallville, 802.11b.

The TEW-222CF

The TEW-222CF is a CompactFlash Type 1 card, which is about 1.7 mm thinner than a CompactFlash Type II card. The TEW-222CF wireless NIC operates in the unlicensed 2.4 MHz ISM (Industrial Scientific Medical) band. That puts the poor NIC in competition with microwave ovens and 2.4 MHz wireless telephones. The TEW-222CF NIC complicated modulation schemes (CCK (Complimentary Code Keying), DQPSK (Differential Quadrature Phase Shift Keying) and DBPSK (Differential Binary Phase Shift Keying)) help avoid much of the interference from other devices using or disrupting the frequencies the TEW-222CF NIC wants to use. The aforementioned modulation techniques are also responsible for the increased data rate of 802.11b over the original 802.11 specification. DBPSK delivers 1 Mbps data rates, DQPSK is used for 2 Mbps data rates and CCK can provide 5.5 Mbps and 11 Mbps data rates. I'm purposely not going to delve into the technical details of how Direct Sequence Spared Spectrum (DSSS) and its modulation schemes work as it is not essential to our getting an AirDrop on the air. If you're interested in the nitty gritty of 802.11b RF, just get a broadband connection to the internet and read to your heart's delight. There's LOTS of "how the radio works" stuff on the big network. You'll also find that I've included all that I can on the 802.11b RF/modulation subject that relates directly to AirDrops on of the CDROM that accompanies this book.

I believe authors should go out of their way to try to get the most pertinent information they can about their book's major subject matter. So, before they stopped marketing the TEW-222CF, I contacted the TRENDnet engineering department in an attempt to get more detailed information on the TEW-222CF. They basically ignored me until I went away. In fact, now that I'm thinking back on it all, while I was searching and begging for 802.11b information for this book I got ignored by 802.11b vendors a lot. That's all I'm going to say about that. A "commercial" view of the TEW-222CF is shown in Photo 4.1.

Photo 4.1: You're looking at a stock TEW-222CF. About the only difference you can see is that if you place it side by side with the other CompactFlash NICs, it is a bit taller than the other CompactFlash NICs I tested. Like all 802.11b CompactFlash NICs, the TEW-222CF uses DSSS technology. The original 802.11 specification described wireless devices that used FHSS (frequency-hopping spread spectrum) technology. I'll bet you don't know who invented frequency-hopping technology?

Never Ignore an Inquisitive Author with Hand Tools

In this chapter, I had planned on describing the different 802.11b CompactFlash NICs I experimented with while writing the AirDrop 802.11b drivers. However, I soon realized that the only thing I would be describing to you would be the color and size of each individual 802.11b CompactFlash NIC's enclosure.

By chance the very first 802.11b CompactFlash NIC I was going to talk about has been withdrawn from the 802.11b marketplace. You can still purchase the D-Link DCF-660W, but you will find a DCF-660W end-of-life statement on the D-Link web site. Since the D-Link 802.11b CompactFlash NIC is for all intents and purposes dead, I figured it was of no use to post a photo of the D-Link NIC here. However, this is a good opportunity to "open up" an 802.11b CompactFlash NIC and oogle at its innards.

I gathered up a small screwdriver and a set of small electronic pliers and set about removing the front and back metal shields and plastic outer covering of the DCF-660W. The easiest part to remove was the bulbous piece of plastic that I had imagined made room the Compact-Flash card's antenna. As you can see in Photo 4.2, I was right.

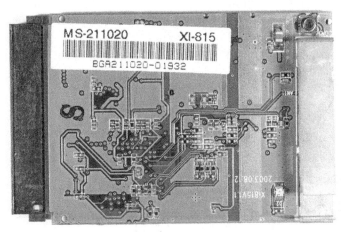

Photo 4.2: The strip of metal you see to the far right of the printed circuit board is called a PIFA (Patched Inverse "F" Antenna). If an 802.11b CompactFlash NIC has a bulge at the opposite end of the CompactFlash connector, it's probably protecting a PIFA. I tested this theory by looking at the Netgear MA701 data sheet. The MA701 has a bulge and the data sheet says it is covering a PIFA. Photo 4.3 is the flip side of the DCF-660W 802.11b CompactFlash NIC. Behold the DCF-660W's PRISM 2.5 chipset and supporting cast.

To expose all of the DCF-660W 802.11b CompactFlash NIC's ICs, I removed all of the shielding tops from the RF section of the DCF-660W printed circuit board you see in Photo 4.3.

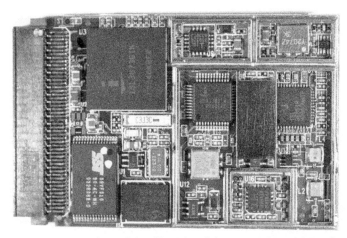

Photo 4.3: This is a nose-to-IC shot of the flip side of the DCF-660W 802.11b CompactFlash NIC. Behold the DCF-660W's PRISM 2.5 chipset and supporting cast.

Everything 802.11b started out in the open with Intersil and is all now being declared "TOP SECRET" by the new owners of the PRISM chipset. So, instead of doing it the scientific way (with a valid data sheet and moral scientific methods), I was forced to positively identify the DCF-660W circuitry as PRISM 2.5 the "Yo Ho!" patch-on-my-eye pirate's way by comparing the Intersil IC part numbers on the DCF-660W printed circuit board to an Intersil press release for PRISM 2.5. The Intersil IC numbers led me to a document on the Intersil FTP server that confirmed without a doubt that the DCF-660W is a PRISM 2.5 baby. You can match up the components in Photo 4.3 with their PRISM component counterparts depicted in Figure 4.1.

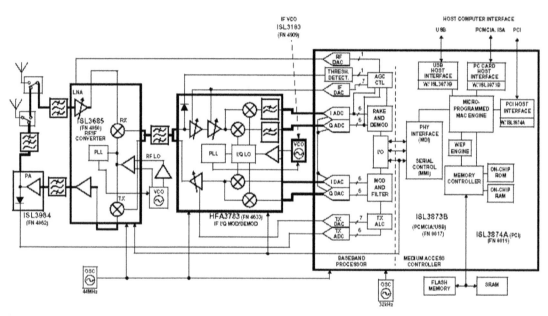

Figure 4.1: The finite details of how all of this works is not important to us as we can simply apply the PRISM technology to get our AirDrop's on a LAN. Hedy Lamarr, an MGM beauty queen and actress, invented frequency hopping technology for use by torpedoes during WWII. Unlike the inventor of Ethernet, Hedy never made a penny on her invention. The WWII Allied military machine didn't take her invention seriously and it was never used during WWII. Imagine that. In fact, Hedy's brainchild was not put to practical use by the military until the 1950's after her patent had expired. However, frequency hopping technology was used by the military during the Cuban missile crisis in the 1960's. By this time it was called "spread spectrum" technology.

The largest IC of the 4-IC PRISM 2.5 chipset gang in the upper left corner of Photo 4.3 is the ISL3873B integrated Baseband Processor (BBP) and Medium Access Controller (MAC). The ISL3873B includes both a PCMCIA interface and a USB interface. The PC-MCIA interface is logically identical to a CompactFlash interface and this bodes well for us in the AirDrop modules. Supporting the ISL3873B are a 1Mbit (128K × 8) Flash IC (SST 39VF010), which is located in the bottom left corner of the printed circuit board, and an un-identified chuck of what looks to be 55 nS SRAM sitting directly to the right of the Flash IC. Although I can't verify it, the SRAM part number, 62S16128BU-55LLT, leans the SRAM IC towards being a 128K device. This 128K guess is also based on information gleaned from the ISL3873A data sheet. The ISL3873B data sheet has all of a sudden become "TOP SECRET" and cannot be found on the Internet. Note that in Figure 4.1 the "FN" numbers in parentheses are the Intersil data sheet numbers. None of the "FN" data sheets in Figure 4.1 are obtainable via the web. So, I'll use hard facts from here and there, intuition and wild scientific educated guesses to attempt to give you some idea about the electronics that make up the 802.11b CompactFlash NICs I tested and used to develop the AirDrop 802.11b drivers.

Working Figure 4.1 right to left, the next PRISM building block we encounter is the ISL3183 IF (intermediate frequency) VCO, which is called a companion VCO in the PRISM 2.5 Overview document. The ISL3183 seems to be an integral part of the HFA3783, which performs IF conversion and I/Q baseband modulator and demodulator duties. In Photo 4.3, the ISL3183 is the topmost IC directly to the right of the large ISL3873B BBP/MAC IC. The HFA3783 is located directly below the ISL3183 and to the right of the large ISL3873B BBP/MAC IC. The output of the HFA3783 is then fed to the ISL3685 RF/IF Converter, which is also aided by an external VCO. You'll find the ISL3685 to the right of the HFA3783 on the other side of the rectangular EMI shield. Incoming RF is passed through the ISL3685 back to the input of the HFA3783 and finally makes its way back to the ISL3873B. Since 802.11b operates in half-duplex mode (no simultaneous transmission and reception), only one common IF filter is required as both transmit and receive operations use the same physical path. A power amplifier/detector, the ISL3984, puts the modulated RF from the output of the ISL3685 RF/IF Converter out to the antenna. The ISL3984 is situated directly below the rectangular EMI shield under the HFA3783 and ISL3685. Figure 4.1 also implies that some of the transmitted signal is fed back to the ISL3873B. Since the feedback path from the ISL3984 to the ISL3873B terminates at the input of an analog-to-digital converter, my best guess is that the ISL3873B is monitoring the RF transmission power level and controlling something in the PRISM radio chain as a result. PRISM radio element clocking is provided by a 44 MHz master clock oscillator, which feeds the ISL3685, the HFA3783 and the ISL3873B. The ISL3873B also uses a 32.768 KHz watch crystal to keep time and allow the ISL3873B electronics to sleep and wake up at a predetermined time interval. The 44 MHz oscillator is seen as the relatively large "can" directly above the SRAM IC and to the right of the Flash IC. The 32.768 KHz watch crystal is the relatively larger rectangular object directly below the large ISL3873B BBP/MAC IC.

Although the latest and greatest PRISM chipset data sheets aren't publicly available, we can still use the clues that are present in the PRISM stuff that is available to formulate the AirDrop drivers. Now that I've unwrapped the DCF-660W, I'm really anxious to see why the TRENDnet engineers ignored me. With that..

Unwrapping the TEW-222CF

A naked TEW-222CF stands before you in Photo 4.4. The PIFA, common to the DCF-660W and the Netgear MA701, is replaced by a chip antenna on the TEW-222CF printed circuit board. The TEW-222CF's chip antenna is the rectangular object at the far upper right edge of the TEW-222CF printed circuit board. As with the DCF-660W, in Photo 4.4 I've removed the TEW-222CF's metal RF shielding to expose the TEW-222CF's PRISM chipset.

An ISL3873B resides over the non-RF portion of the printed circuit board just as it did with the DCF-660W. The ISL3873B is the largest IC you see in Photo 4.4. A 44 MHz oscillator and a 32.768 KHz watch crystal are nearby supposedly performing the same duties I described earlier. The small "can" directly to the upper right of the ISL3873B is the 44 MHz oscillator. The 32.768 KHz watch crystal is placed below the ISL3873B to the immediate right of the SRAM IC. The TEW-222CF's SRAM IC, a W26B02B-55LE, is positively identified as a 55nS, 128K × 16 part from Winbond and that pretty much says the DCF-660W's SRAM is also a 55 ns, 128K × 16 part. The TEW-222CF's Flash IC, a 128K × 8, 45 nS SST39VF010, is identical to the Flash IC found on the DCF-660W printed circuit board. The SST39VF010 is positioned directly below the Winbond SRAM IC and the 32.768 KHz watch crystal.

It appears that the TEW-222CF's HFA3783B, which was not RF shielded, is serviced by a special PRISM 2 VCO from Delta Electronics, the 748UT63, instead of the ISL3183, which is nowhere to be found on the TEW-222CF printed circuit board. You can see the Delta Electronics 748UT63 as a larger "can" directly right of the smaller 44 MHz oscillator. Recall that the ISL3183 is not considered part of the core PRISM chipset as it is referred to as a companion in the data sheet. The discovery of the 748UT63 specification gives up a clue about the properties of the ISL3183. The 748UT63 is a VCO that operates from 741 MHz to 755 MHz with a center frequency of 748 MHz. It would be safe to assume the ISL3183 specs aren't far from the 748UT63 specs since the 748UT63 seems to replace the ISL3183 in the TEW-222CF.

The second PRISM 2 VCO, a Delta VCX2074S6, looks to be performing the same duty with the ISL3685 RF/IF Converter as the DCF-660W's unknown manufacturer's Y2074Z VCO. In Photo 4.4, the Delta VCX2074S6 is the large can positioned at the bottom far left of the TEW-222CF's printed circuit board. The DCF-660W's Y207AZ VCO "can" is found at the far top right of the DCF-660W's printed circuit board in Photo 4.3 The Delta VCX2074S6 specification states that the VCX2074S6 operates between 2.023 GHz and 2.125 GHz. Using some intuition, I'd say that the DCF-660W's Y2074Z has the same specs as the Delta part and operates at a center frequency of 2.074 GHz.

The HFA3783, which performs IF conversion and I/Q baseband modulator and demodulator duties, is located almost in the center of the TEW-222CF printed circuit board to the left of the shielding strip that runs the vertical length of the TEW-222CF printed circuit board. On the other side of the vertical shielding strip and directly to the right of the the HFA3783, you'll find the TEW-222CF's ISL3685 RF/IF Converter IC.

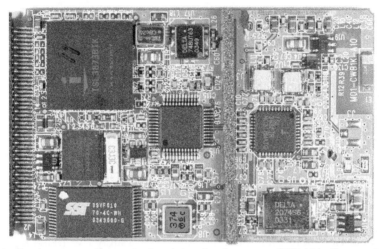

Photo 4.4: Don't worry...I'm not about to disrobe every 802.11b CompactFlash NIC I get my hands on. If there's a way to find out what's inside one without hand tools, we'll find it. All of this PRISM IC spelunking I'm doing has a purpose. Every little bit of PRISM hardware knowledge you can get will come in handy when it comes time to understand the code behind the AirDrop 802.11b drivers. Note the TEW-222CF's lack of elaborate RF shielding compared to the DCF-660W. Judging from the conductive strips on the top and bottom of the TEW-222CF printed circuit board, it seems that the TEW-222CF uses the CompactFlash card's metal enclosure panels as shielding. Another clue is given up in this shot. Make a mental note of the silk screen label to the left of the TEW-222CF's chip antenna, "M01-CWB1K-A10."

The TEW-222CF's ISL3984 power amplifier/detector, shown in the bottom right of Photo 4.5, is located alone on the opposite side of the TEW-222CF printed circuit board.

Now I can see why the DCF-660W costs more than the TEW-222CF in the retail market. The TEW-222CF and DCF-660W PRISM chipsets are identical. So, the cost difference has to be partly in the manufacturing of the elaborate shielding and PIFA I found inside the DCF-660W. I'm sure the rest of the cost difference lies in the marketing philosophy of each respective manufacturer. The cost factor really has no significance to us as we just want to know how to make the 802.11b CompactFlash NICs work. Besides, we can't waste time trying to outguess marketing types and I've got one more NIC to tear apart..

Photo 4.5: The presence of conductive shielding strips on both sides of the TEW-222CF's printed circuit board is a dead give away that the metal parts of the TEW-222CF's enclosure were used for EMI shielding.

An Undercover Look at the Zonet ZCF1100

I'm going to start paying more attention to press releases. Lately I've found that if you really read them, they can tell you bunches of good stuff about the product they're touting. For instance, the Zonet ZCF1100 you see in Photo 4.6 does not contain the expected PRISM 2.5 ISL3873B BBP/MAC IC. Instead, an ISL3871 (largest IC in the photo) is in that space. So, I did the good old web search on "ISL3871". My search resulted in a press release describing the PRISM 3 chipset. Here's what I found out.

When I began my look at various 802.11b CompactFlash NIC innards, there was one relatively large multi-pinned component on both the DCF-660W and TEW-222CF printed circuit boards that I could not positively identify. PRISM 3 is a Zero-IF (ZIF) architecture, which simply means that the intermediate frequency (IF) stage that is common in most every radio made has been eliminated and replaced with direct conversion circuitry. Normally, radio IF stages require IF transformers or filters somewhere in the circuit. I recall from somewhere in the volumes of information I've read about PRISM 2.5 and PRISM 3 that the PRISM IF stage uses a SAW filter. The only SAW filter clue I have is that the unidentified IF component on the DCF-660W is actually marked "EPCOS 3677." If you peek back at Photo 4.3, the EPCOS 3677 is a large "can" positioned below the HFA3783 and to the lower left of the rectangular EMI shield. A web search using "saw filter epcos" yielded an EPCOS page stating that the B3677, as it is called, is a WLAN-DSSS IF device with a center frequency of 374 MHz. HMMMM…374 was marked on the TEW-222CF's mystery "can". Look for the TEW-222CF mystery "can" at the bottom of the TEW-222CF printed circuit board to the right of the SST39VF010 Flash IC. Two birds, one stone. The UFO on the DCF-660W and TEW-222CF printed circuit boards is a WLAN-DSSS IF SAW filter with a center frequency of 374 MHz. If you look closely at Photo 4.6, the multi-pinned SAW filter that graces the PRISM 2.5 DCF-660W and TEW-222CF printed circuit boards is absent from the PRISM 3 Zonet ZCF1100 printed circuit board.

Photo 4.6: As Gomer would say, "Surprise, Surprise, Surprise!" The Zonet ZCF1100 threw us a curve ball. Note the Zonet ZCF1100's printed circuit board trace antenna.

The PRISM 3 press release goes on to say that the PRISM 3 chipset is a 2-IC set consisting of the ISL3871 integrated Baseband Processor/Medium Access Controller (BPP/MAC) and the ISL3684 Direct Down Conversion (DDC) Transceiver. The ISL3984, the power amplifier detector, is considered optional as is the ISL3084, a 5 GHz VCO. The ISL3871 is the largest IC on the Zonet ZCF1100 printed circuit board. The ISL3684 Direct Down Conversion (DDC) Transceiver is positioned near the center of the Zonet ZCF1100 printed circuit board to the lower right of the ISL3871 BBP/MAC IC. The ISL3084 5 GHz VCO is the only IC above the ISL3684 Direct Down Conversion (DDC) Transceiver. The ISL3984 power amplifier detector IC is mounted just to the upper right of the ISL3684 IC.

The Zonet ZCF1100's ISL3871 is supported by a 55 nS, 2 Mbit (128K × 16) Cypress CY62137CV33 SRAM and a 70 nS PMC PM39LV010-70 1 Mbit Flash IC. In Photo 4.6, the CY62137CV33 SRAM IC is directly below the ISL3871 BBP/MAC IC and the PMC PM39LV010-70 1 Mbit Flash IC is directly below the SRAM IC. Nothing new here except the use of a 16-bit SRAM instead of an 8-bit SRAM and a slower 70 nS Flash part instead of the 55 nS part used in the other 802.11b CompactFlash NIC designs. The standard PRISM 2.5/3 44 MHz master clock oscillator and the 32.768 KHz sleepy time crystal are present as expected. The 32.768 KHz watch crystal is mounted directly to the right of the CY62137CV33 SRAM IC and the 44 MHz oscillator "can" is just to the upper left of the ISL3084 5 GHz VCO.

The ISL3684 DDC Transceiver receives high frequency radio waves and directly down-converts them to a baseband signal. During transmission, the ISL3684 DDC Transceiver directly up-converts the baseband signal to high frequency RF for transmission. This direct coupling of the ISL3684 DDC Transceiver and the ISL3871 BPP/MAC eliminates the legacy PRISM 2 IF stage and thus, the SAW filter, the HFA3783 IF I/Q Modulator/Demodulator IC and the ISL3183 IF VCO.

I really wasn't expecting to find any PRISM 3 devices in my flock of 802.11b Compact-Flash NICs. I'm sitting here looking at the Linksys WCF12. It's about the same size as the Zonet ZCF1100. Hmmm…I'll bet it is PRISM 3 too. OK.. I give. Let's tear into just one more.

What's Behind Door Number 4

It's a brand new PRISM 3 chipset! All of the standard PRISM 3 goodies we found on the Zonet ZCF1100 printed circuit board are mounted on the Linksys WCF12 printed circuit board. There's no reason to describe them all again. Just marvel at the shot of the Linksys WCF12 in Photo 4.7.

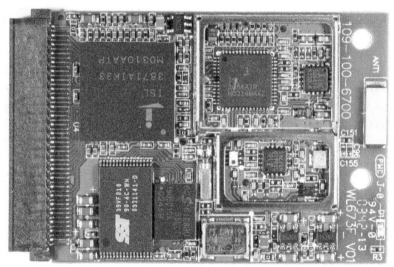

Photo 4.7: Photographing these miniature component works of art is akin to looking down upon a city from 30,000 feet. This is a shot of the Linksys WCF12.

The largest IC in Photo 4.7 is the ISL3871 BBP/MAC IC. Directly below the ISL3871 we find the SST39VF010 Flash IC. Moving your eyes directly to the right of the SST39VF010 Flash IC reveals the SRAM IC. The 44 MHz oscillator "can" sits just to the lower right of the SRAM IC and the 32 KHz watch crystal is positioned directly to the SRAM IC's upper right. The ISL3684 DDC Transceiver lives to the right of the ISL3871 BPP/MAC and the ISL3084 resides to the right of the ISL3084. Centered below the ISL3084 and ISL3684 is the ISL3984 power amplifier/detector IC. The WCF12's chip antenna rests at the far right edge of the Linksys WCF12 printed circuit board.

RF, Witchcraft, Pointy Hats, Ghouls, Goblins...Same Thing

I've always believed RF engineers make a pact with the devil before receiving their diplomas. I darned near "sold my soul" once working for NPR (National Public Radio) at WLRH in Huntsville, Alabama. I remember submitting my demo tape to the station manager, George Dickerson, way back in 1976. George took me in as a news guy at the fledgling station. I was to become one of the fourteen original "personalities" at WLRH. With the help of Chief Engineer Don McComb (he wore a really pointed hat as he was a television RF engineer as well), I ended up with a First Class Radio license and as the chief engineer of WLRH for a short time, after Don left the station, before leaving broadcast radio forever myself. You can get a peek at George, Don and me at http://www.wlrh.org/About/history.asp. GEEZ...I am old! I'm in a "history" section of a website.

Stop laughing and think about what you've seen in this chapter: Tiny components in wafer thin enclosures executing mathematically complex algorithms that result in radio waves that can accurately and quickly carry information for hundreds of feet indoors and tens of hundreds of feet outdoors.

The good news is that none of us have to sell our souls (or work as an RF engineer for NPR) to use this devilish technology. We've learned lots about what PRISM is and the electronic devices behind it by simply observing and following up on clues from our observations. Now you know why the AirDrop hardware is so simple. The key to 802.11b connectivity is in the PRISM chipset within the 802.11b CompactFlash NIC.

Just in case you're wondering what the rest of the naked NICs look like with their clothes on, the Zonet ZCF1100 is shown in Photo 4.8 and the Linksys WCF12 can be seen in Photo 4.9. Sorry, I destroyed the DCF-660W before I took its mug shot.

Photo 4.8: This is the smallest card that I tested and certified.

Photo 4.9: This is the most expensive card that I tested and certified. However, it is the most widely available CompactFlash NIC in the bunch.

In the chapters that follow, we will approach the creation of the AirDrop 802.11b drivers in the same manner as we did with the PRISM NICs. We'll observe, deduce, search for answers and just plain guess about things we can't figure out.

Since you're still laughing at my double-knit suit, what great guitar was introduced in the year that I was born (1954) and is still one of the most popular guitars today?

Talking With 802.11b CompactFlash NICs

The mere fact that a CompactFlash card can be inserted into a PCMCIA adapter sleeve and used as if it were a PCMCIA card implies that the CompactFlash and PCMCIA specifications are almost identical. In fact, the official CompactFlash specification states that CompactFlash cards provide complete PCMCIA-ATA functionality and compatibility including PCMCIA True IDE capability. While we're discussing CompactFlash card specifications, I'll clear up a much-asked question about what voltages CompactFlash cards can tolerate. The Compact-Flash specification states that CompactFlash cards can operate with either a +5VDC or +3.3 VDC power supply. The AirDrop modules use the +3.3 VDC power rail. Hmmm…The Com-pactFlash card dual-voltage statement makes me believe that one or all of those unidentified 5-pin SOT-23-5 devices on the 802.11b CompactFlash NICs I "opened up" is a low-dropout voltage regulator. With the exception of the DCF-660W, all of the other 802.11b Compact-Flash NIC's have three such devices on their printed circuit board. The DCF-660W has two SOT-23-5 packages and one unpopulated SOT-23-5 site. If we take a look at the PRISM chipset data sheets that we can get our hands on and make the assumption that the newer parts operate in a similar manner, my theory is proven as the ISL3873A's operating voltage range is +2.7 VDC to +3.6 VDC.

We won't be discussing PCMCIA or PCMCIA 802.11b NICs to any great detail within the pages of this book. However, I don't like to talk about things and not show them to you. After all, you're reading this book to obtain 802.11b knowledge without having to go look for it here and there. So, I figured you'd want to see what that CompactFlash-to-PCMCIA adapter sleeve looks like. I've supplied a shot of the SanDisk 802.11b CompactFlash NIC-to-PCMCIA adapter sleeve in Photo 5.1.

If you're already familiar with PCMCIA and CompactFlash card services, you're ahead of the game. If you're not, read on. I'll tell you all you need to know about CompactFlash cards as they relate to the AirDrop.

Photo 5.1: The SanDisk CompactFlash-to-PCMCIA adapter sleeve shown here adapts the 50-pin 802.11b CompactFlash NIC to the 68-pin PCMCIA format. This adapter allows a CompactFlash Type 1 card to be fitted in a Type II PCMCIA card slot and used just as a standard PCMCIA card would be.

Physically Connecting a Microcontroller to a CompactFlash Card

Schematic 5.1 is a connection-oriented view of the CompactFlash card connector used by both the AirDrop-A and AirDrop-P. Note that the CompactFlash card is natively a 16-bit device as implied by CompactFlash card data I/O pins *D0-D15*. However, the CompactFlash standard allows for the use of 8-bit devices, like our PIC and AVR microcontrollers, and provides a scheme to read and write 16-bit data to and from the CompactFlash card in easily managed 8-bit chunks from the lower eight CompactFlash card data I/O lines *D0-D7*. If you decide that an 8-bit microcontroller is crunching your style, you can easily covert to a 16-bit microcontroller as CompactFlash cards, with the right logic levels on some key configuration I/O pins, are always ready for either 8-bit or 16-bit operation.

In our CompactFlash card connector mini schematic (Schematic 5.1) note that Compact-Flash data I/O pins *D08-D15* are not connected and all of the CompactFlash address input pins (*A0-10*) are used. This is a standard 8-bit CompactFlash card connector arrangement, which allows us to address and access all of the CompactFlash card internals in 8-bit mode. Some of the other CompactFlash card connector's I/O pins aren't intuitively obvious. So, let's go through them all and define them.

The pins are interwoven on the AirDrop modules' CompactFlash card connectors. Com-pactFlash card connector pin 1 is followed by pin 26, pin 2 is followed by pin 27 and so forth. That puts the first and the last CompactFlash card connector pins at GND (ground) potential. With the exception of the active low *CE1 and *OE pins (the * denotes active low), the odd-numbered CompactFlash card I/O pins are self-explanatory. Assigning each CompactFlash card connector address input pin to a microcontroller I/O pin assures us that we can reach any address space or memory area inside the CompactFlash card. Photo 5.2 is close-up look at the AirDrop modules' CompactFlash card connector.

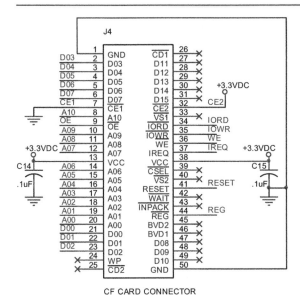

CF CARD CONNECTOR

Schematic 5.1: The CompactFlash card connector is a well-engineered interface. The circuit you see in Schematic 5.1 can be used to interface to both 802.11b CompactFlash NICs and standard CompactFlash memory cards. CompactFlash memory cards? What a puzzled look you just gave me. Yes, I said CompactFlash memory cards.

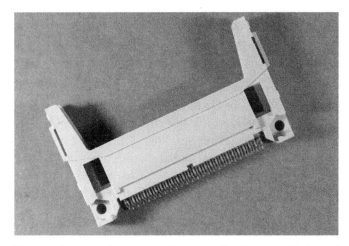

Photo 5.2: I could have chosen a cheaper and less massive CompactFlash card connector but the AirDrop modules are designed to be development platforms as well as working 802.11b stations. I figure you'll be plugging and unplugging all sorts of 802.11b CompactFlash NICs into and out of this CompactFlash card connector. So, the CompactFlash card connector must be capable of taking any punishment you might dish out in a development environment. The only gotcha to this connector is that it will only accommodate the thinner Type 1 CompactFlash cards. However, I've ripped off the upper panels of this connector and it takes Type II cards quite well then.

We really can't talk about input pin *CE1* without involving input pin *CE2*, which are both active low CompactFlash card input signals. These CompactFlash card input signals are used both to select the CompactFlash card and to indicate to the CompactFlash card whether a byte (8-bit) or a word (16-bit) access operation is being performed. CompactFlash card connector pin 20 (*A0*) is also involved when the AirDrop modules' firmware operates against the CompactFlash card's internals in 8-bit mode. To add a bit more confusion to the mix, the CompactFlash card connector *REG* input pin (CompactFlash card connector pin 37) also plays a part in the 8-bit mode CompactFlash card memory area access scheme. With the Air-Drop module design, I'm going to take the complexity out of the *CE1*-*CE2*-*A0*-*REG* pin function matrix right away by eliminating the CompactFlash card's 16-bit functionality at the hardware level.

Note that the *CE1* CompactFlash card input pin (CompactFlash card pin 7) is tied to ground and the CompactFlash card input pin *CE2* (CompactFlash card pin 32) is tied to +3.3VDC in Schematic 5.1. This arrangement of the CompactFlash card's *CE1* and *CE2* input pins' logic levels negates the use of the CompactFlash card's data I/O lines *D8-D15*. The CompactFlash card standby mode is also eliminated as *CE1*, which is permanently tied to ground, is incapable of being driven to a logical high level. Although the CompactFlash card's 16-bit mode and standby mode have been blasted out of existence, this give-and-take of the aforementioned CompactFlash card's 16-bit capabilities supports the first rule of embedded programming; "Nothing is free". However, tying *CE1* and *CE2* puts the 802.11b CompactFlash NIC into permanent 8-bit mode and we can perform all of the 8-bit 802.11b operations necessary to run an AirDrop without ever having to worry about the logical states of *CE1* and *CE2*.

Table 5.1 is a graphical representation of the logic levels on the CompactFlash card's *A0*, *CE1*, *CE2*, *REG* inputs and read/write control input pins that are required to read the CompactFlash card's Card Information Structure (CIS) memory area and read and write the CompactFlash card's configuration registers if that becomes necessary. The Compact-Flash card's *REG* input pin (CompactFlash card pin 44) should be driven logically low for CompactFlash card Attribute Memory accesses and CompactFlash card I/O cycles when an active I/O address is on the CompactFlash card's address lines. The *REG* input pin should be pulled logically high for Common Memory accesses. The AirDrop driver code pulls the Com-pactFlash card's *REG* input pin low for all 802.11b CompactFlash NIC operations. Thus, there are no Common Memory accesses performed by the AirDrop-A or AirDrop-P 802.11b CompactFlash NIC drivers against the 802.11b CompactFlash NIC. Attaching the Compact-Flash card's *REG* input pin to a microcontroller I/O pin allows the flexibility of interfacing the AirDrop module's support electronics to a CompactFlash card that contains a Common Memory area.

CompactFlash+ Card Register and Memory Area Decoding							
*CE1	*CE2	*REG	*OE	*WE	A10-A1	A0	Selected Memory Area
0	1	0	0	1	X	0	Configuration Registers Read
0	1	0	1	0	X	0	Configuration Registers Write
0	1	0	0	1	X	0	Card Information Structure Read
Legend:							
Active Low signals are preceded by a '*'							
x' = "don't care"							

Table 5.1: The CompactFlash card standby and 16-bit operations have been trashed, but in the case of the AirDrop modules, we don't need 16-bit mode or standby operation. The fact is that we gain an advantage both physically and logically by permanently squashing standby mode and 16-bit mode. Physically, there are two less microcontroller I/O pins tied up by the CompactFlash card interface. Logically, we never have to be concerned about flipping bits in our AirDrop driver code that may accidentally put us into 16-bit mode or standby mode.

Assuming that we may just see some Common Memory accesses (I did say the words CompactFlash memory card earlier), and to stabilize the facts behind *OE* and *WE* usage, I've drawn up Table 5.2, which shows the *OE* and *WE* input logic needed to access a CompactFlash Storage Card's or a memory-mode enabled CompactFlash+ card's Common Memory area.

Common Memory Functions								
Function Code	*CE1	*CE2	*REG	*OE	*WE	A0	D15-D8	D7-D0
Byte Read (8-bits)	0	1	1	0	1	0	High Z	Even Byte
Byte Read (8-bits)	0	1	1	0	1	1	High Z	Odd Byte
Byte Write (8-bits)	0	1	1	1	0	0	x	Even Byte
Byte Write (8-bits)	0	1	1	1	0	1	x	Odd Byte
Legend:								
Active Low signals are preceded by a '*'								
x' = "don't care"								

*Table 5.2: Here's where having the CompactFlash card's *REG input pin tied to a microcontroller I/O pin pays off. To be able to read a CompactFlash Storage card or a CompactFlash+ card in memory mode all we have to do is use the microcontroller I/O pin to drive the CompactFlash card's *REG input pin logically high. Otherwise, the song remains the same.*

I'm getting a bit ahead of schedule and talking out of school here. Later, you will see the importance of being able to read a CompactFlash card's CIS memory information. Another very important CompactFlash card register we will use in our AirDrop driver code is the Configuration Option Register (COR). Both of these terms, CIS and COR, will be discussed in detail when we start picking apart the AirDrop 802.11b driver source code. A look at the logic levels needed to operate on the CompactFlash card's COR is shown in Table 5.3.

CompactFlash+ Card Configuration Registers Decoding										
*CE1	*CE2	*REG	*OE	*WE	A4	A3	A2	A1	A0	Selected Register
0	x	0	0	1	0	0	0	0	0	Configuration Option Reg Read
0	x	0	1	0	0	0	0	0	0	Configuration Option Reg Write
Legend:										
Active Low signals are preceded by a '*'										
Address lines A5-A10 are "don't care"										
x' = don't care										

Table 5.3: Note that the *CE1, *CE2 and *REG input logic levels never change while the AirDrop 802.11b driver firmware is performing either the COR read or COR write operation. Although *CE2 is a "don't care" in this instance, remember that the AirDrop CompactFlash card connector's *CE2 pin is permanently tied logically high to the +3.3VDC voltage rail.

As you can see in Table 5.1, the CompactFlash card's *OE* (Output Enable) input pin (CompactFlash card connector pin 9) is used to read the CompactFlash card's CIS memory area. The CompactFlash card connector *WE* (Write Enable) input pin (CompactFlash card pin 36) is driven low only when writing to the CompactFlash card's configuration registers. These active low CompactFlash card signals, *OE* and *WE*, are also used to read and write data to and from a CompactFlash Storage Card or a CompactFlash+ card in memory mode. The *OE* and *WE* pins are used to read the CompactFlash card CIS memory area and write to the CompactFlash card configuration registers when the CompactFlash card is configured for I/O mode. As you will see later when we examine the AirDrop 802.11b source code in detail, the AirDrop 802.11b CompactFlash NIC runs in I/O mode. An example of using the *OE* and *WE* signals for the manipulation of registers, in particular the CompactFlash card's COR, is logically demonstrated in Table 5.3.

Once the CompactFlash card is put into I/O mode, it looks more and more like a PCMCIA card is plugged into the CompactFlash card connector. Take a look at Table 5.4

PCMCIA Mode I/O Functions								
Function Code	*CE1	*CE2	*REG	*IORD	*IOWR	A0	D15-D8	D7-D0
Byte Input (8-bits)	0	1	0	0	1	0	High Z	Even Byte
Byte Input (8-bits)	0	1	0	0	1	1	High Z	Odd Byte
Byte Output (8-bits)	0	1	0	1	0	0	x	Even Byte
Byte Output (8-bits)	0	1	0	1	0	1	x	Odd Byte
Legend:								
Active Low signals are preceded by a '*'								
x' = don't care								

Table 5.4: The CompactFlash card *REG input pin has been returned to its normally low logic level and the *OE and *WE inputs have been dumped for the new *IORD and *IOWR I/O mode read/write control signals. Again, the CompactFlash card's *CE1 and *CE2 inputs don't have to be considered.

Using *IORD* and *IOWR* allows 16-bits of data to be clocked into or out of a CompactFlash card using even and odd 8-bit bytes. The CompactFlash card's *A0* address line determines which byte is even (*A0*=0) and which byte is odd (*A0*=1). You'll see the CompactFlash card's *IORD* and *IOWR* control signals again when we start talking to the AirDrop's 802.11b CompactFlash NIC.

The next chapter will take us into the bowels of the 802.11b CompactFlash NIC's CIS memory area. The CIS tells all about a CompactFlash card if you know how to read and decipher the CIS code. Fortunately, we can simply ignore 99% of the CIS information we glean from the 802.11b CompactFlash NIC as the operation of the AirDrop 802.11b drivers only require the 802.11b CompactFlash NIC's COR address from the CIS.

Musical Overtones

We've got some unfinished musical business from Chapter 4 that needs to be tabled and put to rest. That renowned guitar I was alluding to is Fender's Stratocaster, which in 2004 celebrated its 50th birthday. And, to answer the question you've just asked yourself, YES, I have a couple of Stratocasters including a 50th Anniversary model. Buddy Holly brought the Strat into the Rock and Roll limelight during his short professional musical career and if you've never heard of Eric Clapton's "Blackie," one of the world's most desirable Stratocasters, you are living under a rock.

For those of you that missed that one, here's a chance to redeem yourselves. What equally as famous Fender guitar was originally called the "Broadcaster," and why was it renamed? Chomp on that one while you're taking a tour of Tuple City in Chapter 6.

Touring the Card Information Structure

All of the source code you're about to encounter was written to help you understand the basic concepts behind writing a microcontroller-based driver for 802.11b CompactFlash NICs containing PRISM chipsets. You won't find any fancy coding techniques in this book's AirDrop driver source code. However, the AirDrop driver source code you will see in the pages that follow and on the CD-ROM performs the most important task that can be asked of any firmware. It works. I'm sure that there are areas of the code that can be optimized or improved. My feeling is that once you've grasped the fundamentals of the AirDrop driver code, you can then apply your personal coding methods and techniques and any degree of optimization you desire. I realize that some of you reading right now are seasoned microcontroller C coders and some of you aren't. So, I wrote the AirDrop driver code with beginning coders in mind. If you claim to be a "professional" microcontroller coder, you'll be able to easily follow along and absorb all of the fundamental concepts offered by the AirDrop driver source code. Because the AirDrop 802.11b driver source code doesn't contain "advanced" or hard-to-figure-out C constructs, the same "you can easily understand it" goes for you rookie C coders in the audience as well. I didn't write the AirDrop driver code or this book to showcase my C coding abilities. My goal is to show *ALL* of you how to easily enable 802.11b technology using inexpensive off-the-shelf microcontrollers and 802.11b CompactFlash NICs.

As we make our way through the AirDrop 802.11b driver source code, I'll show you code snippets, which are incomplete physical code segments but complete logical code segments of the whole of the AirDrop 802.11b driver source code. To aid you in understanding each code snippet's purpose, I'll gather all of the supporting functions and definitions used in the code snippet and include them in the code snippet's dialog. All of the source code you will see from here on out will be taken from the AirDrop-P and is based on the HI-TECH PICC-18 C compiler. If I stray from that convention, I'll point out where the source code in the code snippet originates since the AirDrop-A and AirDrop-P 802.11b CompactFlash NIC driver source code is identical in most areas. The complete AirDrop 802.11b driver source code for both the AirDrop-P and AirDrop-A is included on the CD-ROM that accompanies this book.

The AirDrop 802.11b driver is simply a combination of various CompactFlash card I/O routines coupled with PRISM commands. Driving a PRISM-based CompactFlash card is actually very easy to do and I don't intend to add any complication to the process. Whether you're a pro C coder or a rookie to all of microcontrollerdom, that "ground training" you started in Chapter 1 is just about over and shortly you'll be ready to climb into the AirDrop cockpit.

Talking in Tuples

No, tuples are not beings that live in "Middle Earth". However, they do exist at the beginning of a CompactFlash card's Attribute Memory area and can be seen immediately after the CompactFlash card performs a successful reset operation. Once you've entered Tuple City, it's not hard to find your way around, but don't ask for directions if you can't speak the language and don't drink the water.

Pick up your 802.11b CompactFlash NIC and stare it straight in the antenna. The only thing that 802.11b CompactFlash NIC in your hand knows is what was programmed into it by the manufacturer, and torturing the 802.11b CompactFlash NIC by applying high voltages to its I/O pins won't make it talk. It's totally up to us to coax the 802.11b CompactFlash NIC into telling us what it holds in its read-only Attribute Memory area. The way we initially learn about the contents of any CompactFlash card is by retrieving and examining the CompactFlash card's tuples, which are found in the CompactFlash card's CIS memory area.

A tuple is technically a "set of *x* amount of elements". In the case of the AirDrop, a tuple is actually a collection of bytes that describes a physical or electrical element of the CompactFlash card in which the tuple resides. An AirDrop 802.11b CompactFlash NIC tuple consists of a tuple code byte followed by a tuple link byte followed by the tuple data byte(s). The tuple code byte tells us what kind of tuple we are examining (or about to examine). The tuple link byte tells us how many bytes are in the tuple structure. Knowing how many bytes we must read from a particular tuple structure also tells us where the next tuple structure begins in the CompactFlash card address space. As you have already ascertained, the tuple data bytes contain some of the information about the 802.11b CompactFlash NIC we are looking for to use in our AirDrop driver code. Each individual tuple structure ends with a hexadecimal 0xFF, which is considered as part of that particular tuple's byte count. However, as you will see when we begin to examine the tuples in detail, some tuples use the linked list concept without the inclusion of the 0xFF tuple ending byte sequence. Tuples that we are immediately interested in are called Layer 1 tuples. I've identified the Layer 1 tuples we're likely to see in Table 6.1.

Layer 1 Tuples		
Tuple Code	**Tuple Name**	**Tuple Description**
0x00	NULL	
0x01	DEVICE	Device Information for Common Memory
0x02-0x05	RESERVED	
0x06	LONGLINK_MFC	Long-link for Multi-Function Card
0x07-0x0F	RESERVED	
0x10	CHECKSUM	Checksum Control
0x11	LONGLINK_A	Long-Link Control to Attribute Memory
0x12	LONGLINK_C	Long-Link Control to Common Memory
0x13	LINKTARGET	Link Target
0x14	NO_LINK	End of Current Chain

Layer 1 Tuples		
Tuple Code	**Tuple Name**	**Tuple Description**
0x15	VERS_1	Identifies PCMCIA Compliance Level of CIS
0x16	ALTSTR	Alternate Language String
0x17	DEVICE_A	Device Information to Attribute Memory
0x18	JEDEC_C	JEDEC Programming Algorithm for Tuple 0x01
0x19	JEDEC_A	JEDEC Programming Algorithm for tuple 0x17
0x1A	CONFIG	Configuration Tuple
0x1B	CFTABLE_ENTRY	Configuration Table Entry
0x1C	DEVICE_OC	Other Conditions Device Information (Common Memory)
0x1D	DEVICE_OA	Other Conditions Device Information (Attribute Memory)
0x1E	DEVICEGEO	Device Geometry (Common Memory)
0x1F	DEVICEGEO_A	Device Geometry (Attribute Memory)
0x20	MANFID	PCMCIA Manufacturers Identification
0x21	FUNCID	Function Identification
0x22	FUNCE	Function Extension
0xFF	END	Termination Tuple

Table 6.1: It would stand to reason that we could simply walk the TEW-222CF tuple chain and key on the tuple codes to get all of the raw CIS information we need. That's the easy part. Correctly interpreting the raw CIS data that we get can be a little tricky.

Tuple data is purposely arranged on even byte boundaries. The even-byte-only alignment makes things done with tuples easy for 8-bit setups like the AirDrop. That means that the very first tuple, a code tuple, is located at CIS address 0 (zero). The first link tuple resides at CIS address 0x02 and the first of any data bytes can be found at CIS address 0x04. The even-byte tuple arrangement for the TEW-222CF initial tuple entry is illustrated in Table 6.2.

Typical Tuple Layout			
Type	Addr	Data	Description
END	0x08	0xFF	End of Tuple
	0x07	x	
DATA	0x06	0x00	
	0x05	x	
DATA	0x04	0x00	
	0x03	x	
LINK	0x02	0x03	3 Bytes Follow
	0x01	x	
CODE	0x00	0x01	Device Tuple

Table 6.2: This is pretty simple stuff. The tuple chain is no more than a simple linked list. The odd bytes, notated with an 'x' in the table, contain nothing and are not read by the AirDrop driver code.

Now that you know how to walk the tuple chain, let's put some code behind our new-found tuple knowledge and inspect the TEW-222CF's CIS memory area.

First Steps with the AirDrop-P

Let's begin our tuple journey by taking care of some necessary business that relates to the PIC microcontroller's USART and I/O pins. Listing 6.1 is a code snippet from the *main* function of the AirDrop-P 802.11b driver. At this point, we're only concerned with the calls *init_USART1* and *init_cf_card*.

Code Snippet 6.1
****

```
void main(void)
{
        unsigned int i,temp,evstat_data;
        char rc;

        init_USART1();
        init_cf_card();
```

****
Code Snippet 6.1: The USART code used in the AirDrop driver modules is the very same code used by the Easy Ethernet devices I described in excruciating detail in my previous book "Networking and Internetworking with Microcontrollers". So, I'm not going to reinvent the wheel and hash out the minute details of the USART initialization process and the USART interrupt structure again here.

The only difference in the AirDrop-P USART code and the Easy Ethernet W USART code is the PIC microcontroller that it is written for. I'll take you through USART code the changes quickly here. However, if you need the "how-does-it-work" detail, reference the RS-232 chapters in my book "Networking and Internetworking with Microcontrollers". So, let's get started and assume that the *init_USART1* function has been called and begin at the top of the AirDrop-P's *init_USART1* function code.

Code Snippet 6.2
****

```
#define BAUD1    57600        //desired baud rate
#define FOSC     20000000     //oscillator frequency
#define DIVIDER1   ((unsigned int)(FOSC/(16 * BAUD1) -1))
//******************************************************************
//*    Init USART Function
//******************************************************************
void init_USART1(void)
{
```

```
SPBRG1 = DIVIDER1;        //load baud rate divisor
TRISC7 = 1;               //receive pin
TRISC6 = 0;               //transmit pin
TXSTA1 = 0x04;            //high speed baud rate
RCSTA1 = 0x80;            //enable serial port and serial port pins
USART_RxTail = 0x00;      //flush receive buffer
USART_RxHead = 0x00;
USART_TxTail = 0x00;      //flush transmit buffer
USART_TxHead = 0x00;
RC1IP = 1;                //receive interrupt = high priority
TX1IP = 1;                //transmit interrupt = high priority
RC1IE = 1;                //enable receive interrupt
PEIE = 1;                 //enable all unmasked peripheral interrupts
GIE = 1;                  //enable all unmasked interrupts
CREN1 = 1;                //enable USART1 receiver
TX1IE = 0;                //disable USART1 transmit interrupt
TXEN1 = 1;                //transmitter enabled
}
```

****

Code Snippet 6.2: This is all standard PIC microcontroller USART stuff with a little interrupt spice thrown in. This code works pretty much everywhere with just about any microcontroller that has its own internal USART and is not specific to the AirDrop.

The code shown in Code Snippet 6.2 sets the USART1 baud rate divisor, clears the receive buffer area and the transmit buffer area, prioritizes the transmit interrupt and the receive interrupt and enables the USART1 transmitter and receiver circuitry. If you've ever used a microcontroller USART before, there's absolutely nothing new here.

I've taken some shortcuts with the various PIC USART configuration bits and the PIC USART interrupt handling settings. So, to clear things up, in Table 6.3 I've assembled the various PIC configuration register layouts that all of the bits in Code Snippet 6.2 belong to.

TSXTA1: TRANSMIT STATUS AND CONTROL REGISTER 1							
CSRC	TX9	TXEN	SYNC	SENDB	BRGH	TMRT	TX9D
0	0	1	0	0	1	0	0

RCSTA1: RECEIVE STATUS AND CONTROL REGISTER 1							
SPEN	RX9	SREN	CREN	ADDEN	FERR	OERR	RX9D
1	0	0	1	0	0	0	0

IPR1: PERIPHERAL INTERRUPT PRIORITY REGISTER 1							
PSPIP	ADIP	RC1IP	TX1IP	SSPIP	CCP1IP	TMR2IP	TMR1IP
1	1	1	1	1	1	1	1

PIE1: PERIPHERAL INTERRUPT ENABLE REGISTER 1							
PSPIE	ADIE	RC1IE	TX1IE	SSPIE	CCP1IE	TMR2IE	TMR1IE
0	0	1	0	0	0	0	0

INTCON REGISTER							
GIE/GIEH	PEIE/GIEL	TMR0IE	INT0IE	RBIE	TMR0IF	INT0IF	RBIF
1	1	0	0	0	0	0	0

Table 6.3: I've relied on power-up defaults to place the configuration bits that I didn't twiddle at the correct logic level. All of my bit short cuts are made possible by a predefined compiler-specific PIC18LF8621 include file.

The USART interrupt routines are intended for use with character operations, which entail the use of the byte-oriented *sendchr* and *recvchar* functions found in the AirDrop driver code. The use of the USART interrupt-driven character functions is optional and you can still use the standard *printf* function to send character and formatted string data to the AirDrop's RS-232 port.

Once the USART is up, the next call goes out to the 802.11b CompactFlash NIC. So, let's move on and take a closer look at the bytes contained within the TEW-222CF 802.11b CompactFlash NIC's CIS.

Walking the Tuple Chain

The C statements in Code Snippet 6.3 define the local variables that will be used in the 802.11b CompactFlash NIC initialization procedure. The C statements in Code Snippet 6.3 also reset the 802.11b CompactFlash NIC electronics and initialize the input/output states of the PIC microcontroller's I/O ports.

Code Snippet 6.3
****

```
//**********************************************************
//*    PORT Definitions
//**********************************************************
#define AOUT        LATA
#define BOPEN       PORTB
#define COPEN       PORTC
#define data_in     PORTD
#define data_out    LATD
#define EOPEN       PORTE
#define PCTL        LATF
#define addr_hi     LATG
#define addr_lo     LATH
#define PCOMM       LATJ
//**********************************************************
//*    CF I/O Definitions
//**********************************************************
#define RSET        0x10
#define set_RSET    PCTL |= RSET
#define clr_RSET    PCTL &= ~RSET
//**********************************************************
//*    Flags
//**********************************************************
```

```
#define cisflag 0x04
#define bcisflag        flags & cisflag
#define setf_cis        flags |= cisflag
#define clrf_cis        flags &= ~cisflag
//**************************************************************
//*    PORT TRIS Definitions
//**************************************************************
#define TO_NIC       TRISD = 0x00
#define FROM_NIC     TRISD = 0xFF
#define SETUP_A      TRISA = 0x00;          \
                     LATA = 0xFF;
#define SETUP_C      TRISC = 0x80;
#define SETUP_F      TRISF = 0x20;          \
                     LATF = 0xFB;
#define SETUP_G      TRISG = 0x00;          \
                     LATG = 0xF0;
                     //CE1 = 0;
                     //CE2 = 1;
#define SETUP_H      TRISH = 0x00;          \
                     LATH = 0x00;
#define SETUP_J      TRISJ = 0xEF;          \
                     LATJ = 0xFF;

void init_cf_card(void)
{
 char cor_addr_lo,cor_addr_hi,cor_data;
 char cis_device;
 unsigned int x;
 unsigned int cis_addr;    // attribute memory index (EVEN ADDRS ONLY)
 char code_tuple;          // tuple type
 char link_tuple;          // number of bytes in tuple structure
 char tuple_link_cnt;      // tuple byte count
 char tuple_data;          // current tuple contents
 char funce_cnt;
 char mac_cnt;
 char rc;

set_RSET;                  //reset CF card
SETUP_A;                   //configure I/O ports
SETUP_C;
SETUP_F;
SETUP_G;
SETUP_H;
SETUP_J;
FROM_NIC;                  //set to receive from NIC
clr_RSET;                  //take CF card out of reset
clrf_cis;                  //clear cis flag
rc = 0;                    //clear result code

****************************
```

Code Snippet 6.3: No UFM's (Unidentified Flying Microcontrollers) here. This section of code initializes the 802.11b CompactFlash NIC and at the same time sets up the I/O direction of the PIC microcontroller's I/O ports. Again, this is standard PIC microcontroller fare and if you've ever done anything with a PIC, you've written this kind of code. The local variables needed by this function are defined at the beginning of the init_cf_card function. To keep things as simple as possible, I've tried to give all of the AirDrop variables logical names that relate closely to their function. So far, this 802.11b driver stuff is pretty easy, huh?

If you're new to coding C and to the PIC microcontroller, the code in Code Snippet 6.3 is nothing that you can't handle. All we're doing here is toggling the 802.11b CompactFlash NIC's *RESET* pin and giving the reset process ample time to complete by setting up the PIC microcontroller's I/O lines during the 802.11b CompactFlash NIC's reset time. We'll use the Fred-defined *cisflag* bit and *rc* variable to signal the availability of the CIS data. So, we must make sure that the *cisflag* and *rc* variable are set to a known state before we continue with the rest of the code in the *init_cf_card* function.

Following the 802.11b CompactFlash NIC reset, the first real code that works on the PRISM chipset complement inside the 802.11b CompactFlash NIC is a read operation aimed at retrieving the data contained within the first byte of the CIS memory area. In Code Snippet 6.4, we give the 802.11b CompactFlash NIC one second to cough up the first CIS tuple entry. Once we see the first legitimate CIS tuple code, that's a signal that the rest of the 802.11b CompactFlash NIC's CIS tuples are also available to us. Referencing Table 6.1 and Table 6.2, we know that the very first CIS byte should be 0x01, which identifies the Device tuple.

Code Snippet 6.4
****

```
//**********************************************************
//*    Macros
//**********************************************************
#define make8(var,offset)   ((unsigned int)var >> (offset * 8)) & 0x00FF
//**********************************************************
//*    Flags
//**********************************************************
#define cisflag 0x04
#define bcisflag        flags & cisflag
#define setf_cis        flags |= cisflag
#define clrf_cis        flags &= ~cisflag
//**********************************************************
//*    PORT Definitions
//**********************************************************
#define data_in        PORTD
#define PCTL           LATF
#define addr_hi        LATG
#define addr_lo        LATH
//**********************************************************
//*    CF I/O Definitions
//**********************************************************
```

```
#define OE              0x01
#define set_OE          PCTL |= OE
#define clr_OE          PCTL &= ~OE
//****************************************************************
//*   CIS
//****************************************************************
#define rd_code_tuple       rd_cf_reg

****************************
void wr_cf_addr(unsigned int addr)
{
 addr_hi = (make8(addr,1));
 addr_lo = addr & 0x00FF;
}

char rd_cf_reg(unsigned int reg_addr)
{
 char data,i;
 wr_cf_addr(reg_addr);
 clr_OE;
 i=1;
 while(--i);                //read access delay
 data = data_in;
 set_OE;
 return(data);
}
****************************

x = 500;              //wait for cis functionality to come up
do{
 delay_ms(2);
 if(--x == 0)
  {
   rc = 1;
   break;
  }
 code_tuple = rd_code_tuple(0x0000);
 if(code_tuple == 1)     //1 denotes the Device Tuple
  setf_cis;
}while(!(bcisflag));
if(rc)
{
 printf("\r\nCIS Entry Failed..");
 printf("\r\nDriver Halted..");
 while(1);
}

****************************
```

Code Snippet 6.4: Following a successful reset, one second is more than enough time to allow the 802.11b CompactFlash NIC's internal electronics to get their act together and offer up the first CIS tuple byte. Attempting to read the initial tuple is our very first encounter with the actual inner workings of the 802.11b CompactFlash NIC.

In Code Snippet 6.4, we continually read the CIS memory area at address 0x0000 for one second. The *rd_code_tuple* function calls the *wr_cf_addr* function, which places the CIS memory area address we wish to read on the 802.11b CompactFlash NIC's address lines. The 802.11b CompactFlash NIC is not in I/O mode at this time and we are attempting to read Attribute Memory. Therefore, we must utilize the **OE* and **WE* CompactFlash card read/write control lines to gain access to the CIS memory area. After setting up the CompactFlash card address lines and asserting the CompactFlash card's **OE* pin, we implement a very short read access delay before reading in the CIS data. Once we have the data (or think we have the data), the OE pin is driven high to end the read cycle.

If a CIS read results in the variable *code_tuple* being equal to 0x01, the *cisflag* bit is set and the reading/timing loop is exited with a result code of 0x00. If one second flies by and the *code_tuple* variable is anything other than 0x01, the AirDrop driver assumes that there is no valid CIS information to be had, and after signaling the user via the AirDrop's serial port of the bad situation, the AirDrop driver code goes into a never-ending do-nothing loop. If the CIS is invalid or the 802.11b CompactFlash NIC isn't plugged into the CompactFlash card connector, there's no reason to continue anyway.

If, in fact, we retrieve the "golden" 0x01 into the *code_tuple* variable, we're ready to start parsing the CIS tuples. Before we roll the AirDrop-P out to the runway and prepare for flight, let's do some catch-up with the AirDrop-A driver source code.

Same Tuples, Different Microcontroller

Code Snippet 6.1 kicks off the AirDrop-P 802.11b driver and the same can be said for the Air-Drop-A. The interrupt routines and transmit/receive buffers for the AirDrop-P and AirDrop-A are also logically identical with the only differences being in the manufacturer names of the USART registers.

As you can see in Code Snippet 6.5, the AirDrop-A USART initialization is a bit less bulky due to the inclusion of the interrupt bit settings inside the AVR USART register value declarations and a pair of global interrupt enable/disable functions.

Code Snippet 6.5
****

```
AirDrop-A Driver Source Code
*****************************

#pragma interrupt_handler USART_RX_interrupt:iv_USART1_RX
#pragma interrupt_handler USART_TX_interrupt:iv_USART1_UDRE

void init_usart1(void)
{
 CLI();                     //disable interrupts
 UCSR1B = 0x00;             //disable while setting baud rate
 UCSR1A = 0x00;
 UCSR1C = 0x06;             //8-bits, no parity
```

```
UBRR1L = 0x07;              //set baud rate lo
UBRR1H = 0x00;              //set baud rate hi
UCSR1B = 0x98;

USART_RxTail = 0x00;        //flush receive buffer
USART_RxHead = 0x00;
USART_TxTail = 0x00;        //flush transmit buffer
USART_TxHead = 0x00;
SEI();                      //enable interrupts
}
```

****
Code Snippet 6.5: Again, this is pretty simple stuff. The code shown here disables the global interrupt mechanism, sets a baud rate of 57600 Kbps with communications parameters of 8 data bits, no parity bit, one stop bit, turns on the receive interrupt, flushes the transmit and receive buffers and reenables the global interrupt mechanism.

Table 6.4 lays out the AVR USART bit settings.

UCSR1A: USART Control and Status Register A							
RXC1	TXC1	UDRE1	FE1	DOR1	UPE1	U2X1	MPCM1
0	0	0	0	0	0	0	0

UCSR1B: USART Control and Status Register B							
RXCIE1	TXCIE1	UDRIE1	RXEN1	TXEN1	UCSZ12	RXB81	TXB81
1	0	0	1	1	0	0	0

UCSR1C: USART Control and Status Register C								
–	UNSEL1	UPM11	UPM10	USBS1	UCSZ11	UCSZ10	UCPOL1
0	0	0	0	0	1	1	0

Table 6.4: Data sheets are free and there's no good reason to repeat their contents in the text of this book you just paid for. If you need greater detail, all of the AVR acronyms in this table are explained fully in the ATmega128L data sheet.

As with AirDrop-P USART operation, you can find the details of AirDrop-A USART operation by reading the RS-232 section of my *Networking and Internetworking with Microcontrollers* book. The AirDrop-A USART interrupt routines are also optional. If you decide not to ever use them, you can eliminate them from your final AirDrop-A 802.11b driver firmware.

OK...The AVR USART is ready to go and the call is made to *init_cf_card*. Everything flows within the AirDrop-A's AVR in a similar fashion to the AirDrop 802.11b driver firmware flow that is rushing through the internal electronics of the AirDrop-P's PIC. I have kept consistent variable and function naming conventions for both to the AirDrop modules. Only the manufacturer register and I/O pin names are different. The similarity of the AirDrop-A and AirDrop-P 802.11b driver source code is evident in Code Snippet 6.6.

Code Snippet 6.6
****

```
AirDrop-A Driver Source Code
****************************

//************************************************************
//*    PORT Definitions
//************************************************************
#define addr_lo      PORTA
#define PREG         PORTB
#define data_out     PORTC
#define data_in      PINC
#define PCOMM        PORTD
#define POPEN        PORTE
#define addr_hi      PORTF
#define PCTL         PORTG
//************************************************************
//*    CF I/O Definitions
//************************************************************
#define RSET         0x10
#define set_RSET     PCTL |= RSET
#define clr_RSET     PCTL &= ~RSET
//************************************************************
//*    Flags
//************************************************************
#define cisflag 0x04
#define bcisflag     flags & cisflag
#define setf_cis     flags |= cisflag
#define clrf_cis     flags &= ~cisflag
//************************************************************
//*    PORT TRIS Definitions
//************************************************************
#define TO_NIC DDRC = 0xFF;
#define FROM_NIC     DDRC = 0x00; \
                     data_out = 0xFF;
#define SETUP_A      DDRA = 0xFF; \
                     addr_lo = 0x00;
#define SETUP_B      DDRB = 0x40; \
                     PORTB = 0xBF;
#define SETUP_C      DDRC = 0x00; \
                     data_out = 0xFF;
#define SETUP_D      DDRD = 0x0A;
#define SETUP_E      DDRE = 0x02;
#define SETUP_F      DDRF = 0x07; \
                     addr_hi = 0xF8;
#define SETUP_G      DDRG = 0xFF; \
                     PCTL = 0xFF;

void init_cf_card(void)
{
```

```
char cor_addr_lo,cor_addr_hi,cor_data;
char cis_device;
unsigned int x;
unsigned int cis_addr;     // attribute memory index (EVEN ADDRS ONLY)
char code_tuple;           // tuple type
char link_tuple;           // number of bytes in tuple structure
char tuple_link_cnt;       // tuple byte count
char tuple_data;           // current tuple contents
char funce_cnt;
char mac_cnt;
char rc;
CLI();                     //disable all interrupts
set_RSET;                  //reset CF card
SETUP_A;                   //configure I/O ports
SETUP_B;
SETUP_C;
SETUP_D;
SETUP_E;
SETUP_F;
SETUP_G;
FROM_NIC;                  //set to receive from NIC
clr_RSET;                  //take CF card out of reset
SEI();                     //enable interrupts
clrf_cis;                  //clear cis flag
rc = 0;                    //clear result code
```

****

Code Snippet 6.6: There's not much difference between the PIC functions and the AVR code in this snippet. Notice, however, that I had to disable global interrupts while I setup the AVR's I/O ports during the CompactFlash card reset time.

The only difference in the AirDrop-A 802.11b driver and the AirDrop-P 802.11b driver up to this point is the addition of a pair of AVR disable/enable interrupt calls. Other than that, logically the AirDrop-A and AirDrop-P code segments are identical. In the AVR 802.11b driver source code that takes us up to the point of beginning to read the tuples, there is only one deviation from the AirDrop-P 802.11b driver source code, a masking of the upper three CompactFlash card address lines in the *wr_cf_addr* function. The logical AND mask operation is illustrated in Code Snippet 6.7.

Code Snippet 6.7
****

```
AirDrop-A Driver Source Code
****************************

#define nop              NOP()
//**************************************************************
//*    Macros
//**************************************************************
#define make8(var,offset)   ((unsigned int)var >> (offset * 8)) & 0x00FF
```

```
//*************************************************************
//*     Flags
//*************************************************************
#define cisflag      0x04
#define bcisflag     flags & cisflag
#define setf_cis     flags |= cisflag
#define clrf_cis     flags &= ~cisflag
//*************************************************************
//*     PORT Definitions
//*************************************************************
#define addr_lo      PORTA
#define data_in      PINC
#define addr_hi      PORTF
#define PCTL         PORTG
//*************************************************************
//*     CF I/O Definitions
//*************************************************************
#define OE           0x01
#define set_OE       PCTL |= OE
#define clr_OE       PCTL &= ~OE

void wr_cf_addr(unsigned int addr)
{
 addr_hi = (make8(addr,1)) & 0x07;
 addr_lo = addr & 0x00FF;
}

char rd_cf_reg(unsigned int reg_addr)
{
 char data,i;
 wr_cf_addr(reg_addr);
 clr_OE;
 i=1;
 while(--i);
 data = data_in;
 set_OE;
 NOP();
 return(data);
}
****************************

x = 500;              //wait for cis functionality to come up
do{
 delay_ms(2);
 if(--x == 0)
 {
  rc = 1;
  break;
 }
 code_tuple = rd_code_tuple(0x0000);
 if(code_tuple == 1)     //1 denotes the Device Tuple
  setf_cis;
```

```
}while(!(bcisflag));
if(rc)
{
 printf("\r\nCIS Entry Failed..");
 printf("\r\nDriver Halted..");
 while(1);
}
```

****

Code Snippet 6.7: For very obvious reasons, the AirDrop-A's AVR is not pin compatible with the PIC18LF8621. Note that in the AirDrop-A's wr_cf_addr function that we logically AND the high byte of the address to make sure we compensate for the open I/O pin, PF3.

OK…The Device tuple has been identified and the PIC pilots and AVR aviators are side by side on the runway. We know what the CIS landmarks look like as they were put before us in Table 6.1. So, let's write some code to survey the CIS and look for the tuple code landmarks before committing to writing a formal CIS parsing routine.

CIS Reconnaissance

The idea here is to get a dump of all of the TEW-222CF's tuples and try to navigate through the dump using the tuple codes shown in Table 6.1. I've written some code in Code Snippet 6.8 to do just that. Once we've identified the actual tuples that exist within the TEW-222CF's CIS memory area we can then write some more specific CIS parsing code that keys on the tuple codes we discovered.

Code Snippet 6.8
****

```
#define END_TUPLE            0xFF
//*******************************************************
//*    CIS
//*******************************************************
#define rd_code_tuple       rd_cf_reg
#define rd_link_tuple       rd_cf_reg
//*******************************************************
//*    PORT Definitions
//*******************************************************
#define data_in      PORTD
#define PCTL         LATF
#define addr_hi       LATG
#define addr_lo       LATH
//*******************************************************
//*    CF I/O Definitions
//*******************************************************
#define OE           0x01
#define set_OE       PCTL |= OE
#define clr_OE       PCTL &= ~OE
//*********************************************************
//*    CIS
```

```
//*****************************************************************
#define rd_code_tuple        rd_cf_reg

*****************************
void wr_cf_addr(unsigned int addr)
{
 addr_hi = (make8(addr,1));
 addr_lo = addr & 0x00FF;
}


char rd_cf_reg(unsigned int reg_addr)
{
 char data,i;
 wr_cf_addr(reg_addr);
 clr_OE;
 i=1;
 while(--i);                //read access delay
 data = data_in;
 set_OE;
 return(data);
}
*****************************

//***********************************************************
//*    CIS Tuple Collection Utility
//***********************************************************
unsigned int index;
unsigned int cis_addr;       // attribute memory index (EVEN ADDRS ONLY)
char code_tuple;             // tuple type
char link_tuple;             // number of bytes in tuple structure
char tuple_link_cnt;         // tuple byte count
char tuple_data[1024];       //tuple dump buffer

cis_addr   = 0;              // CIS starts at location 0 of Attribute memory
index = 0;                   //pointer into tuple dump buffer
do{
     code_tuple = rd_code_tuple(cis_addr);       //read code tuple
     tuple_data[index++] = code_tuple;           //store code tuple in buffer
     cis_addr+=2;                                 //increment addr to link tuple
     link_tuple = rd_link_tuple(cis_addr);       //read link tuple
     tuple_data[index++] = link_tuple;           //store link tuple in buffer
     cis_addr+=2;                                 //increment addr to data tuple

   for(tuple_link_cnt=0;tuple_link_cnt<link_tuple;++tuple_link_cnt)
   {
     tuple_data[index++] = rd_tuple_data(cis_addr);
     cis_addr+=2;
   }
}while(code_tuple != END_TUPLE);


*****************************
```

Code Snippet 6.8: This little bit of code captures all of the tuple bytes and places them in a buffer called tuple_data.

The CIS Tuple Collection Utility code shown in Code Snippet 6.8 reads every tuple byte and stores them in a buffer called *tuple_data*. Note that the CIS Tuple Collection Utility code only works against even bytes in the CIS memory area. The code tuple is read first and stored in the buffer. The tuple buffer pointer (*index*) is then incremented to point to the next byte in the tuple buffer. The cis_*addr*, which addresses tuples in the CIS memory area is incremented by two to point to the tuple element, which always resides on an even-byte boundary. After reading and storing the link tuple, the data tuple area is accessed, read and stored in the tuple buffer according to the byte value contained within the link tuple. This process of addressing, reading and storing the tuples continues until the end tuple is encountered. The results of running the CIS Tuple Collection Utility code can be seen in the CIS memory area hexadecimal dump, Hex Dump 6.1.

Hex Dump 6.1

```
****************************
TRENDNET TEW-222CF CIS DUMP
CIS DATA BEGINS AT 0x02FA
 Address   00 01 02 03 04 05 06 07 08 09 0A 0B 0C 0D 0E 0F      ASCII

  02F0     00 00 00 00 00 00 00 00 00 00 01 03 00 00 FF 17  ........ ........
  0300     04 67 5A 08 FF 1D 05 01 67 5A 08 FF 15 27 05 00  .gZ..... gZ...'..
  0310     57 69 72 65 6C 65 73 73 00 4C 41 4E 20 41 64 61  Wireless .LAN Ada
  0320     70 74 65 72 00 56 65 72 73 69 6F 6E 20 30 31 2E  pter.Ver sion 01.
  0330     30 32 00 00 FF 20 04 56 01 02 00 21 02 06 00 22  02... .V ...!..."
  0340     02 01 07 22 05 02 40 42 0F 00 22 05 02 80 84 1E  ..."..@B ..".....
  0350     00 22 05 02 60 EC 53 00 22 05 02 C0 D8 A7 00 22  ."..`.S. ".....
  0360     02 03 07 22 08 04 06 00 E0 98 BD 61 8A 22 02 05  ..."..... ...a."..
  0370     01 1A 07 03 01 E0 03 00 00 01 1B 0F C1 01 19 76  ........ .......v
  0380     C5 4B D5 19 36 36 05 46 7F FF FF FF FF 00 00 00  .K..66.F ........
****************************
```
Hex Dump 6.1: You can easily see the DEVICE tuple code (0x01) at address 0x02FA. Note that following the VERS_1 tuple (0x15) the end of tuple structures is no longer identified with 0xFF.

The DEVICE tuple is all about Common Memory. Judging by the data tuples' values of 0x00, I'd say there is nothing to be said about Common Memory as far as the TEW-222CF is concerned. The zero values also imply that Common Memory is not something that resides within the TEW-222CF's electronics and if it actually does exist, it is never used.

The next code tuple we encounter begins at address 0x02FF. The 0x17 at this memory location equates to the code tuple called DEVICE_A. The DEVICE_A tuple contains three bytes of Attribute Memory information. After a couple of days of looking at those three bytes, I finally think I matched them up to some Linux code I had seen. The 0x67 at address 0x0301 in Hex Dump 6.1 says that there is an SRAM device involved with the Attribute Memory. However, the write speed bits (0bxxxxx111) of the 0x67 byte didn't match up to anything I could find in the Linux documentation I was studying.

I had been working on other places within the CIS using this really neat code snippet from the *cistpl.c – 16-bit PCMCIA Card Information Structure Parser* written by David A. Hinds. David's code that we are interested in is shown in Code Snippet 6.9.

Code Snippet 6.9

****

```
39 static const u_char mantissa[] = {
40     10, 12, 13, 15, 20, 25, 30, 35,
41     40, 45, 50, 55, 60, 70, 80, 90
42 };
43
44 static const u_int exponent[] = {
45     1, 10, 100, 1000, 10000, 100000, 1000000, 10000000
46 };
47
48 /* Convert an extended speed byte to a time in nanoseconds */
49 #define SPEED_CVT(v) \
50     (mantissa[(((v)>>3)&15)-1] * exponent[(v)&7] / 10)
51 /* Convert a power byte to a current in 0.1 microamps */
52 #define POWER_CVT(v) \
53     (mantissa[((v)>>3)&15] * exponent[(v)&7] / 10)
```

****

Code Snippet 6.9: This little bit of code really helped me get a handle on converting the CIS bytes to voltages, speeds and currents. It also answers the question as to what constitutes a mantissa and exponent in the CIS tuple data. Thanks, David.

The flow in the *cistpl.c* document swerved off to do a macro calculation when it saw the least significant three bits of the first DEVICE_A data byte all set to one. So, after seeing that tearing apart the bits in the 0x67 byte were getting me nowhere, I decided to chase the bits in the byte that followed at address 0x0302, which is 0x5A. Here's what I came up with.

I'll apply what I can of the macros David wrote in his Linux code. I assumed I was looking at a speed parameter in the 0x5A byte. And, since I know it is related to SRAM, I may be looking for 0x5A to translate to 55 nS. Recall that earlier, in Chapter 4, we took the TEW-222CF 802.11b CompactFlash NIC apart and positively identified a 55 nS SRAM on the TEW-222CF printed circuit board. With those assumptions in mind, I broke the 0x5A byte down into its binary (0b01011010) form. If I shave off the least significant bits of 0x5A by shifting right three times (0b00001011) and logically AND the shifted result with 15 (0b00001011 AND 0b00001111), I end up with 0b00001011, or 0x0B. Subtracting 1 from the shifted and logically ANDed result yields 0b00001010, or 0x0A. The value of 50 decimal lies at index 0x0A in the *mantissa* array shown in Code Snippet 6.9.

The next step involved calculating the *exponent* index value. Logically ANDing 0b01011010 with 0b00000111 yields 0b00000010, or simply 2 decimal. That puts my *exponent* value at 100, which is the third element (0,1,2) of the *exponent* array. Multiplying 50, my *mantissa*, by my exponent of 100 and then dividing by 10 results in 500 time units. We must convert the time units to nanoseconds. That is done by multiplying the 500 time units by .000000001, or 1 nS, which gives us a 500 nS CompactFlash card write cycle time. The host will always assume a 300 nS CompactFlash card read access cycle time at this point. As it turns out, this has very little to do with the speed of the SRAM as the relatively large read and write cycle times obviously take other internal delay factors into account. The final byte of tuple data, 0x08, is a device size parameter. The least significant three bits represent the number of 512-byte units of address space. In the case of the TEW-222CF, that's 8 units of 512, or 2K of address space. The address space figure means nothing to us as far as the AirDrop driver is concerned but it's nice to know what it represents.

The next code tuple, 0x1D, at address 0x0305 signifies the beginning of the DEVICE_OA tuple structure. This tuple structure is basically the same as the DEVICD_A tuple with CompactFlash card voltage information thrown in. The 0x01 at address 0x0307 in Hex Dump 6.1 alludes to wait function use of devices in the Attribute Memory address space but does not specify +3.3 VDC operation.

From the DEVICE_OA tuple forward there is very little information we can put to use in our AirDrop driver code until we encounter tuple 0x1A, the CONFIG tuple, and code tuple 0x22, the FUNCE tuple structure. Bytes 4–7 of the CONFIG tuple contain the COR (Configuration Option Register) base address. We must know the COR base address value in order to enable the 802.11b CompactFlash NIC. The COR base value begins at address 0x0375 in Hex Dump 6.1 and is arranged with the low order bytes first. Thus, the 4-bytes E0 03 00 00 hexadecimal translate to a COR base value of 0x000003E0.

There are multiple FUNCE tuple structures. The only one that could mean anything to the AirDrop driver is the FUNCE that contains the 802.11b CompactFlash NIC's MAC address. Although the 802.11b CompactFlash NIC's MAC address can be programmatically obtained from this tuple, it is safer to enable the 802.11b CompactFlash NIC and then obtain the 802.11b CompactFlash NIC's MAC address via a command. Fifty percent of the 802.11b CompactFlash NICs I tested did not always put the correct MAC address in the tuple. In the case of the TEW-222CF, the correct 6-byte MAC address can be found in Hex Dump 6.1 beginning at address 0x0367. The 0x06 at address 0x0366 defines the number of bytes in the MAC address that follows.

Tuple 0x1B, the CFTABLE_ENTRY tuple, is very interesting. The fifth byte of the CFTABLE_ENTRY tuple structure, 0x76, breaks down as follows:

Power Descriptor Byte

x	PWRDWN I	PEAK I	AVG I	STATIC I	MAX V	MIN V	NOM V
0	1	1	1	0	1	1	0

Where:

x	= NOT USED
PWRDWN I	= POWER DOWN CURRENT
PEAK I	= PEAK CURRENT
AVG I	= AVERAGE CURRENT
STATIC I	= STATIC CURRENT
MAX V	= MAXIMUM OPERATING VOLTAGE
MIN V	= MINIMUM OPERATING VOLTAGE
NOM V	= NOMINAL OPERATING VOLTAGE

The values of the aforementioned voltage and current parameters are located in the bytes that immediately follow this descriptor byte beginning with the parameter represented by the least significant bit, NOM V.

Since the nominal operating voltage bit is zero, the fun begins with bytes 6 and 7 of the CFTABLE_ENTRY tuple structure, which is supposedly the minimum operating voltage of the 802.11b CompactFlash NIC. Here's where we once again turn to David A. Hinds' cool little Linux macros. Byte 6 contains our mantissa and exponent values, which allow us to translate the hexadecimal coding into meaningful data a human can understand. Byte 6 of the CFTABLE_ENTRY tuple is 0xC5 or 0b11000101. Applying the shift right (0b00011000) and mask operation (0b00011000 AND 0b00001111) leaves the *mantissa* index value at 0b00001000, or 8 decimal. The *exponent* index value is 0b101, or 5 decimal. Byte 7 represents hundredths of a volt and here happens to have a value of 0x4B, or 75 decimal, which equates to .75 volts. Here's where we jump off of David's boat for a bit and throw the *mantissa* and *exponent* arrays overboard. Instead of using David's *exponent* array, we use the exponent value as a multiplier of the mantissa and then divide the whole thing by 10 and add the fractional second byte to the result. Let's do the simple math:

mantissa	= (0b11000101 >> 3) & 0b00001111
	= 0b00011000 & 0b00001111
	= 0b00001000 or 0x08 or 8
exponent	= 0b11000101 & 0b00000111
	= 0b101 or 0x05 or 5
minimum voltage	= ((mantissa * exponent) / 10) + fractional byte
	= ((8 * 5) / 10) + (75 / 100)
	= 4.75 V

You can't trust everything you see in the CIS. The TEW-222CF's CIS is telling us that the TEW-222CF cannot operate at the 3.3-volt level. That's a bogus statement.

Let's apply our newly modified voltage conversion math on the next set of bytes, which represent the maximum operating voltage of the 802.11b CompactFlash NIC. The byte 0xD5 (0b11010101) after shifting and masking gives us a mantissa of 0b1010 (0b00011010 AND 0b00001111), or 10 decimal, with an exponent value of 0b101, or 5 decimal. Byte 9 is the fractional voltage byte and has a value of 0x19, or 25 decimal. Applying our voltage formula:

mantissa	= (0b11010101 >> 3) & 0b00001111
	= 0b00011010 & 0b00001111
	= 0b00001010 or 0x0A or 10
exponent	= 0b11010101 & 0b00000111
	= 0b101 or 0x05 or 5

maximum voltage = ((mantissa * exponent) / 10) + fractional byte
= ((10 * 5) / 10) + (25 / 100)
= 5.25V

The maximum value matches up with the TEW-222CF's marketing data sheet figure. As it turns out, in the two voltage conversion examples we've looked at we could have used David's *mantissa* array to provide our (mantissa * exponent) number. However, our little modified voltage conversion formula worked out just fine.

Let's jump back into David's boat and recover those water-logged *mantissa* and *exponent* arrays. The next byte of the CFTABLE_ENTRY tuple we want to look at is byte 10, which represents the 802.11b CompactFlash NIC's average current. Byte 11 is the 802.11b CompactFlash NIC's peak current and byte 12 represents the 802.11b CompactFlash NIC's power down current. David's macros should work here without modification. Let's do the math:

average current = (mantissa[((0x36)>>3)&15] * exponent[(0x36)&7] / 10)
= (mantissa[6] * exponent[6] / 10)
= (30 * 1000000 / 10)
= 3000000 * .1 uA
= .300 A or 300 mA

OK…So far, so good. Byte 11 of the CFTABLE_ENTRY tuple, the peak current byte, has the same value of 0x36 and thus results in a peak current value of 300 mA as well:

peak current = (mantissa[((0x36)>>3)&15] * exponent[(0x36)&7] / 10)
= (mantissa[6] * exponent[6] / 10)
= (30 * 1000000 / 10)
= 3000000 * .1 uA
= .300 A or 300 mA

Let's do the math and calculate the 802.11b CompactFlash NIC's power down current:

power down current = (mantissa[((0x05)>>3)&15] * exponent[(0x05)&7] / 10)
= (mantissa[0] * exponent[5] / 10)
= (10 * 100000 / 10)
= 100000 * .1 uA
= .01 A or 10 mA

Using the TEW-222CF marketing data sheet as a guide, it seems that the peak current figure is in the ballpark. The TEW-222CF minimum operating voltage is bogus and that is re-iterated in the TEW-222CF marketing data sheet specs. Also, the power down current is stated to be 50 mA in the TEW-222CF data sheet and shows as 10 mA in the CIS. Hmmm…

Dumping Linksys WCF12 Tuples

This CIS stuff is pretty interesting. So, let's examine a hexadecimal dump of the Linksys WCF12's CIS Memory area and see if we can spot any consistencies or inconsistencies.

Hex Dump 6.2

```
*****************************
LINKSYS WCF12 CIS DUMP
CIS DATA BEGINS AT 0x2FA
Address   00 01 02 03 04 05 06 07 08 09 0A 0B 0C 0D 0E 0F       ASCII

02F0    00 00 00 00 00 00 00 00 00 00 01 03 00 00 FF 17  ........ ........
0300    04 67 5A 08 FF 1D 05 03 67 5A 08 FF 15 28 01 00  .gZ..... gZ...(..
0310    4C 69 6E 6B 73 79 73 00 57 69 72 65 6C 65 73 73  Linksys. Wireless
0320    20 43 6F 6D 70 61 63 74 46 6C 61 73 68 20 43 61   Compact Flash Ca
0330    72 64 00 00 00 FF 20 04 8A 02 73 06 21 02 06 00  rd.... . ..s.!...
0340    22 02 01 07 22 05 02 40 42 0F 00 22 05 02 80 84  "..."..@ B.."....
0350    1E 00 22 05 02 60 EC 53 00 22 05 02 C0 D8 A7 00  .."..`.S ."......
0360    22 02 03 07 22 08 04 06 00 0C 41 DD 68 5E 22 02  "..."... ..A.h^".
0370    05 01 1A 07 03 01 E0 03 00 00 01 1B 10 C1 01 19  ........ ........
0380    77 B5 1E 35 B5 3C 36 36 05 46 FF FF FF FF FF 00  w..5.<66 .F......
*****************************
```

Hex Dump 6.2: I plugged in a Linksys WCF12 802.11b CompactFlash NIC into an AirDrop-P and ran the same CIS Tuple Collection Utility code from Code Snippet 6.8 to generate the bytes in this hexadecimal dump.

The WCF12's write access timing matches that of the TEW-222CF. However, the Attribute Memory voltage byte at address 0x0307, 0x03, signifies +3.3 VDC operation used with the wait function. I know that the WCF12 runs at +3.3 VDC. So, I must assume the least significant bit of the 0x03 byte signifies +3.3 VDC operation.

The CONFIG tuple, 0x1A, at address 0x0372 shows us that the COR base address is 0x000003E0 just like the TEW-222CF. The COR base address information is the only data the AirDrop driver needs from the CIS Memory area. However, let's have some fun and look at the WCF12's voltage and current parameters. Here's a look at the power parameter descriptor byte, 0x77 at address 0x0380, for the WCF12:

Power Descriptor Byte

x	PWRDWN I	PEAK I	AVG I	STATIC I	MAX V	MIN V	NOM V
0	1	1	1	0	1	1	1

We'll start by computing the WCF12's nominal voltage from the bytes at address 0x0381 and 0x0382, 0xB5 and 0x1E:

mantissa
$$= (0b10110101 >> 3) \,\&\, 0b00001111$$
$$= 0b00010110 \,\&\, 0b00001111$$
$$= 0b00000110 \text{ or } 0x06 \text{ or } 6$$

exponent
$$= 0b11000101 \,\&\, 0b00000111$$
$$= 0b101 \text{ or } 0x05 \text{ or } 5$$

nominal voltage
$$= ((\text{mantissa} * \text{exponent}) / 10) + \text{fractional byte}$$
$$= ((6 * 5) / 10) + (30 / 100)$$
$$= 3.30V$$

That makes sense. However, at first glance the 0x35 byte at address 0x0383 doesn't. This should be two bytes that we can translate into a minimum voltage value. According to the rules of our voltage conversion formula, the hundredths-of-a-volt second byte, 0xB5, can't exceed the value of 99 decimal. So, the 0x35/0xB5 combination cannot be a valid voltage byte set. Even coupling the renegade 0x35 with the previous 0xB5 makes no sense as the minimum voltage should fall below the nominal voltage. Since 0x35 is larger than 0x1E, that ain't gonna happen. What if there's no fractional voltage value to be added? You would think 0x35 followed by 0x00 would be appropriate. Let's assume this is a valid minimum voltage byte and run it through the voltage conversion formula:

mantissa	= (0b00110101 >> 3) & 0b00001111
	= 0b00000110 & 0b00001111
	= 0b00000110 or 0x06 or 6
exponent	= 0b11000101 & 0b00000111
	= 0b101 or 0x05 or 5
minimum voltage	= ((mantissa * exponent) / 10) + fractional byte
	= ((6 * 5) / 10) + (0 / 100)
	= 3.00V

Well, how about that! I wonder if the most significant bit being set signifies a two-byte voltage value? Hmmm…Now things fall into place nicely. The most significant bit is set in the byte 0xB5 at address 0x0384. That must mean that the maximum operating voltage value is obtained by coupling the next two bytes, 0xB5 and 0x3C:

mantissa	= (0b10110101 >> 3) & 0b00001111
	= 0b00010110 & 0b00001111
	= 0b00000110 or 0x06 or 6
exponent	= 0b11000101 & 0b00000111
	= 0b101 or 0x05 or 5
maximum voltage	= ((mantissa * exponent) / 10) + fractional byte
	= ((6 * 5) / 10) + (60 / 100)
	= 3.60V

The last three bytes, 0x36, 0x36, 0x05, represent the Linksys WCF12's average current, peak current and power down current, respectively:

average current	= (mantissa[((0x36)>>3)&15] * exponent[(0x36)&7] / 10)
	= (mantissa[6] * exponent[6] / 10)
	= (30 * 1000000 / 10)
	= 3000000 * .1 uA
	= .300A or 300 mA
peak current	= (mantissa[((0x36)>>3)&15] * exponent[(0x36)&7] / 10)
	= (mantissa[6] * exponent[6] / 10)

$$= (30 * 1000000 / 10)$$
$$= 3000000 * .1 \text{ uA}$$
$$= .300\text{A or } 300 \text{ mA}$$

power down current $\quad$ = (mantissa[((0x05)>>3)&15] * exponent[(0x05)&7] / 10)

$$= (\text{mantissa}[0] * \text{exponent}[5] / 10)$$
$$= (10 * 100000 / 10)$$
$$= 100000 * .1 \text{ uA}$$
$$= .01\text{A or } 10 \text{ mA}$$

The Linksys WCF12 marketing data sheet states that the operating voltage is +3.3 VDC and the current consumption is 250 mA. I suspect the 250 mA figure is receive operating current. In any event, our computed figures are totally on for the Linksys 802.11b CompactFlash NIC. Notice also that the Linksys WCF12 reported a correct MAC address with the six bytes beginning at address 0x0368 in Hex Dump 6.2.

Dumping Netgear MA701 Tuples

We're on a roll. So, let's dump some Netgear MA701 tuples and see if we can decrypt the tuple data. I've supplied the raw material in Hex Dump 6.3.

Hex Dump 6.3
****
```
NETGEAR MA701 CIS DUMP
CIS DATA BEGINS AT 0x02FA
Address   00 01 02 03 04 05 06 07 08 09 0A 0B 0C 0D 0E 0F       ASCII

02F0    00 00 00 00 00 00 00 00 00 00 01 03 00 00 FF 17   ........ ........
0300    04 67 5A 08 FF 1D 05 03 67 5A 08 FF 15 23 05 00   .gZ..... gZ...#..
0310    4E 45 54 47 45 41 52 00 4D 41 37 30 31 20 57 69   NETGEAR. MA701 Wi
0320    72 65 6C 65 73 73 20 43 46 20 43 61 72 64 00 00   reless C F Card..
0330    FF 20 04 01 D6 02 00 21 02 06 00 22 02 01 07 22   . .....! ..."..."
0340    05 02 40 42 0F 00 22 05 02 80 84 1E 00 22 02 03   ..@B..". ......"..
0350    07 22 08 04 06 00 00 00 00 00 00 22 02 05 01 1A   ."...... ..."....
0360    07 03 01 E0 03 00 00 01 1B 10 C1 01 19 77 B5 1E   ........ .....w..
0370    35 B5 3C 36 3E 06 46 7F FF FF FF 00 00 00 00 00   5.<6>.F. ........
0380    00 00 00 00 00 00 00 00 00 00 00 00 00 00 00 00   ........ ........
```
****

Hex Dump 6.3: The Netgear MA701 CIS is begging to be different. Note the zeros in the MAC address fields and the slightly different current values.

Right off, I checked for a valid MAC address. As you can see, beginning at address 0x0355 there are six bytes of 0x00 where a MAC address should be. My MA701's MAC address of *00095BB3F945* is nowhere to be found in Hex Dump 6.3.

The write access speeds are identical to what we have seen in the previous dumps and the COR base address is defined as 0x000003E0 beginning at address 0x0363 in Hex Dump 6.3. The power parameter descriptor byte at address 0x036D of Hex Dump 6.3, 0x77, tells us that the nominal voltage value is included in the CIS voltage bytes.

Power Descriptor Byte

x	PWRDWN I	PEAK I	AVG I	STATIC I	MAX V	MIN V	NOM V
0	1	1	1	0	1	1	1

The MA701 voltage bytes look just like the Linksys WCF12 voltage bytes. Let's not make any assumptions and do the math anyway. Let's compute the MA701's nominal voltage from the bytes at address 0x036E and 0x036F, 0xB5 and 0x1E:

mantissa	= (0b10110101 >> 3) & 0b00001111
	= 0b00010110 & 0b00001111
	= 0b00000110 or 0x06 or 6
exponent	= 0b11000101 & 0b00000111
	= 0b101 or 0x05 or 5
nominal voltage	= ((mantissa * exponent) / 10) + fractional byte
	= ((6 * 5) / 10) + (30 / 100)
	= 3.30V

The next voltage value according to the MA701's power descriptor byte is the minimum voltage, which is represented by 0x35 at address 0x0370 of Hex Dump 6.3. Note the most significant bit of the minimum voltage value byte is not set. We know what to do:

mantissa	= (0b00110101 >> 3) & 0b00001111
	= 0b00000110 & 0b00001111
	= 0b00000110 or 0x06 or 6
exponent	= 0b11000101 & 0b00000111
	= 0b101 or 0x05 or 5
minimum voltage	= ((mantissa * exponent) / 10) + fractional byte
	= ((6 * 5) / 10) + (0 / 100)
	= 3.00V

Just as before, everything continues to fall into place. The most significant bit of the byte at address 0x0371 of Hex Dump 6.3 is set. So, next up are the maximum voltage bytes beginning at address 0x0371 of Hex Dump 6.3, 0xB5 and 0x3C:

mantissa	= (0b10110101 >> 3) & 0b00001111
	= 0b00010110 & 0b00001111
	= 0b00000110 or 0x06 or 6
exponent	= 0b11000101 & 0b00000111
	= 0b101 or 0x05 or 5
maximum voltage	= ((mantissa * exponent) / 10) + fractional byte
	= ((6 * 5) / 10) + (60 / 100)
	= 3.60V

Things look a bit different here as the last three bytes, 0x36, 0x3E, 0x06, represent differing current values than we've seen previously. Here's a rundown of the MA701's average current, peak current and power down current:

average current = (mantissa[((0x36)>>3)&15] * exponent[(0x36)&7] / 10)
= (mantissa[6] * exponent[6] / 10)
= (30 * 1000000 / 10)
= 3000000 * .1 uA
= .300A or 300 mA

peak current = (mantissa[((0x3E)>>3)&15] * exponent[(0x3E)&7] / 10)
= (mantissa[7] * exponent[6] / 10)
= (35 * 1000000 / 10)
= 3500000 * .1 uA
= .350A or 350 mA

power down current = (mantissa[((0x06)>>3)&15] * exponent[(0x06)&7] / 10)
= (mantissa[0] * exponent[6] / 10)
= (10 * 1000000 / 10)
= 1000000 * .1uA
= .1A or 100mA

The power down current value stuck out as I thought that was quite high considering what we have already seen. Checking the MA701 marketing data sheet confirms our calculations. The MA701 marketing data sheet shows a power down current of 80 mA with a peak (transmit) current of 350 mA. Receive current was posted as less than 250 mA, which confirms our average current calculation.

So far our calculations are on the mark, which means our 802.11b CompactFlash NIC voltage, speed and current formulas work. We've got one more 802.11b CompactFlash NIC to dump. The Zonet ZCF100.

Dumping Zonet ZCF100 Tuples

This should be interesting. Take a look at the VERS_1 data area of Hex Dump 6.4. See the word "Eval-RevA". Hmmm…This 802.11b CompactFlash NIC has been on the market since June 2003. However, with that written in the CIS, we'll keep our eyes on this rascal.

You're probably pretty good with CIS dumps by now. So, I won't be redundant with the speed/voltage/current math unless something deserves the attention. Starting at the top, we have the same 500 nS write access time bytes in the DEVICE and DEVICE_OA tuples, which are 0x5A at addresses 0x0303 and 0x0309. The COR base address of 0x000003E0 is also present within the CONFIG tuple at address 0x037D of Hex Dump 6.4. The Zonet ZCF1100 MAC address in the CIS tuple dump shown in Hex Dump 6.4 is not correct. My Zonet ZCF1100 sports a MAC address of *00026F09CC27C*.

Let's jump down to the CFTABLE_ENTRY tuple and check out the voltage bytes. The power parameter descriptor byte at address 0x0387 of Hex Dump 6.4, 0x77, tells us we can plan to see nominal, minimum, and maximum operating voltage bytes plus average, peak and power down current bytes. Comparing the voltage and current bytes with the Linksys WCF12 CIS dump reveals no discrepancies. The Zonet ZCF1100 marketing data sheet does not give current consumption numbers. However, we know our voltage/speed/current calculations are accurate. If we believe what the Zonet ZCF1100 CIS is telling us, the ZCF1100 is much like the Linksys WCF12 at the PRISM chipset level. In fact, the Linksys WCF12 and the Zonet ZCF1100 both are built around PRISM 3 chipsets.

Hex Dump 6.4

```
*****************************
ZONET ZCF100 CIS DUMP
CIS DATA BEGINS AT 0x2FA
 Address  00 01 02 03 04 05 06 07 08 09 0A 0B 0C 0D 0E 0F      ASCII

   02F0   00 00 00 00 00 00 00 00 00 00 01 03 00 00 FF 17   ........ ........
   0300   04 67 5A 08 FF 1D 05 03 67 5A 08 FF 15 2F 01 00   .gZ..... gZ.../..
   0310   57 4C 41 4E 00 31 31 4D 62 70 73 5F 50 43 2D 43   WLAN.11M bps_PC-C
   0320   61 72 64 5F 33 2E 30 00 49 53 4C 33 37 31 30 30   ard_3.0. ISL37100
   0330   50 00 45 76 61 6C 2D 52 65 76 41 00 FF 20 04 0B   P.Eval-R evA.. ..
   0340   00 00 71 21 02 06 00 22 02 01 07 22 05 02 40 42   ..q!..." ...".. @B
   0350   0F 00 22 05 02 80 84 1E 00 22 05 02 60 EC 53 00   .."..... ."..`.S.
   0360   22 05 02 C0 D8 A7 00 22 02 03 07 22 08 04 06 00   "......" ..."....
   0370   02 6F 01 5B AB 22 02 05 01 1A 07 03 01 E0 03 00   .o.[.".. ........
   0380   00 01 1B 10 C1 01 19 77 B5 1E 35 B5 3C 36 36 05   .......w ..5.<66.
   0390   46 FF FF FF FF FF 00 00 00 00 00 00 00 00 00 00   F....... ........
*****************************
```

Hex Dump 6.4: As you would expect, I have a few Zonet ZCF1100 802.11b CompactFlash NICs laying around. I plugged more than one into an AirDrop-P and dumped the CIS. All of the Zonet ZCF1100 CIS hexadecimal dumps were identical byte by byte to Hex Dump 6.4.

The good news is that the data we really need from the CIS for the AirDrop driver code (the COR base address) is being given to us correctly by all of the 802.11b CompactFlash NICs I tested. We really don't care about minimum voltages or peak currents as they don't play into the AirDrop 802.11b CompactFlash NIC driver firmware flow. For machines running Windows and Linux, the bytes contained within the 802.11b CompactFlash NIC's CIS Memory area are very important as they tell the operating system what it needs to know about the 802.11b CompactFlash NIC. In out case, we only need to know how to kick-start the 802.11b CompactFlash NIC as our AirDrop driver does not need to be as sophisticated as its Linux and Windows cousins. I thought it would be neat to explore the CIS stuff anyway.

Code Snippet 6.10 is the resultant code that was spawned by the Code Tuple Collection Utility shown in Code Snippet 6.8. I threw in some comments as to what is where in Tupleville. I also added case statements for CIS tuples we ignore. The case statements allow you to enter your own production or debug code concerning that particular tuple. For instance, you can add *printf* statements to send a tuple's contents to the AirDrop RS-232 port. Code Snippet

6.10 ends with the capture of the COR base address value, which is the only CIS information we use in the AirDrop 802.11b CompactFlash NIC drivers.

Code Snippet 6.10

```
****************************

//*************************************************************
//*    Macros
//*************************************************************
#define make16(varhigh,varlow)      (((unsigned int)varhigh & 0xFF)* 0x100) +
((unsigned int)varlow & 0x00FF)
//*************************************************************
//*    Flags
//*************************************************************
#define cisflag 0x04
#define bcisflag        flags & cisflag
#define setf_cis        flags |= cisflag
#define clrf_cis        flags &= ~cisflag
//*************************************************************
//*    PORT Definitions
//*************************************************************
#define addr_lo         PORTA
#define data_in         PINC
#define addr_hi         PORTF
#define PCTL            PORTG
//*************************************************************
//*    CF I/O Definitions
//*************************************************************
#define OE              0x01
#define set_OE          PCTL |= OE
#define clr_OE          PCTL &= ~OE


****************************
void wr_cf_addr(unsigned int addr)
{
 addr_hi = (make8(addr,1)) & 0x07;
 addr_lo = addr & 0x00FF;
}

char rd_cf_reg(unsigned int reg_addr)
{
 char data,i;
 wr_cf_addr(reg_addr);
 clr_OE;
 i=1;
 while(--i);
 data = data_in;
 set_OE;
 NOP();
 return(data);
}
```

```
****************************

char macaddrc[6];
char cor_addr_lo,cor_addr_hi,cor_data;
char cis_device;
unsigned int x,index;
unsigned int cis_addr;      //attribute memory index (EVEN ADDRS ONLY)
char code_tuple;            //tuple type
char link_tuple;            //number of bytes in tuple structure
char tuple_link_cnt;        //tuple byte count
char tuple_data;            //current tuple contents
char funce_cnt;
char mac_cnt;
char rc;

cis_device  = 0;
cor_addr    = 0;
funce_cnt = 0;                          // reset to the first CF_CISTPL_FUNCE
mac_cnt   = 0;
cis_addr    = 0;                        // CIS starts at location 0 of Attribute
memory
clrf_cis;
do{
    code_tuple = rd_code_tuple(cis_addr);        //read code tuple
    cis_addr+=2;                                 //increment addr to link tuple
    link_tuple = rd_link_tuple(cis_addr);        //read link tuple
    cis_addr+=2;                                 //increment addr to data tuple

    if(code_tuple != END_TUPLE)
    {
     for(tuple_link_cnt=0;tuple_link_cnt<link_tuple;++tuple_link_cnt)
     {
       tuple_data = rd_tuple_data(cis_addr);   //read tuple data bytes
       cis_addr+=2;

       switch(code_tuple)
       {
         case DEVICE_TUPLE:
             break;
         case DEVICE_A_TUPLE:
             break;
         case DEVICE_OC_TUPLE:
             break;
         case DEVICE_OA_TUPLE:
             break;
         case VERS_1_TUPLE:
             break;
         case MANFID_TUPLE:
             break;
         case FUNCID_TUPLE:
           if(tuple_link_cnt == 0)
```

73

```
            cis_device = tuple_data;        // device type 0x06 is Network
Adapter
        break;
          case FUNCE_TUPLE:
            if(funce_cnt == 4)              // funce[4] holds the MAC address
             if(tuple_link_cnt>1)           // MAC is body [2..7] in funce[4]
              macaddrc[mac_cnt++] = tuple_data; // store the MAC
            if(tuple_link_cnt == (link_tuple-1))// we are done with this
funce
            funce_cnt++;                    // increment to indicate the next
funce tuple
           break;
          case CONFIG_TUPLE:                       // GET THE COR ADDRESS
            if(tuple_link_cnt == 2)
            cor_addr_lo = tuple_data;      // copying low-byte
            if(tuple_link_cnt == 3)        // shift, then copy hi-byte
          {
             cor_addr_hi = tuple_data;
       cor_addr = make16(cor_addr_hi,cor_addr_lo);
          }
          break;
        case CFTABLE_ENTRY_TUPLE:
          break;
        default:
          break;
      }// end switch
    }// end for loop tuple_link_cnt
  }//if(code_tuple != END_TUPLE)
      else
      {
        setf_cis;
      }
}while(!(bcisflag));
```

****

Code Snippet 6.10: The AirDrop-A AVR source code is identical to what you see here. Again, the only differences in the AVR and PIC code in this snippet are the microcontroller-specific names and locations.

In actuality, from what we've seen every 802.11b CompactFlash NIC I tested used a COR base address of 0x000003E0 and we could eliminate the CIS parsing done with the code in Code Snippet 6.10. However, that's not a safe thing to do. By actually searching out the COR base address, we move one step further towards knowing that we can positively put the 802.11b CompactFlash NIC on the air.

Enabling the 802.11b CompactFlash NIC

At this point, the 802.11b CompactFlash NIC is not yet been put into I/O mode. However, we're done with the CIS and need to move to I/O mode so we can issue commands. The following set of C statements in Sub Snippet 6.1 is all it takes to put the 802.11b CompactFlash NIC into I/O mode:

Sub Snippet 6.1

```
*****************************************************************
cor_data = rd_cor_data(cor_addr);        //get the current bit mask
delay_ms(4);
cor_data |= io_en_mask;                   //add I/O enable bit
wr_cor_data(cor_data,cor_addr);           //write the modified bit mask
*****************************************************************
```

Once the write to the COR is complete, the NIC enters I/O mode. Now instead of using the *OE* and *WE* signals to communicate with the 802.11b CompactFlash NIC, we will begin using the *IORD* and *IOWR* signals. A look at all of the code things necessary to accomplish the 802.11b CompactFlash NIC's move to I/O mode are shown in Code Snippet 6.11.

Code Snippet 6.11

```
****************************

//**********************************************************
//*    CIS
//**********************************************************
#define io_en_mask          0x01
#define rd_cor_data         rd_cf_reg
#define wr_cor_data         wr_cf_reg
//**********************************************************
//*    PORT Definitions
//**********************************************************
#define data_in         PORTD
#define data_out        LATD
#define addr_hi         LATG
#define addr_lo         LATH
//**********************************************************
//*    PORT TRIS Definitions
//**********************************************************
#define TO_NIC          TRISD = 0x00
#define FROM_NIC            TRISD = 0xFF
//**********************************************************
//*    CF I/O Definitions
//**********************************************************
#define OE              0x01
#define WE              0x40
#define set_OE          PCTL |= OE
#define clr_OE          PCTL &= ~OE
#define set_WE          PCTL |= WE
#define clr_WE          PCTL &= ~WE
****************************

void wr_cf_addr(unsigned int addr)
{
 addr_hi = (make8(addr,1));
 addr_lo = addr & 0x00FF;
}
```

```
void wr_cf_reg(unsigned int reg_data,unsigned int reg_addr)
{
 TO_NIC;
 wr_cf_addr(reg_addr);
 data_out = reg_data;
 clr_WE;
 delay_ms(2);
 set_WE;
 FROM_NIC;
}

char rd_cf_reg(unsigned int reg_addr)
{
 char data,i;
 wr_cf_addr(reg_addr);
 clr_OE;
 i=1;
 while(--i);                  //read access delay
 data = data_in;
 set_OE;
 return(data);
}
*****************************

cor_data = rd_cor_data(cor_addr);        //get the current bit mask
delay_ms(4);
cor_data |= io_en_mask;                  //add I/O enable bit to mask
wr_cor_data(cor_data,cor_addr);          //write the modified bit mask

*****************************
```

Code Snippet 6.11: We're on our way. Once the 802.11b CompactFlash NIC enters I/O mode the PRISM chipset comes to life.

The SanDisk manual calls the COR byte we just read and wrote back to address 0x03E0 the WLAN Configuration Option Register. Although the layout of the SanDisk COR may differ from the other 802.11b CompactFlash NICs we have looked at, writing a 1 to the least significant bit performs the same function, enables I/O mode, in all of the CompactFlash card NICs I've tested thus far. Writing a 1 to the most significant bit of the COR will also generate a reset in the 802.11b CompactFlash NIC's I've tested. That tells me that the SanDisk COR bit layout is most likely a standard with all of the 802.11b CompactFlash NICs. I've provided a look at the SanDisk COR bit layout below.

SanDisk WLAN Configuration Option Register Bit Definitions							
D7	D6	D5	D4	D3	D2	D1	D0
SRE-SET	LevlREQ	0	0	0	Enable IRQ	0	Enable I/O

As you can see in the SanDisk bit layout, the SanDisk COR allows the programmer/user to issue a soft reset the NIC, configure interrupts and enable I/O mode. The soft reset code I tested is shown in Code Snippet 6.12.

Code Snippet 6.12

****

```
cor_data = rd_cor_data(cor_addr);      //get the current bit mask
delay_ms(4);
cor_data |= 0x80;                      //add soft reset bit
wr_cor_data(cor_data,cor_addr);        //write the modified bit mask
```

****

Code Snippet 6.12: I tested the soft reset function by first power-on resetting the 802.11b CompactFlash NIC and then enabling I/O functionality. I then simply executed the code in this snippet. Note that I can still read and write Attribute Memory using the OE and WE signals even while in I/O mode.

The Value of Parsing the CIS

As I've mentioned earlier, the TEW-222CF is out of production. However, I did manage to find an exact replacement for the TEW-222CF by comparing CIS dumps of various Compact-Flash NICs I tested during the writing of this book. The Xterasys CWB1K you see in Photo 6.1 is an exact duplicate of the TEW-222CF down to the CIS tuples. Recall that I had you make a mental note of a label we saw next to the TEW-222CF's chip antenna. Remember what it said? Here's the line taken from the caption of Photo 4.4 in Chapter 4:

> *Make a mental note of the silkscreen label to the left of the TEW-222CF's chip antenna, "M01-CWB1K-A10".*

Hmmm…In addition, in its CIS, the Xterasys CWB1K also states that it cannot run at +3.3 volts just like the TEW-222CF. Hmmm...Can you say clone?

Photo 6.1: I had a feeling about this 802.11b CompactFlash NIC when I first saw a picture of it. It even looks like a TEW-222CF, LEDs and all. The CIS tuple chain of the Xterasys CWB1K is a byte-for-byte copy of the TEW-222CF tuple chain.

Well, we won't have to tear the Xterasys CWB1K CompactFlash NIC apart, now will we?

Full Throttle

At the beginning of this chapter, we taxied our AirDrops out to the runway and waited for permission to take off. We got take-off permission once we started reading bytes from the CIS

Memory area. Our throttle up and take-off roll began when we parsed the CIS for the COR base address. Right now, the 802.11b CompactFlash NIC is in I/O mode and we're moving pretty fast. So, pull back on the stick and let's rotate and take off!

More Musical Overtones

While you're gaining altitude and "cleaning up" (that's pilot slang for retracting the landing gear), here's the answer to that guitar question I posed at the end of Chapter 5. The Fender "Broadcaster" name was dropped because Gretsch already had a drum kit called "Broadkaster" it had been touting for years. As a result, Fender's Broadcaster became what we all know today as the Telecaster. In the interim, Leo Fender took the Broadcaster decals off the new Broadcaster guitars he sold while waiting for the new Telecaster decals. Collectors call the decal-less guitars *No-casters*.

Learning to Talk to
802.11b CompactFlash NICs

Unless you've been privy to some previous 802.11b and PRISM experience, right now the words FID, RID and BAP most likely mean very little to you. However, those little three-letter acronyms are all over any 802.11b driver source code you're ever likely to see. The PRISM acronyms are also carefully placed in PRISM chipset data sheets as well. Fortunately, careful observation of the available PRISM chipset data sheets gives us an insight as to what those acronyms mean and what they are to the AirDrop driver modules and 802.11b. With that, let's do some PRISM chipset spelunking.

What the 802.11b NIC Does for Us

If you've scanned the massive amount of Linux 802.11b source on the Internet, you've probably seen references to downloading code to PRISM chipsets. We won't be doing any of that. However, just the notion of being able to push bytes into an 802.11b NIC implies that firmware controls the beast. In fact, I came across a couple of internet documents that detailed the

downloading of code to a PRISM-equipped NIC. The concrete was set when I read the second paragraph of the ISL3873A data sheet. It stated that firmware implements the full IEEE 802.11 wireless LAN MAC protocol. That one sentence containing the words "firmware implements" is our 802.11b emancipation proclamation.

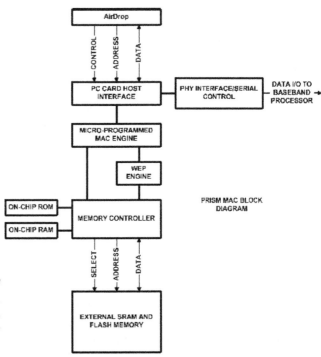

Figure 7.1: If I didn't know better, I'd say this was a typical microcontroller-based subsystem. What do you suppose "Micro-Programmed MAC Engine" means? Hmmm...

Figure 7.1 is a graphical depiction of the MAC area of the ISL3873A. Lots of things we'll talk about later when we start a detailed examination of the AirDrop driver source code are performed within the bold-lined blocks of Figure 7.1. One of the more obvious clues we can pick up from the block diagram shown in Figure 7.1 is the inclusion of a WEP engine functional block.

Yet another three-letter wireless LAN acronym, WEP, or wired equivalency privacy, is supposed to provide an equivalent level of security for wireless LANs that exists by nature in wired LAN environments. All we have to do to use WEP with an AirDrop module is provide a WEP key and enable the WEP Engine. I'll show you how that's done in a later chapter of this book.

WEP encryption doesn't exist above the data link layer. That means we don't have to worry with it in our AirDrop driver or AirDrop application firmware. WEP only works on the data transmitted by the radio. WEP's inner workings are transparent to us as the PRISM chipset takes care of coding and decoding the WEP-encrypted radio link. Piece of cake..

Other "automatic" functions that are performed by a PRISM-equipped NIC include the support of BSS and IBSS operation, beacon monitoring and, fragmentation/defragmentation. Active scanning is also provided by the NIC once the microcontroller and driver firmware initialize the NIC. What does all of that mean? As Ricky would say "Lucy, you've got some esplanin' to do!"

Some 802.11b Language Basics

If you're totally new to wireless LAN technology and haven't been chapter hopping, you're up to speed on what you need to know about PRISM chipsets that operate under the 802.11b banner. In addition to the AirDrop and PRISM hardware knowledge you have, there are a few more things about 802.11b that you must know. First, will your AirDrop be participating in an infrastructure network or will it operate in ad hoc mode? An infrastructure (BSS or Basic Service Set) network implies that some type of wireless access point (AP) is actively operating in the network. Your AP can stand alone as a wireless access point or it can double as a gateway to the Internet via your DSL or Broadband connection. An example of a home wireless access point is the basic Netgear MR814, which is shown in Photo 7.1. Ad hoc (IBBS or Independent Basic Service Set) networks are simple point-to-point wireless connections between wireless stations that do not have an access point in their network structure. The PRISM chipsets we've discussed can handle IBBS networks with just a few bytes of encouragement. The AirDrop 802.11b driver allows the use of IBSS mode. However, thus far, I haven't come up with a very good reason to couple AirDrop modules in an IBSS.

802.11b can be described as a MAC (Medium Access Control), PHY (Physical) and LLC (Link Layer Control) combination that operates in the unlicensed 2.4GHz ISM (Industrial, Scientific, Medical) band at a maximum speed of 11Mbps using standards set forth by the IEEE. Whew! 802.11b devices operate in the ISM S-Band, which extends from 2.4GHz to 2.5GHz. Operation in this band puts the typical 802.11b device in competition with radio emissions wafted by other devices in that bandwidth such as microwave ovens and cord-

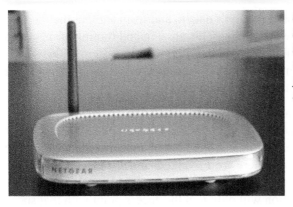

Photo 7.1: This unassuming little box is the "Air Boss" of the wireless LAN. Nothing flows through a BSS without this guy knowing about it. If you don't believe that, just turn it off and see for yourself.

less telephones. To reduce the interference factor and raise data delivery reliability, 802.11b devices don't use collision detection schemes (CSMA/CD or Carrier Sense Multiple Access/ Collision Detection) like those found in wired 802.3 networks. Instead, a collision avoidance method (CSMA/CA or Carrier Sense Multiple Access/Collision Avoidance) is implemented. This means that the 802.11b device listens to the communications ether before attempting to transmit unlike 802.3 devices that collide and then back off for a period of time before trying to gain access to the communications channel again. To make the 802.11b CSMA/CA scheme work, a set of timing rules is implemented that allow an 802.11b station to cleanly enter the communications channel, transmit its message and release the ether to the next 802.11b station that needs to send a message. This timing is automatic and is taken care of by the PRISM chipset firmware. If an 802.11b station hears traffic, it does not attempt a transmission and performs an exponentially-timed backoff procedure. To further enhance data delivery reliability, every transmitted message must be acknowledged by the receiver. If communications conditions degrade, the PRISM chipset automatically throttles back the transmission speed from 11 Mbps to 5.5 Mbps to 2 Mbps. When things get better, the NIC's transmission speed is automatically restored to an acceptable maximum level.

The 802.2 LLC layer is common to both the wired 802.3 MAC and 802.3 PHY OSI layers and the wireless 802.11 MAC and 802.11b PHY OSI layers. This kinship in the upper OSI layers allows the 802.11b LAN to play easily with an 802.3 LAN at the LLC layer and above.

If your wireless network includes an access point, your AirDrop is operating in infrastructure mode. 802.11b devices that communicate peer-to-peer are said to be operating in ad-hoc mode. You may see these modes described as IBSS (Independent Basic Service Set) for ad-hoc networks or BSS (Infrastructure Basic Service Set) for infrastructure networks. A BSS (Basic Service Set) is simply a number of 802.11b stations that are communicating with each other using ad-hoc mode or infrastructure mode. The word *infrastructure* in the BSS network mode description is implied and never used in the infrastructure mode abbreviation to avoid confusion.

In addition to managing some of the network traffic, access points enhance the mobility of stations in a wireless network. Usually, an access point acts as a bridge between a wireless station and a wired system. When multiple access points are used in a network, each access point must be able to pass the mobile station's data to and from another mobile station, another access point or a wired network station. This passing and routing of messages is performed using what is called a distribution system. A distribution system in this sense is a logical component of 802.11 that simply routes messages to their destinations no matter which access point in the network the mobile 802.11b station is using.

An SSID (Service Set Identity) and a channel number are used to distinguish access points. The SSID is usually a human-readable description that can be up to 32 characters in length. An 802.11b station will scan the communications ether channel by channel to detect an access point to join. In the United States, there are 11 valid channels that can be used for 802.11b. Most European countries can use 13 of the 14 available 802.11b channels. If a particular SSID is specified, the 802.11b station will only join the access point with that specific SSID. Otherwise, the 802.11b station can be configured to join the strongest signaled access point it can find. An 802.11b station can simply listen to the ether (passive scanning) for a Beacon from an access point or probe the communications ether (active scanning) to detect an in-range access point. A Beacon is transmitted by all access points and delivers all of the information that is necessary for an 802.11b station to determine if it can enter a BSS via that particular access point.

Once all of the available access points are identified by an 802.11b station, the 802.11b station can then introduce itself and join a BBS via the selected access point. Before the 802.11b station can participate in the BSS, it must be authenticated by the access point. There are several ways to authenticate an 802.11b station. One method is called open-system authentication and is really not authentication at all. Using open-system authentication the access point simply allows the requesting 802.11b station to come on in. If your 802.11b station is WEP-enabled (Wired Equivalent Privacy), the access point could invoke shared-key authentication. You can also instruct the access point to only authenticate certain station MAC addresses. This authentication process is called address filtering.

No matter how the 802.11b station gets authenticated, once the authentication process is complete the station can then associate with the access point. Association allows the 802.11b station to use the access point to gain access to the distribution system and thus gain access to the network.

In an ad hoc system each 802.11b station that wants to communicate peer-to-peer must be setup to the same channel and SSID. No access point is involved with stations configured for ad-hoc operation. And, since there is no access point involved, there is no authentication or association of the ad-hoc stations.

We are fortunate that the 802.11b CompactFlash NIC does so many things automatically for us, but to use those automatic features we must be able to listen to what the NIC is telling us and issue commands to the NIC to realize any transfer of data between wireless LAN stations. You've successfully got your AirDrop aircraft in the air and you're cruising along just fine. Now is the time for you to get familiar with some of the controls.

The 802.11b CompactFlash NIC I/O Drivers

Aircraft controls allow the pilot to maneuver his or her aircraft. In the case of the AirDrop, I/O functions are used to steer the 802.11b CompactFlash NIC. Code Snippet 7.1 represents the final C statements in the *init_cf_card* call.

Code Snippet 7.1

```
****************************
//* COMMAND/STATUS
#define  Command_Register        0x0000
#define  Param0_Register         0x0002
//* EVENT
#define  EvStat_Register         0x0030
#define  EvAck_Register          0x0034
//* COMMANDS
#define  Initialize_Cmd          0x0000
//* BIT MASKS
#define  CmdCode_Mask            0x003F
#define  CmdBusy_Bit_Mask        0x8000
#define  EvStat_Tick_Bit_Mask    0x8000
#define  EvStat_WTErr_Bit_Mask   0x4000
#define  EvStat_InfDrop_Bit_Mask 0x2000
#define  EvStat_Info_Bit_Mask    0x0080
#define  EvStat_DTIM_Bit_Mask    0x0020
#define  EvStat_Cmd_Bit_Mask     0x0010
#define  EvStat_Alloc_Bit_Mask   0x0008
#define  EvStat_TxExc_Bit_Mask   0x0004
#define  EvStat_Tx_Bit_Mask      0x0002
#define  EvStat_Rx_Bit_Mask      0x0001
#define  Status_Result_Mask      0x7F00


#define nop                NOP()
//**********************************************************
//*   Macros
//**********************************************************
#define make8(var,offset)   ((unsigned int)var >> (offset * 8)) & 0x00FF
#define make16(varhigh,varlow)     (((unsigned int)varhigh & 0xFF)* 0x100) +
((unsigned int)varlow & 0x00FF)
/* Macros to access bytes within words */
#define LOW_BYTE(x)  ((unsigned char)(x)&0xFF)
#define HIGH_BYTE(x) ((unsigned char)(x>>8)&0xFF)
//**********************************************************
//*   PORT Definitions
//**********************************************************
#define data_in      PORTD
#define data_out     LATD
#define addr_hi      LATG
#define addr_lo      LATH
//**********************************************************
//*   PORT TRIS Definitions
//**********************************************************
```

```
#define TO_NICTRISD = 0x00
#define FROM_NIC      TRISD = 0xFF
//*************************************************************
//*    CF I/O Definitions
//*************************************************************
#define OE                0x01
#define IORD              0x02
#define WE                0x40
#define IOWR              0x80
#define set_OE            PCTL |= OE
#define clr_OE            PCTL &= ~OE
#define set_IORD          PCTL |= IORD
#define clr_IORD          PCTL &= ~IORD
#define set_IOWR          PCTL |= IOWR
#define clr_IOWR          PCTL &= ~IOWR

void wr_cf_addr(unsigned int addr)
{
        addr_hi = (make8(addr,1));
        addr_lo = addr & 0x00FF;
}

unsigned int rd_cf_io16(unsigned int addr)
{
        char data_lo,data_hi,i;
        unsigned int data16;

        wr_cf_addr(addr);
        clr_IORD;
        i=3;
        while(--i);
        data_lo=data_in;
        set_IORD;
        NOP();

        wr_cf_addr(addr+1);
        clr_IORD;
        i=3;
        while(--i);
        data_hi=data_in;
        set_IORD;
        NOP();
        data16 = make16(data_hi,data_lo);
        return(data16);

}

void wr_cf_io16(unsigned int data16,unsigned int addr)
{
        char i;

        wr_cf_addr(addr);
```

```
        data_out = LOW_BYTE(data16);
        TO_NIC;
        clr_IOWR;
        i=1;
        while(--i);
        set_IOWR;
        NOP();
        data_out = 0xFF;

        wr_cf_addr(addr+1);
        data_out = HIGH_BYTE(data16);
        clr_IOWR;
        i=1;
        while(--i);
        set_IOWR;
        NOP();
        data_out = 0xFF;
        FROM_NIC;
}

char    send_command(unsigned int cmd, unsigned int parm0)
{
        char cmd_code,rc;
        unsigned int cmd_status,evstat_data;

        do{
            cmd_data = rd_cf_io16(Command_Register);
          }while(cmd_data &  CmdBusy_Bit_Mask);

        wr_cf_io16(parm0, Param0_Register);
        wr_cf_io16(cmd, Command_Register);

        do{
            cmd_data = rd_cf_io16(Command_Register);
          }while(cmd_data &  CmdBusy_Bit_Mask);

        do{
                evstat_data = rd_cf_io16(EvStat_Register);
          }while(!(evstat_data & EvStat_Cmd_Bit_Mask));

        cmd_status = rd_cf_io16( Status_Register);
        rc = cmd_status &  Status_Result_Mask;
        wr_cf_io16( EvStat_Cmd_Bit_Mask, EvAck_Register);

        switch(rc)
        {
                case 0x00:
                        rc = 0;                 //command completed OK
                        break;
                case 0x01:
                        rc = 1;                 //Card failure
                        break;
```

```
                case 0x05:
                        rc = 2;                 //no buffer space
                        break;
                case 0x7F:
                        rc = 3;                 //command error
                        break;
        }
        return(rc);
}
****************************

void init_cf_card(void)
{
 char rc;

 rc =   send_command( Initialize_Cmd, 0);
 switch(rc)
  {
    case 0:
     printf("\r\nAirDrop-P Initialized");
     break;
    case 1:
     printf("\r\nAirDrop-P Initialization Failure");
     break;
    case 2:
     printf("\r\nNo buffer space");
     break;
    case 3:
     printf("\r\nCommand error");
     break;
    default:
     printf("\r\nResult Code = %x",rc);
     break;
  }
}
****************************
```

Code Snippet 7.1: Initializing the 802.11b CompactFlash NIC is akin to arming the weaponry of an aircraft. Flipping the NIC's initialization switch brings everything "automatic" we've talked about to life.

There are lots of new concepts being introduced here. So, let's take the code apart line by line beginning with Sub Snippet 7.1.

Sub Snippet 7.1
```
************************************************************
rc =   send_command( Initialize_Cmd, 0);
************************************************************
```

The *send_command* function's purpose is pretty obvious. In this instance, we are sending a command to initialize the 802.11b CompactFlash NIC. There are other commands that may be issued with the *send_command* function and I've listed them in Sub Snippet 7.2.

Sub Snippet 7.2

```
***************************************************************
//* COMMANDS
#define   Initialize_Cmd           0x0000
#define   EnableMAC_Cmd            0x0001
#define   Disable_Cmd             0x0002
#define   Diagnose_Cmd            0x0003
#define   Allocate_Cmd            0x000A
#define   Transmit_Cmd            0x000B
#define   TransmitReclaim_Cmd     0x010B
#define   Inquire_Cmd             0x0011
#define   Access_Cmd             0x0021
***************************************************************
```

We won't use all of the commands I've listed here in the AirDrop driver code but the complete set of commands are there for you if you want to explore them. I'll talk about each command we use in detail as we encounter that command in the AirDrop driver source code. For now, let's concentrate on the initialization of the 802.11b CompactFlash NIC.

Sub Snippet 7.3 lays out the local variables and then, by way of a *do-while* loop, checks for the busy bit inside the *Command_Register* using the *CmdBusy_Bit_Mask*.

Sub Snippet 7.3

```
***************************************************************
char   send_command(unsigned int cmd, unsigned int parm0)
{
       char cmd_code,rc;
       unsigned int cmd_status,evstat_data;

       do{
          cmd_data = rd_cf_io16(Command_Register);
          }while(cmd_data &  CmdBusy_Bit_Mask);
***************************************************************
```

The heart of the *do-while* loop is the *rd_cf_io16* function in Sub Snippet 7.4, which retrieves a 16-bit value (*cmd_data*) from a designated address (*Command_Register*). You're already familiar with the *wr_cf_addr* function, which in this case is writing an address of 0x0000 to the 802.11b CompactFlash NIC address lines. To perform a CompactFlash card I/O read operation we must take *IORD* low for 2.2 µS, read the CompactFlash card data bus and return *IORD* to a logical high state. Earlier we found that the 802.11b CompactFlash NIC SRAM is rated at 55 nS and one would conclude that 2.2 µS is way too long a time to take to read a 55 nS SRAM. Remember that we're not talking directly to the SRAM. We're talking to the PRISM chipset's firmware, which controls and accesses the 55 nS SRAM.

Two CompactFlash card I/O read operations are required. One read is performed to get the low byte of the 16-bit value and another to retrieve the high byte of the 16-bit value. Note that the base address is incremented by one to access the high byte value. The *NOPs* that follow the *set_IORD* macros are there to ensure any hold time requirements are met. The extra time afforded by the *NOPs* also gives the microcontroller I/O ports some time to settle. Once the high and low bytes are gathered, they are merged to form the 16-bit data value *cmd_data*.

Sub Snippet 7.4

```
*************************************************************
unsigned int rd_cf_io16(unsigned int addr)
{
      char data_lo,data_hi,i;
      unsigned int data16;

      wr_cf_addr(addr);
      clr_IORD;
      i=3;
      while(--i);
      data_lo=data_in;
      set_IORD;
      NOP();

      wr_cf_addr(addr+1);
      clr_IORD;
      i=3;
      while(--i);
      data_hi=data_in;
      set_IORD;
      NOP();
      data16 = make16(data_hi,data_lo);
      return(data16);
}
*************************************************************
```

When the busy bit is clear the busy bit *do-while* loop is exited and the command sequence can be written to the 802.11b CompactFlash NIC. If present, the command parameter is written first followed by the actual command bit sequence as shown in Sub Snippet 7.5.

Sub Snippet 7.5

```
*************************************************************
      wr_cf_io16(parm0, Param0_Register);
      wr_cf_io16(cmd, Command_Register);
*************************************************************
```

The *wr_cf_io16* function is used to write a 16-bit value to the 802.11b CompactFlash NIC. As with the CompactFlash card I/O read operation, the desired address is sent to the 802.11b CompactFlash NIC's address lines before the actual data is put on the 802.11b CompactFlash NIC's data bus. The *LOW_BYTE* and *HIGH_BYTE* macros are very similar to the *make8* macro. However, the *LOW_BYTE* and *HIGH_BYTE* macros are more efficient as far as code size and execution time is concerned.

A 16-bit CompactFlash card I/O write operation is performed by executing the *TO_NIC* macro, which puts the microcontroller data bus I/O in output mode. To write the low byte of the 16-bit value, the *IOWR* microcontroller I/O pin is taken to a logic low level for 800 nS before it is returned to a logical high state. Recall that we used values gleaned from the CIS to calculate a maximum write access time of 500 nS. Again 800 nS is half again as much time needed to perform the 500 nS write operation. If we had used *NOPs* (200 nS each) the closest we could come to the ideal 500 nS write pulse is 600 nS. As always, the first rule of embedded programming applies. "Nothing is free."

To avoid any I/O contention and satisfy any hold time requirements, a *NOP* is executed following the return of the *IOWR* I/O pin to a logical high state, which is then followed by the writing of all ones to the microcontroller's data I/O port. The base address is then incremented by one and the high byte of the 16-bit value is placed on the 802.11b CompactFlash NIC's data bus. An *IOWR* write pulse is generated in the same fashion as it was in the low byte write sequence and the data bus is allowed to settle. Before leaving the *wr_cf_io16* function in Sub Snippet 7.6, the microcontroller's data bus I/O port is put into input mode with the execution of the *FROM_NIC* macro.

Sub Snippet 7.6
```
*************************************************************
void wr_cf_io16(unsigned int data16,unsigned int addr)
{
        char i;

        wr_cf_addr(addr);
        data_out = LOW_BYTE(data16);
        TO_NIC;
        clr_IOWR;
        i=1;
        while(--i);
        set_IOWR;
        NOP();
        data_out = 0xFF;

        wr_cf_addr(addr+1);
        data_out = HIGH_BYTE(data16);
        clr_IOWR;
        i=1;
        while(--i);
        set_IOWR;
        NOP();
        data_out = 0xFF;
        FROM_NIC;
}
*************************************************************
```

The 802.11b CompactFlash NIC will go into a busy state while executing the command. We must wait for the busy bit to clear before proceeding. This is done by once again polling the *Command_Register* and checking the most significant bit (the busy bit) of the *Command_Register* for a value of 1 (1 = busy). The polling of the busy bit is done with the code you see in Sub Snippet 7.7.

Sub Snippet 7.7
****
```
      do{
          cmd_data = rd_cf_io16(Command_Register);
          }while(cmd_data &  CmdBusy_Bit_Mask);
```
****

When the busy bit finally clears, a command complete event will be generated. The command complete event is signaled by the setting of a bit in the *Event Status Register*. As you can see in Sub Snippet 7.8, we look for the command complete bit by reading the *EvStat_ Register* and applying a bit mask that represents the command complete bit, which happens to be the *EvStat_Cmd_Bit_Mask* (0x0010).

Sub Snippet 7.8
****
```
      do{
             evstat_data = rd_cf_io16(EvStat_Register);
         }while(!(evstat_data & EvStat_Cmd_Bit_Mask));
```
****

Once the command complete bit materializes the 802.11b CompactFlash NIC will post a command status word in the Status Register. What we want to see is a return code of zero (rc = 0) after applying the *Status_Result_Mask* to the *cmd_status* word as we've done in Sub Snippet 7.9.

Sub Snippet 7.9
****
```
      cmd_status = rd_cf_io16( Status_Register);
      rc = cmd_status &  Status_Result_Mask;
      wr_cf_io16( EvStat_Cmd_Bit_Mask, EvAck_Register);
```
****

All events must be acknowledged by the AirDrop driver firmware. In this case, we must acknowledge the command complete event. Since the *EvStat_Cmd_Bit_Mask* value represents the command complete bit's location in the *Event Acknowledge Register*, all we have to do is write the *EvStat_Cmd_Bit_Mask* value to the *Event Acknowledge Register* to acknowledge the pending command complete event.

In Sub Snippet 7.10, the return code (rc) is parsed and returned to the calling function, which in this case is *send_command*.

Sub Snippet 7.10
****
```
      switch(rc)
      {
          case 0x00:
                 rc = 0;                //command completed OK
                 break;
          case 0x01:
                 rc = 1;                //Card failure
                 break;
```

```
            case 0x05:
                    rc = 2;                  //no buffer space
                    break;
            case 0x7F:
                    rc = 3;                  //command error
                    break;
    }
    return(rc);
}
```
****

Another *switch* structure then translates the returned return code value into a human-readable message. What we want to see coming out of the AirDrop RS-232 port is "AirDrop-P Initialized". When we get that positive message, the 802.11b CompactFlash NIC is ready to receive commands and punch holes in electromagnetic space. Other undesirable messages are shown in Sub Snippet 7.11.

Sub Snippet 7.11
****
```
rc =  send_command( Initialize_Cmd, 0);
 switch(rc)
  {
    case 0:
     printf("\r\nAirDrop-P Initialized");
     break;
    case 1:
     printf("\r\nAirDrop-P Initialization Failure");
     break;
    case 2:
     printf("\r\nNo buffer space");
     break;
    case 3:
     printf("\r\nCommand error");
     break;
    default:
     printf("\r\nResult Code = %x",rc);
    break;
  }
```
****

OK, we're ready to rock! At this point, if all went well (we'll assume that it did), the 802.11b CompactFlash NIC is initialized and ready to be told what we want to do next. The 802.11b CompactFlash NIC is in limbo land and has not uttered a word of RF yet. All of the 802.11b CompactFlash NICs exhibit a flashing LED condition at this point with the exception of the Zonet ZCF100 NIC. The Zonet flashed the LED once and appears to die. We'll see if it's just playing possum' in the next chapter.

Speaking of playing possum', what famous Bluegrass duo shared possum' pie regularly with Granny and Ellie May? And, if you had to show up at the mansion in a formal kind of way, how would you address Granny?

CHAPTER **8**

Setting Up An
AirDrop Wireless Network

Thus far, this wireless LAN stuff isn't as hairy as it's been cranked up to be. In the previous chapters we've built some really simple 802.11b hardware and we've taken an extraordinarily "in depth" look at the electronics inside a typical 802.11b CompactFlash NIC. We've even taken some swipes at coding the AirDrop 802.11b driver.

In this chapter, we'll set up a BSS on Channel 1. Our little BSS will consist of a standard desktop personal computer equipped with a Netgear 802.11b PCI NIC, a laptop equipped with built-in 802.11b capability and an AirDrop-P. I'll use a Netgear MR814v2 as the AP for our BSS. Once we get all of the wireless LAN hardware up, we'll use a wireless LAN network Sniffer to follow the 802.11b action as if unfolds. A family portrait of all of the AirDrop BSS hardware is shown in Photo 8.1.

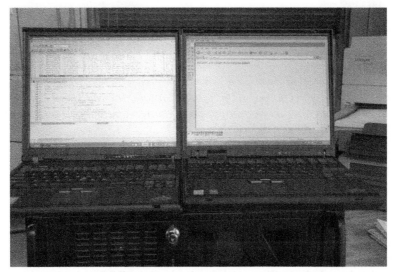

Photo 8.1: The desktop personal computer, TEST, is contained within the rackmount case that the laptops are perched on. The laptop loaded with the sniffer application, SNOOP, has a built-in wireless LAN interface but the Netasyst W Sniffer driver requires an approved PCMCIA 802.11b NIC, in this case a Symbol LA4121, which you see shoved into the laptop's PCMCIA slot. The second laptop, SNOOPER, is 802.11b capable and is another wireless station I can use to communicate with the AirDrop module.

Setting Up the AP

This is a piece of cake. I started by resetting the Netgear AP to its default settings and logging on to the Netgear AP's browser-based management interface using the desktop personal computer with a wired Ethernet connection. Since we won't be going out onto the "Big Wire" via and ISP connection, I ignored the Netgear AP's Basic Settings window and tooled right over to the Netgear AP's Wireless Settings window you see in Screen Capture 8.1.

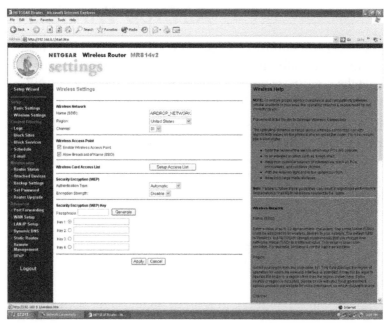

Screen Capture 8.1: As you can see in this screen capture, the MR814v2 has lots of knobs that you can twist. Right now, all we want to do is setup an operating Channel and establish a network name via the AP's SSID.

I decided to set our little BSS's SSID to AIRDROP_NETWORK. I wanted an SSID that we could easily identify in the wireless LAN traces and I believe the name I have chosen will stand out in our Sniffer captures. If you've read any of the many "how to set up a wireless LAN" articles or books, you already know that adjacent 802.11b Channels can overlap each other and cause interference. The rule is to separate APs operating in the same vicinity by five 802.11b Channels. The EDTP Florida Room Lab is currently running on Channel 11. There is also another Florida Room server-based wireless LAN running on Channel 6. So, to keep things between the lines, I chose Channel 1 for the AirDrop BSS we're creating. It's not a perfect world and there will be times where physical 802.11b networks will overlap. When that happens, the SSID earns it keep. Wireless LAN stations with hard-coded SSIDs will always look for and associate with an AP with a matching SSID.

For now, we'll leave everything else in the Netgear AP just as it is. That way our laptop and desktop personal computers can be given IP addresses via the AP's DHCP (Dynamic

Host Configuration Protocol) engine. The AirDrop 802.11b hardware does not participate in the DHCP exchanges and will be hard-coded with an IP address of 192.168.0.151. This puts the AirDrop module out of range of the default AP DHCP IP address lease range, which is 192.168.0.2 to 192.168.0.51 on the Netgear AP.

Something's in the Air

I applied the SSID and Channel changes to the Netgear AP. Shortly thereafter, I checked the Netgear AP's status and saw two IP address leases had been distributed. As you can see in Screen Capture 8.2, IP addresses 192.168.0.3 was assigned to the desktop named TEST, and SNOOPER; the 802.11b laptop, received 192.168.0.11. The Netgear AP uses the first IP address of 192.168.0.1.

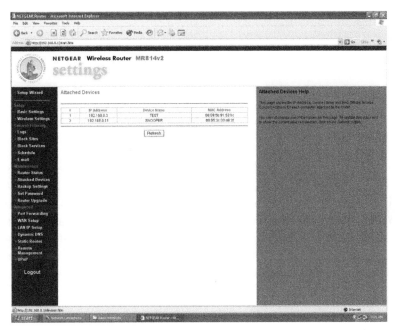

Screen Capture 8.2: The Device Names in this screen capture are the Windows NetBIOS names that were given to the desktop personal computer and the 802.11b-enabled station laptop. Note that the station laptop and desktop personal computer's Netgear 802.11b PCI adapter MAC addresses are also given here. The laptop running the sniffer application will not show up here as its Symbol PCMCIA NIC is under total control of the Netasyst Sniffer driver and does not perform any transmissions.

A typical Netasyst Sniffer capture is shown in Sniffer Capture 8.3. This looks great on a 19-inch color monitor, but lacks a bit in the smaller pages of this book. So, to make it a bit easier to follow the Netasyst Sniffer capture data, I'll format the capture data in a manner similar to the way I'm presenting the AirDrop driver source code.

Sniffer Capture 8.3: These Netasyst Sniffer screen captures can get really busy. So, when it makes sense, I'll supplement the Netasyst Sniffer graphical capture data with Netasyst Sniffer text content.

The 802.11b CompactFlash NIC mounted on the AirDrop-P hasn't uttered one single RF word. However, the Netgear AP is screaming. At regular intervals, the Netgear AP transmits a Beacon frame. Beacons advertise the AP's capabilities and present BSS identification and BSS rules to a wireless station looking to join the AP's BSS. The actual Netasyst Sniffer shot of Beacons flying about is represented in Sniffer Capture 8.4. The content of one of the Beacon frames is presented to you in Sniffer Text 8.1. Let's tear apart one of the Beacon frames and see what our Netgear AP is hollering about.

Sniffer Text 8.1

```
DLC: ----- DLC Header -----
      DLC:
      DLC: Frame 1 arrived at  15:34:11.0342; frame size is 76 (004C hex)
bytes.
      DLC: Signal level              = 100%
      DLC: Channel                   = 1
      DLC: Data rate                 = 2 ( 1.0 Megabits per second)
      DLC:
      DLC: Frame Control Field #1 = 80
      DLC:                 .... ..00 = 0x0 Protocol Version
      DLC:                 .... 00.. = 0x0 Management Frame
      DLC:                 1000 .... = 0x8 Beacon (Subtype)
      DLC: Frame Control Field #2 = 00
      DLC:                 .... ...0 = Not to Distribution System
      DLC:                 .... ..0. = Not from Distribution System
      DLC:                 .... .0.. = Last fragment
```

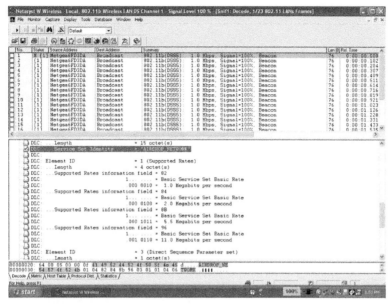

Sniffer Capture 8.4: Sometimes it is impossible to show you everything in the Sniffer Capture as the Netasyst Sniffer display must be scrolled to access the full content of the capture. So, I'll take you through screens like this and highlight related text in the capture's written content and the associated hexadecimal representation.

Sniffer Text 8.1 (continued)

```
DLC:                    .... 0... = Not retry
DLC:                    ...0 .... = Active Mode
DLC:                    ..0. .... = No more data
DLC:                    .0.. .... = Wired Equivalent Privacy is off
DLC:                    0... .... = Not ordered
DLC: Duration                     = 0 (in microseconds)
DLC: Destination Address          = BROADCAST FFFFFFFFFFFF, Broadcast
DLC: Source Address               = Station Netgea6FD3DA
DLC: Basic Service Set ID         = Netgea6FD3DA
DLC: Sequence Control             = 0x63E0
DLC: ...Sequence Number           = 0x63E (1598)
DLC: ...Fragment Number           = 0x0   (0)
DLC: Timestamp                    = 7093760389 (in microseconds)
DLC: Beacon Interval              = 100
DLC: Capability information field #1 = 05
DLC:                    .... ...1 = Extended Service Set is on
DLC:                    .... ..0. = Independent Basic Service Set
                                    is off
DLC:                    .... 01.. = Point coordinator at Access
                                    Point for delivery and polling
DLC:                    ...0 .... = No privacy
DLC:                    ..0. .... = Short Preamble option is not
                                    allowed
DLC:                    .0.. .... = Packet Binary Convolutional
```

```
                                        Coding Modulation mode option
                                        is not allowed
        DLC:                        0... .... = Channel agility is not in use
        DLC: Capability information field #2 = 00
        DLC:                        0000 0000 = Reserved
        DLC:
        DLC: Element ID                 = 0 (Service Set Identifier)
        DLC: ...Length                  = 15 octet(s)
        DLC: ...Service Set Identity    = "AIRDROP_NETWORK"
        DLC:
        DLC: Element ID                 = 1 (Supported Rates)
        DLC: ...Length                  = 4 octet(s)
        DLC: ...Supported Rates information field = 82
        DLC:                        1... .... = Basic Service Set Basic Rate
        DLC:                        .000 0010 = 1.0 Megabits per second
        DLC: ...Supported Rates information field = 84
        DLC:                        1... .... = Basic Service Set Basic Rate
        DLC:                        .000 0100 = 2.0 Megabits per second
        DLC: ...Supported Rates information field = 8B
        DLC:                        1... .... = Basic Service Set Basic Rate
        DLC:                        .000 1011 = 5.5 Megabits per second
        DLC: ...Supported Rates information field = 96
        DLC:                        1... .... = Basic Service Set Basic Rate
        DLC:                        .001 0110 = 11.0 Megabits per second
        DLC:
        DLC: Element ID                 = 3 (Direct Sequence Parameter set)
        DLC: ...Length                  = 1 octet(s)
        DLC: ...dot11CurrentChannelNumber = 1
        DLC:
        DLC: Element ID                 = 4 (Contention Free Parameter set)
        DLC: ...Length                  = 6 octet(s)
        DLC: ...CFP Count               = 0
        DLC: ...CFP Period              = 2
        DLC: ...CFP Maximum Duration    = 0 (in Time Unit)
        DLC: ...CFP Duration Remaining  = 0 (dot11CurrentPattern)
        DLC:
        DLC: Element ID                 = 5 (Traffic Indication Map)
        DLC: ...Length                  = 4 octet(s)
        DLC: ...Delivery Traffic Indication Message Count = 0
        DLC: ...Delivery Traffic Indication Message Period = 1
        DLC: ...Bitmap control field = 00
        DLC:                .... ...0 = Traffic Indicator bit
        DLC:                0000 000. = 0 Bitmap offset
        DLC: ...Partial Virtual Bitmap    = 00
        DLC:
ADDR  HEX                                                    ASCII
0000: 80 00 00 00 ff ff ff ff ff ff 00 09 5b 6f d3 da  | €...ÿÿÿÿÿÿ..[oóÚ
0010: 00 09 5b 6f d3 da e0 63 85 31 d2 a6 01 00 00 00  | ..[oóÚàc…1Ò¦....
0020: 64 00 05 00 00 0f 41 49 52 44 52 4f 50 5f 4e 45  | d.....AIRDROP_NE
0030: 54 57 4f 52 4b 01 04 82 84 8b 96 03 01 01 04 06  | TWORK..,„<-.....
0040: 00 02 00 00 00 00 05 04 00 01 00 00              | ...........
```

Sniffer Text 8.1: There's lots of good information to be had from a Beacon frame, and using a Netasyst Sniffer capture to decipher the Beacon bits beats the heck out of counting bits and bytes while at the same time trying to reference the 802.11b standards documentation.

As you can see in Mini Sniff 8.1, there are four areas of the captured Beacon frame that have meaning to us as it relates to our AirDrop BSS. The first Beacon frame area provides a Channel number, the quality of the signal and the data rate. The Channel, signal strength and data rate information in this area is provided as a feature of the Netasyst Sniffer.

Note that the Beacon is transmitted at 1 Mbps. The reason for this is that the AP doesn't assume a station that can hear its Beacon can operate at any rate greater than 1 Mbps and less than or equal to 11 Mbps. Beaconing at 1 Mbps ensures that any 802.11b station within range can receive and interpret the AP's Beacon frame.

The Frame Control Field #1 identifies the Management frame as a Beacon. Since a Beacon is not moving data to a physical destination inside the BSS and is not sending data from a station within the BSS, the Distribution System is not involved. This is reflected in Frame Control Field #2. We can walk away from this area of the Beacon frame with one more piece of information; WEP is disabled.

Mini Sniff 8.1

```
      DLC: Frame 1 arrived at  15:34:11.0342; frame size is 76 (004C hex)
bytes.
      DLC: Signal level                = 100%
      DLC: Channel                     = 1
      DLC: Data rate                   = 2 ( 1.0 Megabits per second)
      DLC:
      DLC: Frame Control Field #1 = 80
      DLC:                 .... ..00 = 0x0 Protocol Version
      DLC:                 .... 00.. = 0x0 Management Frame
      DLC:                 1000 .... = 0x8 Beacon (Subtype)
      DLC: Frame Control Field #2 = 00
      DLC:                 .... ...0 = Not to Distribution System
      DLC:                 .... ..0. = Not from Distribution System
      DLC:                 .... .0.. = Last fragment
      DLC:                 .... 0... = Not retry
      DLC:                 ...0 .... = Active Mode
      DLC:                 ..0. .... = No more data
      DLC:                 .0.. .... = Wired Equivalent Privacy is off
      DLC:                 0... .... = Not ordered
ADDR  HEX                                             ASCII
0000: 80 00 00 00 ff ff ff ff ff ff 00 09 5b 6f d3 da | €....ÿÿÿÿÿÿ..[oóÚ
0010: 00 09 5b 6f d3 da e0 63 85 31 d2 a6 01 00 00 00 | ..[oóÚàc…1Ò¦....
0020: 64 00 05 00 00 0f 41 49 52 44 52 4f 50 5f 4e 45 | d.....AIRDROP_NE
0030: 54 57 4f 52 4b 01 04 82 84 8b 96 03 01 01 04 06 | TWORK..,„<-.....
0040: 00 02 00 00 00 00 05 04 00 01 00 00             | ...........
```

The second Beacon frame area of interest, which is represented by the text in Mini Sniff 8.2, contains addressing information. The AP's Beacon frame is considered a broadcast frame since it is intended for all in range to hear. Notice that our Netgear AP has taken its own MAC address as its BSSID (Basic Service Set ID). That's normal. Since each MAC address must be unique, using one's own MAC address as its BSSID automatically makes the AP's BSSID unique as well.

Mini Sniff 8.2

```
        DLC: Destination Address            = BROADCAST FFFFFFFFFFFF, Broadcast
        DLC: Source Address                 = Station Netgea6FD3DA
        DLC: Basic Service Set ID           = Netgea6FD3DA
        DLC: Sequence Control               = 0x63E0
        DLC: ...Sequence Number             = 0x63E (1598)
        DLC: ...Fragment Number             = 0x0   (0)
        DLC: Timestamp                      = 7093760389 (in microseconds)
        DLC: Beacon Interval                = 100
ADDR  HEX                                                   ASCII
0000: 80 00 00 00 ff ff ff ff ff ff 00 09 5b 6f d3 da  | €...ÿÿÿÿÿÿ..[oÓÚ
0010: 00 09 5b 6f d3 da e0 63 85 31 d2 a6 01 00 00 00  | ..[oÓÚàc…1ò¦....
0020: 64 00 05 00 00 0f 41 49 52 44 52 4f 50 5f 4e 45  | d.....AIRDROP_NE
0030: 54 57 4f 52 4b 01 04 82 84 8b 96 03 01 01 04 06  | TWORK..,„<-.....
0040: 00 02 00 00 00 00 05 04 00 01 00 00              | ...........
```

The Sequence Control field consists of two bytes that identify fragmented 802.11b frames. Fragmentation and reassembly of frame fragments is automatically taken care of by the 802.11b CompactFlash NIC. Fragmentation can occur when long frames are being transmitted into an environment with lots of interference. The idea is that shorter frames are more likely to make it through the interference without alteration and if they don't, it's more efficient to resend a short frame than a longer one. The four least significant bits of the Sequence Number are dedicated to fragment identification. The twelve most significant bits of the Sequence Number are part of a modulo 4096 counter scheme that is used to identify the fragmented frame. Note that the Netasyst Sniffer has broken the Sequence Control bytes down into the 4-bit and 12-bit patterns I've just described. Don't work yourself into a tizzy about the Sequence Control field in Mini Sniff 8.3, as it's the first and last time you'll hear about it in this book. Some of the stuff I'll talk about is interesting, but not necessary knowledge to understand what the AirDrop 802.11b driver is doing.

Mini Sniff 8.3

```
        DLC: Destination Address            = BROADCAST FFFFFFFFFFFF, Broadcast
        DLC: Source Address                 = Station Netgea6FD3DA
        DLC: Basic Service Set ID           = Netgea6FD3DA
        DLC: Sequence Control               = 0x63E0
        DLC: ...Sequence Number             = 0x63E (1598)
        DLC: ...Fragment Number             = 0x0   (0)
        DLC: Timestamp                      = 7093760389 (in microseconds)
        DLC: Beacon Interval                = 100
ADDR  HEX                                                   ASCII
0000: 80 00 00 00 ff ff ff ff ff ff 00 09 5b 6f d3 da  | €...ÿÿÿÿÿÿ..[oÓÚ
0010: 00 09 5b 6f d3 da e0 63 85 31 d2 a6 01 00 00 00  | ..[oÓÚàc…1ò¦....
0020: 64 00 05 00 00 0f 41 49 52 44 52 4f 50 5f 4e 45  | d.....AIRDROP_NE
0030: 54 57 4f 52 4b 01 04 82 84 8b 96 03 01 01 04 06  | TWORK..,„<-.....
0040: 00 02 00 00 00 00 05 04 00 01 00 00              | ...........
```

The Timestamp, shown highlighted in Mini Sniff 8.4, is a very clever 64-bit way of synchronizing the time within the BSS. Using the Timestamp field an 802.11b station could update its internal clock with the reception of every Beacon from the AP it is associated with. Read the Timestamp hexadecimal bytes in this order: 00 00 00 01 A6 D2 31 85 hexadecimal = 7093760389 decimal.

Mini Sniff 8.4

```
        DLC: Destination Address          = BROADCAST FFFFFFFFFFFF, Broadcast
        DLC: Source Address               = Station Netgea6FD3DA
        DLC: Basic Service Set ID         = Netgea6FD3DA
        DLC: Sequence Control             = 0x63E0
        DLC: ...Sequence Number           = 0x63E (1598)
        DLC: ...Fragment Number           = 0x0   (0)
        DLC: Timestamp                    = 7093760389 (in microseconds)
        DLC: Beacon Interval              = 100
ADDR  HEX                                                   ASCII
0000: 80 00 00 00 ff ff ff ff ff ff 00 09 5b 6f d3 da  | €...ÿÿÿÿÿÿ..[oÓÚ
0010: 00 09 5b 6f d3 da e0 63 85 31 d2 a6 01 00 00 00  | ..[oÓÚàc…1Ò¦....
0020: 64 00 05 00 00 0f 41 49 52 44 52 4f 50 5f 4e 45  | d.....AIRDROP_NE
0030: 54 57 4f 52 4b 01 04 82 84 8b 96 03 01 01 04 06  | TWORK..,„<-.....
0040: 00 02 00 00 00 00 05 04 00 01 00 00              | ...........
```

The granularity of the Timestamp depends on the frequency of the Beacon transmissions and how often a sleeping station wakes up to receive a Beacon and "check the mailbox". The Beacon Interval is a good thing for a sleeping 802.11b station to know because that's how the sleeping station knows when to wake up and check the AP for buffered frames it needs to receive and process. The Beacon Interval is based on a number of TUs, or time units, with each TU being equal to 1024 µS. The Beacon Interval of 100 decimal shown in Mini Sniff 8.5 is typical and that's the default setting for our Netgear AP. Note that all of the values thus far are in little-endian format. Little-endian format simply means that the low-order byte is stored at the lowest memory address, and the high-order byte is stored at the highest memory address. For example, the bytes 64 00 in little-endian format represent the decimal value of 100, not 25,600.

Mini Sniff 8.5

```
        DLC: Destination Address          = BROADCAST FFFFFFFFFFFF, Broadcast
        DLC: Source Address               = Station Netgea6FD3DA
        DLC: Basic Service Set ID         = Netgea6FD3DA
        DLC: Sequence Control             = 0x63E0
        DLC: ...Sequence Number           = 0x63E (1598)
        DLC: ...Fragment Number           = 0x0   (0)
        DLC: Timestamp                    = 7093760389 (in microseconds)
        DLC: Beacon Interval              = 100
ADDR  HEX                                                   ASCII
0000: 80 00 00 00 ff ff ff ff ff ff 00 09 5b 6f d3 da  | €...ÿÿÿÿÿÿ..[oÓÚ
0010: 00 09 5b 6f d3 da e0 63 85 31 d2 a6 01 00 00 00  | ..[oÓÚàc…1Ò¦....
0020: 64 00 05 00 00 0f 41 49 52 44 52 4f 50 5f 4e 45  | d.....AIRDROP_NE
0030: 54 57 4f 52 4b 01 04 82 84 8b 96 03 01 01 04 06  | TWORK..,„<-.....
0040: 00 02 00 00 00 00 05 04 00 01 00 00              | ...........
```

The Netasyst Sniffer is good for more than just capturing wireless LAN frames. The Netasyst Sniffer captures can also tie time to various frame events. To illustrate this, I captured a gaggle of our Netgear AP's Beacons. If you look at the Interval column in Table 8.1, you'll see that all of the numbers are gravitating towards a 0.102.4 pattern. You can also see the Relative Time points clicking in at intervals close to one-tenth of a second each, which supports our Beacon Interval value of 100 decimal time units or .1024 seconds.

The Netgear OUI

00-09-5B (hex)	**Netgear, Inc.**	
00095B (base 16)	**Netgear, Inc.**	
	4500 Great America Parkway	
	Santa Clara CA 95054	
	UNITED STATES	

Source	Destination	Summary				Bytes	Relative Time	Interval
Netgea6FD3DA	Broadcast	802.11b(DSSS): 1.0 Mbps	Signal=100%	Beacon		76	0:00:00.000	0.000.000
Netgea6FD3DA	Broadcast	802.11b(DSSS): 1.0 Mbps	Signal=100%	Beacon		76	0:00:00.102	0.102.384
Netgea6FD3DA	Broadcast	802.11b(DSSS): 1.0 Mbps	Signal=100%	Beacon		76	0:00:00.204	0.102.394
Netgea6FD3DA	Broadcast	802.11b(DSSS): 1.0 Mbps	Signal=100%	Beacon		76	0:00:00.307	0.102.399
Netgea6FD3DA	Broadcast	802.11b(DSSS): 1.0 Mbps	Signal=100%	Beacon		76	0:00:00.409	0.102.472
Netgea6FD3DA	Broadcast	802.11b(DSSS): 1.0 Mbps	Signal=100%	Beacon		76	0:00:00.512	0.102.400
Netgea6FD3DA	Broadcast	802.11b(DSSS): 1.0 Mbps	Signal=100%	Beacon		76	0:00:00.614	0.102.409
Netgea6FD3DA	Broadcast	802.11b(DSSS): 1.0 Mbps	Signal=100%	Beacon		76	0:00:00.716	0.102.417
Netgea6FD3DA	Broadcast	802.11b(DSSS): 1.0 Mbps	Signal=100%	Beacon		76	0:00:00.819	0.102.372
Netgea6FD3DA	Broadcast	802.11b(DSSS): 1.0 Mbps	Signal=100%	Beacon		76	0:00:00.921	0.102.391
Netgea6FD3DA	Broadcast	802.11b(DSSS): 1.0 Mbps	Signal=100%	Beacon		76	0:00:01.024	0.102.388
Netgea6FD3DA	Broadcast	802.11b(DSSS): 1.0 Mbps	Signal=100%	Beacon		76	0:00:01.126	0.102.389
Netgea6FD3DA	Broadcast	802.11b(DSSS): 1.0 Mbps	Signal= 97%	Beacon		76	0:00:01.228	0.102.395
Netgea6FD3DA	Broadcast	802.11b(DSSS): 1.0 Mbps	Signal=100%	Beacon		76	0:00:01.331	0.102.397
Netgea6FD3DA	Broadcast	802.11b(DSSS): 1.0 Mbps	Signal=100%	Beacon		76	0:00:01.433	0.102.414
Netgea6FD3DA	Broadcast	802.11b(DSSS): 1.0 Mbps	Signal=100%	Beacon		76	0:00:01.535	0.102.372
Netgea6FD3DA	Broadcast	802.11b(DSSS): 1.0 Mbps	Signal=100%	Beacon		76	0:00:01.638	0.102.402
Netgea6FD3DA	Broadcast	802.11b(DSSS): 1.0 Mbps	Signal=100%	Beacon		76	0:00:01.740	0.102.393
Netgea6FD3DA	Broadcast	802.11b(DSSS): 1.0 Mbps	Signal= 97%	Beacon		76	0:00:01.843	0.102.392

Table 8.1: The timing columns in this table speak for themselves. It's nice to have the facts to back up our interpretation of the numbers. The Source column is the MAC address of the transmitting device, which is our Netgear AP. The "Netgea" is inserted automatically by the Netasyst Sniffer as it knows how to interpret the manufacturer's OUI (Organizationally Unique Identifier) bytes of the MAC address (00 09 5B).

The next section of our AP's Beacon frame is only two bytes long. Basically, the two bytes broken down below tell the listening 802.11b station what it must do or not do to be able to join the Beaconing AP's BSS. We haven't talked about ESSs (Extended Service Sets) as we won't be implementing one. An ESS is a set of AP's that when properly configured can pass data amongst themselves. An ESS is usually deployed to allow a mobile wireless LAN station to move out of range of one AP and into range of a different AP and still be able to deliver and receive data from its original AP's BSS. In this case, ESS doesn't really follow the definition I've just given to you. If the ESS bit is on and the IBSS bit is off in the Capability information field #1, the network type is infrastructure, or BSS. If the ESS bit is off and the IBBS bit is on, the network is operating in ad-hoc, or IBSS, mode.

The message "Point coordinator at Access Point for delivery and polling" simply says that everything must go through the AP. The rest of the Capability information bits in Mini Sniff 8.6 just say NO and we won't worry about them as they don't change anything we'll do in the AirDrop driver code. The only bit we may see change as we move forward is the privacy bit, which will be turned on by the AP when WEP is required.

Mini Sniff 8.6

```
DLC:  Capability information field #1 = 05
DLC:                    .... ...1 = Extended Service Set is on
DLC:                    .... ..0. = Independent Basic Service Set is off
DLC:                    .... 01.. = Point coordinator at Access Point for
delivery and polling
DLC:                    ...0 .... = No privacy
DLC:                    ..0. .... = Short Preamble option is not allowed
DLC:                    .0.. .... = Packet Binary Convolutional Coding
Modulation mode option is not allowed
DLC:                    0... .... = Channel agility is not in use
DLC:  Capability information field #2 = 00
DLC:                    0000 0000 = Reserved
ADDR  HEX                                              ASCII
0000: 80 00 00 00 ff ff ff ff ff ff 00 09 5b 6f d3 da | €...ÿÿÿÿÿÿ..[oóÚ
0010: 00 09 5b 6f d3 da e0 63 85 31 d2 a6 01 00 00 00 | ..[oóÚàc…1Ò¦....
0020: 64 00 05 00 00 0f 41 49 52 44 52 4f 50 5f 4e 45 | d.....AIRDROP_NE
0030: 54 57 4f 52 4b 01 04 82 84 8b 96 03 01 01 04 06 | TWORK..,„<-.....
0040: 00 02 00 00 00 00 05 04 00 01 00 00              | ...........
```

The final AP Beacon frame segment we'll talk about provides some very important information to a potential 802.11b station partner in the form of Element IDs. For a wireless LAN station to be able to join a particular BSS, the wireless LAN station must know the SSID of the BSS it wants to join. We'll code our AirDrop driver to look for and join the AP with an SSID of "AIRDROP_NETWORK". As you'll see later, the AirDrop wireless LAN station will pick up the Netgear AP's SSID and configure itself accordingly to gain access to the AIRDROP_NETWORK BSS. In addition to getting an SSID, the potential wireless LAN station is told what speeds the AP supports, the transmission method (Direct Sequence Spread Spectrum (DSSS) in our case) and the currently configured Channel. Note that each Element ID is preceded by a byte count, which tells the wireless LAN station how many bytes to parse in that field. For instance, the SSID listed in Mini Sniff 8.7 is 15 octets (bytes) in length. Also note that the most significant bit of each of the Supported Rates determines if the rate is optional or mandatory. The rate is mandatory if the most significant bit of a Supported Rates information field is set. Mandatory in this case means that if the wireless LAN station wants to join this BSS, it must be able to support all of the rates listed in the Supported Rates Element ID.

Mini Sniff 8.7

```
    DLC: Element ID                     = 0 (Service Set Identifier)
    DLC: ...Length                      = 15 octet(s)
    DLC: ...Service Set Identity        = "AIRDROP_NETWORK"
    DLC:
    DLC: Element ID                     = 1 (Supported Rates)
    DLC: ...Length                      = 4 octet(s)
    DLC: ...Supported Rates information field = 82
    DLC:                          1... .... = Basic Service Set Basic Rate
    DLC:                          .000 0010 =  1.0 Megabits per second
    DLC: ...Supported Rates information field = 84
    DLC:                          1... .... = Basic Service Set Basic Rate
    DLC:                          .000 0100 =  2.0 Megabits per second
    DLC: ...Supported Rates information field = 8B
    DLC:                          1... .... = Basic Service Set Basic Rate
    DLC:                          .000 1011 =  5.5 Megabits per second
    DLC: ...Supported Rates information field = 96
    DLC:                          1... .... = Basic Service Set Basic Rate
    DLC:                          .001 0110 = 11.0 Megabits per second
    DLC:
    DLC: Element ID                     = 3 (Direct Sequence Parameter set)
    DLC: ...Length                      = 1 octet(s)
    DLC: ...dot11CurrentChannelNumber   = 1
ADDR  HEX                                               ASCII
0000: 80 00 00 00 ff ff ff ff ff ff 00 09 5b 6f d3 da | €...ÿÿÿÿÿÿ..[oÓÚ
0010: 00 09 5b 6f d3 da e0 63 85 31 d2 a6 01 00 00 00 | ..[oÓÚàc…1Ò¦....
0020: 64 00 05 00 00 0f 41 49 52 44 52 4f 50 5f 4e 45 | d.....AIRDROP_NE
0030: 54 57 4f 52 4b 01 04 82 84 8b 96 03 01 01 04 06 | TWORK..„„<-.....
0040: 00 02 00 00 00 00 05 04 00 01 00 00             | ..........
```

The hardware side of our little AirDrop BSS is in place and the AIRDROP_NETWORK's Netgear AP has some job openings. It just so happens that our AirDrop module is looking for work.

Guitars and Hollywood

Before we move on, I've got an easy one for you. Who was Elvis Presley's first guitarist and what song did they do together first?

If you ever have a chance to visit the "house," eat some possum' stew and take a swim in the cement pond with Lester Flatt and Earl Scruggs, you'd better address Granny as Ms. Daisy Moses. I'm a Tennessee boy and I lived about 20 miles from a place called Bugtussle Hollar, which is a place Jed and Granny mentioned often. Bet you thought they were just makin' that up, huh?

AirDrop Driver Basics

Thus far this 802.11b stuff has been an easy ride and believe it or not, it won't get any more difficult. This chapter will show you what's behind the 802.11b functional building blocks that form the basis of the AirDrop-A and AirDrop-P 802.11b drivers.

When I was a kid (a loooong time ago), Legos were made from wood and instead of constructing plastic motorized robots, I constructed luxurious miniature log cabins. The trick to learning to build little log cabins with Legos was knowing which logs to use on a particular part of the cabin. The same idea holds true for being successful with the components of the AirDrop embedded 802.11b driver. You've already been exposed to some of the AirDrop 802.11b driver "wood" by way of the CompactFlash I/O routines. The AirDrop 802.11b driver concepts that will be presented in this chapter will complete your 802.11b Compact-Flash NIC I/O education and leave you with a big pile of 802.11b "wood" to work with.

When you're finished with this chapter, understanding the rest of the AirDrop 802.11b driver code will be like building Lego log cabins, child's play. With that, let's begin with some definitions of terms you'll see all over the AirDrop 802.11b code and any Linux 802.11b driver code you care to study.

BAP

A BAP, or Buffer Access Path, is the sole method of microcontroller, and ultimately, human interaction with the internal memory management subsystems of the PRISM chipset. Without a BAP, you would have to know exactly where in the PRISM chipset memory subsystem to place or retrieve data and NIC management information. In other words, a BAP provides an automated indirect access path to the PRISM chipset memory complex. For example, you may know the address of a particular register you need to write a value to but you don't have to care about where it resides in memory. Just tell the BAP who you want to talk to and it will take you there. Once you get your AirDrop hardware activated on a live wireless LAN, there will be times when you will have to read and write a dynamically allocated buffer area inside the PRISM memory complex. Using the BAP for data transfers eliminates having to find or keep up with the mobile buffer's location and having to manage the mobile buffer's memory pointers. Basically, all we have to do is provide an address and an offset value to the BAP and it handles the tough stuff behind the scenes. In addition, the BAP will automatically increment the target buffer's buffer pointer as data is transferred between the memory location and the microcontroller.

What's better than a BAP? Two BAPs. According to the ISL3873A data sheet and some passing references in the Linux 802.11b drivers, two BAPs exist within the PRISM chipset system. That means that we could theoretically implement concurrent read and write transfers by reading one BAP and writing to the other BAP. In fact, one of the Linux 802.11b drivers uses one BAP exclusively to perform read operations and one BAP for doing nothing but write operations. The AirDrop 802.11b driver won't adhere to that convention as we'll be checking for the first free BAP before every read or write operation. The pair of BAPs is defined in our AirDrop-A and AirDrop-P 802.11b drivers as shown in Code Snippet 9.1.

Code Snippet 9.1

```
*****************************
//* BAP0
#define   BAP0          0
#define   Select0_Register   0x0018
#define   Offset0_Register   0x001C
#define   Data0_Register     0x0036

//* BAP1
#define   BAP1          1
#define   Select1_Register   0x001A
#define   Offset1_Register   0x001E
#define   Data1_Register     0x0038
*****************************
```

Code Snippet 9.1: A Select Register, an Offset Register and a Data Register are associated with each individual BAP. The Data Register is the data portal that lies between the PRISM chipset and the AirDrop microcontroller.

FID

The PRISM chipset firmware is constantly allocating groups of 128-byte memory blocks into transmit and receive buffers to service incoming and outgoing 802.11b frames. This dynamic allocation of buffers within the PRISM memory subsystem is transparent to both the programmer and the end user. The dynamically allocated buffer chains are built upon linked lists of pointers to noncontiguous 128-byte memory blocks contained within the PRISM chipset memory subsystem. This method of using linked lists of pointers allows the PRISM chipset hardware to assemble buffers quickly from deallocated and unused fragments of the PRISM chipset memory subsystem. In addition to being a very fast and efficient way of allocating buffer memory, the linked list concept allows a BAP to automatically shift from the end of a 128-byte block to the beginning of the next 128-byte memory block in the buffer chain under firmware control and without programmer intervention. The programmer simply reads and writes to what seems to be a single address. The programmer never needs to know where the BAP is pointing to or where the buffer chain resides in the PRISM chipset memory subsystem. If the programmer's code is too big to fit in a buffer, the extra bytes are automatically discarded instead of wrapping around to the beginning of the buffer or overwriting data in an adjacent buffer. Human access to the buffer chains created by the dynamic linking of 128-

byte memory blocks is provided in the form of a FID, or Frame ID. We can actually manually request an allocation of a dynamic buffer area in the AirDrop 802.11b driver. The buffer allocation process returns a FID, which we can then use to access the newly allocated buffer space via a BAP. Yes, it is that simple.

In a nutshell, a FID is simply a unique physical reference to an associated dynamically-allocated, 128-byte-per-block buffer chain. A BAP is a means of accessing the data within the dynamically-allocated buffer chain that a FID identifies. Figure 9.1 shows the relationship between a BAP, a FID and buffer memory.

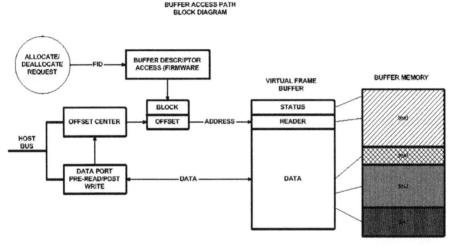

Figure 9.1: All we have to do to gain access to a FID is point a BAP at it, assign an offset into the FID if necessary and read and write at will.

Let's take Figure 9.1 in from left to right. We must first either allocate a buffer area and obtain a FID, or already know what FID we want to access. Once we've chosen one of the two available BAPs, we can introduce and offset into the buffer represented by the FID if we need to. The FID and offset values are used by the BAP mechanism to form an address. As we read or write, the virtual frame buffer takes care of automatically retrieving or writing our data and incrementing the buffer memory pointer for each data transfer.

RID

I've searched the free world of open source documentation to try to arrive at someone calling a RID by its real name. Believe it or not, the acronym RID is all over the Linux 802.11b driver source documentation but it is never called anything else. So, to keep with what must be a secret society convention I'll just call it a RID as well. It's not as important to know what RID stands for as it is to know what a RID is used for. Read through Code Snippet 9.2, which is a list of RID definitions used by the AirDrop 802.11b driver.

Code Snippet 9.2

```
******************************
//* RID CONFIGURATION RECORDS
#define   RID_cfgPortType                   0xFC00
#define   RID_cfgPortType_Length            2
#define   RID_cfgPortType_IBSS              0          //sets IBSS or ad-hoc network operation
#define   RID_cfgPortType_Infrastructure   1          //sets BSS or infrastructure network operation

#define   RID_cfgOwnMACAddress             0xFC01     //our MAC address
#define   RID_cfgDesiredSSID               0xFC02     //the SSID of the BSS we want to join
#define   RID_DesiredSSID_Length           18         //word count of the DesiredSSID RID

#define   RID_cfgOwnChannel                0xFC03     //set the channel
#define   RID_cfgOwnChannel_Length   2

#define   RID_cfgOwnSSID                   0xFC04     //set the SSID
#define   RID_cfgOwnSSID_Length            18

#define   RID_cfgMaxDataLength             0xFC07     //set maximum length of frame body data
#define     RID_cfgMaxDataLength_Length    2
#define   RID_cfgMaxDataLength_Value1500

#define   RID_cfgWEPDefaultKeyID           0xFC23     //which key will do encryption
#define   RID_cfgWEPDefaultKeyID_Length    2
#define   RID_cfgWEPDefaultKeyID_Value0    0x0000     //encryption key0
#define   RID_cfgWEPDefaultKeyID_Value1    0x0001     //encryption key1
#define   RID_cfgWEPDefaultKeyID_Value2    0x0010     //encryption key2
#define   RID_cfgWEPDefaultKeyID_Value3    0x0011     //encryption key3

#define   RID_Default128Key_Length         8          //128-bit encryption with a 13-byte key
#define   RID_Default64Key_Length          4          //64-bit encryption with a 5-byte key and a 24-bit IV

#define   RID_cfgDefaultKey0               0xFC24     //key bytes for key0
#define   RID_cfgDefaultKey1               0xFC25     //key bytes for key1
#define   RID_cfgDefaultKey2               0xFC26     //key bytes for key2
#define   RID_cfgDefaultKey3               0xFC27     //key bytes for key3

#define   RID_cfgWEPFlags                  0xFC28     //turn WEP ON/OFF and decide how much to use IV
#define   RID_cfgWEPFlags_Length           2
#define   RID_cfgWEPFlags_Value_ON   0x0001   //encryption/decryption enabled and IV reuse every frame
#define   RID_cfgWEPFlags_Value_OFF  0x0000   //encryption/decryption disabled

#define   RID_cfgAuthentication            0xFC2A     //choose Open System or Shared Key Authentication
#define   RID_cfgAuthentication_Length     2
#define   RID_cfgAuthentication_Value_OS   0x0001     //Open System authentication
#define   RID_cfgAuthentication_Value_SK   0x0002     //Shared Key authentication
#define   RID_cfgAuthentication_Value_OK   0x0003     //Open System and Shared Key authentication

#define   RID_TxRateControl                0xFC84     //set transmission bit rates
#define   RID_TxRateControl_Length         2
#define   RID_TxRateControl_Value_1        0x0001     //1 Mbps
#define   RID_TxRateControl_Value_2        0x0002     //2 Mbps
#define   RID_TxRateControl_Value_55       0x0004     //5.5 Mbps
#define   RID_TxRateControl_Value_11       0x0008     //11 Mbps
#define   RID_TxRateControl_Value_ALL      0x000F     //11 Mbps
//* RID INFORMATION RECORDS
#define   RID_CommsQuality                 0xFD43     // Request the current communications quality
#define   RID_CommsQuality_Length          4
******************************
```

Code Snippet 9.2: It's pretty obvious that a RID can be used to configure the operation of an 802.11b CompactFlash NIC. A RID is also useful for things like gathering communications quality data information. No matter what you call it, what you see in the Code Snippet is all you really need to know about a RID.

Access to RID data is also accomplished using the services of a BAP. We've got a bunch of work to do with RIDs before we can make our AirDrop module hardware function as a true 802.11b wireless LAN station. And, it just so happens that the next step in the 802.11b CompactFlash NIC setup process is configuring the NIC for BSS network operation. Let's talk about the BSS configuration code you see in Code Snippet 9.3.

Code Snippet 9.3

```
*****************************
//* RID CONFIGURATION RECORDS
#define  RID_cfgPortType                  0xFC00
#define  RID_cfgPortType_Length           2
#define  RID_cfgPortType_IBSS             0       //sets ad-hoc network opera-
tion
#define  RID_cfgPortType_Infrastructure  1       //sets BSS Network operation

//* COMMANDS
#define  Access_Cmd                       0x0021

#define  BAP0Busy_Bit_Mask                0x8000 //busy bit in Offset0 register
#define  BAP0Err_Bit_Mask                 0x4000 //Err bit in Offset0 register
#define  BAP1Busy_Bit_Mask                0x8000 //busy bit in Offset1 register
#define  BAP1Err_Bit_Mask                 0x4000 //Err bit in Offset1 register
#define  AccessWrite_Mask                 0x0100 //Write RID to MAC

#define fidrid_buffer_size                256
unsigned int fidrid_buffer[fidrid_buffer_size];
#define  BSS                              0x01   //infrastructure network

//*************************************************************
//*    PORT Definitions
//*************************************************************
#define data_in      PORTD
#define data_out     LATD
#define addr_hi      LATG
#define addr_lo      LATH
//*************************************************************
//*    PORT TRIS Definitions
//*************************************************************
#define TO_NIC       TRISD = 0x00
#define FROM_NIC     TRISD = 0xFF
//*************************************************************
//*    CF I/O Definitions
//*************************************************************
#define OE           0x01
#define IORD         0x02
#define WE           0x40
#define IOWR         0x80
#define set_OE       PCTL |= OE
#define clr_OE       PCTL &= ~OE
#define set_IORD     PCTL |= IORD
#define clr_IORD     PCTL &= ~IORD
```

```
#define set_IOWR          PCTL |= IOWR
#define clr_IOWR          PCTL &= ~IOWR

void wr_cf_data(char bapnum,char* buffer,unsigned int count)
{
 unsigned int addr,i;
 int *byteptr;
 char data_lo;

 if(bapnum)
  addr = Data1_Register;
 else
  addr = Data0_Register;

 byteptr = (int*)buffer;
 for(i=0;i<(count&0xFFFE);)
  {
   wr_cf_io16(*byteptr++,addr);
   i+=2;
  }
 if(count % 2)
  {
   data_lo = buffer[i++];
   wr_cf_io16(data_lo,addr);
  }
 }

void wr_cf_addr(unsigned int addr)
{
 addr_hi = (make8(addr,1));
 addr_lo = addr & 0x00FF;
}

unsigned int rd_cf_io16(unsigned int addr)
{
 char data_lo,data_hi,i;
 unsigned int data16;

 wr_cf_addr(addr);
 clr_IORD;
 i=3;
 while(--i);
 data_lo=data_in;
 set_IORD;
 NOP();

 wr_cf_addr(addr+1);
 clr_IORD;
 i=3;
 while(--i);
 data_hi=data_in;
 set_IORD;
```

```
 NOP();
 data16 = make16(data_hi,data_lo);
 return(data16);
}

void wr_cf_io16(unsigned int data16,unsigned int addr)
{
 char i;

 wr_cf_addr(addr);
 data_out = LOW_BYTE(data16);
 TO_NIC;
 clr_IOWR;
 i=1;
 while(--i);
 set_IOWR;
 NOP();
 data_out = 0xFF;

 wr_cf_addr(addr+1);
 data_out = HIGH_BYTE(data16);
 clr_IOWR;
 i=1;
 while(--i);
 set_IOWR;
 NOP();
 data_out = 0xFF;
 FROM_NIC;
}

char  send_command(unsigned int cmd, unsigned int parm0)
{
 char cmd_code,rc;
 unsigned int cmd_status,evstat_data;

 do{
     cmd_data = rd_cf_io16(Command_Register);
   }while(cmd_data &  CmdBusy_Bit_Mask);

 wr_cf_io16(parm0, Param0_Register);
 wr_cf_io16(cmd, Command_Register);

 do{
     cmd_data = rd_cf_io16(Command_Register);
   }while(cmd_data &  CmdBusy_Bit_Mask);

 do{
     evstat_data = rd_cf_io16(EvStat_Register);
   }while(!(evstat_data & EvStat_Cmd_Bit_Mask));

 cmd_status = rd_cf_io16( Status_Register);
 rc = cmd_status &  Status_Result_Mask;
```

```
 wr_cf_io16( EvStat_Cmd_Bit_Mask, EvAck_Register);

 switch(rc)
   {
    case 0x00:
     rc = 0;           //command completed OK
    break;
   case 0x01:
    rc = 1;            //Card failure
    break;
  case 0x05:
    rc = 2;            //no buffer space
    break;
  case 0x7F:
    rc = 3;            //command error
    break;
 }
 return(rc);
}

//*********************************************************
//* get a free bap
//*********************************************************
char get_free_bap(void)
{
 unsigned int temp;
 char rc;

 rc = 2;
 temp = rd_cf_io16(Offset1_Register);
 if(!(temp & BAP1Busy_Bit_Mask))
  rc = 1;
 temp = rd_cf_io16(Offset0_Register);
 if(!(temp & BAP0Busy_Bit_Mask))
  rc = 0;
 return(rc);
}

//*********************************************************
//* bap_write
//*********************************************************
char bap_write(unsigned int fidrid,unsigned int offset,char* buffer, unsigned
int count)
{
 unsigned int temp;
 char rc,bapnum;

 bapnum = get_free_bap();
 rc = 0;
 switch(bapnum)
 {
  case 0:
```

```
   wr_cf_io16(fidrid,Select0_Register);
   wr_cf_io16(offset,Offset0_Register);
   do{
       temp = rd_cf_io16(Offset0_Register);
      }while(temp & BAP0Busy_Bit_Mask);
  temp = rd_cf_io16(Offset0_Register);
  if(temp & BAP0Err_Bit_Mask)
   rc=1;
  else
   wr_cf_data(BAP0,buffer,count);
  break;

  case 1:
   wr_cf_io16(fidrid,Select1_Register);
   wr_cf_io16(offset,Offset1_Register);
   do{
       temp = rd_cf_io16(Offset1_Register);
      }while(temp & BAP1Busy_Bit_Mask);
   temp = rd_cf_io16(Offset1_Register);
   if(temp & BAP1Err_Bit_Mask)
    rc=1;
   else
    wr_cf_data(BAP1,buffer,count);
 break;
}
 return(rc);
}


//***********************************************************
//* RID write
//***********************************************************
char rid_write(unsigned int fidrid, unsigned int* buffer, unsigned int count)
{
 char rc;

 rc = bap_write(fidrid, 0x0000, (char*) buffer, (count+1)<<1);
 rc = send_command(Access_Cmd | AccessWrite_Mask, fidrid);
 return(rc);
}


//***********************************************************
//*    AIRDROP CONFIGURATION FUNCTION
//***********************************************************
char airdrop_cfg(unsigned int cmd)
{
 char rc;
 unsigned int i,j,k;

switch(cmd)
{
 case BSS:
  fidrid_buffer[0] = RID_cfgPortType_Length;
```

```
       fidrid_buffer[1] = RID_cfgPortType;
       fidrid_buffer[2] = RID_cfgPortType_Infrastructure;
       rc = rid_write(RID_cfgPortType, fidrid_buffer, RID_cfgPortType_Length);
       break;

*****************************
void main(void)
{
       unsigned int i,temp,evstat_data;
       char rc;

       init_USART1();
       init_cf_card();
       airdrop_cfg(BSS);

*****************************
```

Code Snippet 9.3: The focus is on the **airdrop_cfg** function. The variables in the **airdrop_cfg** function's argument are used to configure many of the 802.11b CompactFlash NIC's operational features.

One of the first things we must do is establish what type of wireless LAN the AirDrop module will be operating within. Since our little AIRDROP_NETWORK contains an AP, it is considered an infrastructure network, or BSS. We tell the 802.11b CompactFlash NIC that it will be a part of a BSS by placing the appropriate PortType RID values in their proper places within the *fidrid_buffer*.

The *fidrid_buffer*, whose beginning elements are shown in Sub Snippet 9.1, is simply a 256-byte array that is allocated to hold the contents or parts and pieces of FIDs and RIDs. The *rid_write* function wants to see the PortType RID elements in the *fidrid_buffer* in the order you see them placed in the *fidrid_buffer* array here.

Sub Snippet 9.1

```
***********************************************************
fidrid_buffer[0] = RID_cfgPortType_Length;
fidrid_buffer[1] = RID_cfgPortType;
fidrid_buffer[2] = RID_cfgPortType_Infrastructure;
***********************************************************
```

When called, the *rid_write* function in Sub Snippet 9.2 immediately calls the *bap_write* function and passes the original *rid_write* function arguments along to the *bap_write* function. The very first thing that the *bap_write* function does is call the *get_free_bap* function, which determines which BAP, BAP0 or BAP1, is free and ready to be used.

Sub Snippet 9.2

```
***********************************************************
rc = rid_write(RID_cfgPortType, fidrid_buffer, RID_cfgPortType_Length);
rc = bap_write(fidrid, 0x0000, (char*) buffer, (count+1)<<1);
bapnum = get_free_bap();
***********************************************************
```

The *(count+1)<<1* statement in the *bap_write* function call turns the word count held in *fidrid_buffer[0]* into a byte count that includes the word in *fidrid_buffer[0]*. Ultimately, the count will be used by the *wr_cf_data* function, which uses bytes in its address counter mechanism. For example, the *RID_cfgPortType_Length* in Sub Snippet 9.3 tells us that there are 2 words in the PortType RID. That's 4 bytes. We also have to write the *RID_cfgPortType_Length* value, which is an additional 2 bytes.

Sub Snippet 9.3
****

```
fidrid_buffer[0] = RID_cfgPortType_Length;           = 0x0002
fidrid_buffer[1] = RID_cfgPortType;                  = 0xFC00
fidrid_buffer[2] = RID_cfgPortType_Infrastructure;   = 0x0001
```
****

Let's do the math:

$$\text{count}+1 = 0x0002 + 1 = 0x0003 = 0b00000011$$
$$0x0003 << 1 = 0x0006 = 0b00000110$$

The *wr_cf_data* function will send 6 bytes or 3 words, which is what we calculated earlier. A byte-by-byte view of the *fidrid_buffer* can be seen in Hex Dump 9.1.

Hex Dump 9.1
****

```
Address        Symbol Name        Value

  00FA       fidrid_buffer
    00FA       [0]                   0x0002
    00FC       [1]                   0xFC00
    00FE       [2]                   0x0001

Address  00 01 02 03 04 05 06 07 08 09 0A 0B 0C 0D 0E 0F      ASCII
  00F0   00 00 00 00 00 00 00 00 00 00 02 00 00 FC 01 00   ........ ........
```
****

Hex Dump 9.1: Here's a look at how the PortType RID bytes sitting in the fidrid_buffer appear. The words in the formatted section of this dump are in HIGH:LOW format for easy pattern recognition.

The *get_free_bap* function, which is partially listed in Sub Snippet 9.4, uses an old friend, the *rd_cf_io16* function, to read both of the BAP Offset Registers, which contain the respective BAP's busy bit. BAP1's Offset Register is read first and BAP1 is tested for a busy condition using the *BAP1Busy_Bit_Mask*. If BAP1 is not busy, the return code is set to 1.

Sub Snippet 9.4
****

```
rc = 2;
temp = rd_cf_io16(Offset1_Register);
if(!(temp & BAP1Busy_Bit_Mask))
  rc = 1;
```

```
temp = rd_cf_io16(Offset0_Register);
if(!(temp & BAP0Busy_Bit_Mask))
 rc = 0;
return(rc);
```
****

BAP0 is then tested for a busy condition and if BAP0 is found not to be busy, the return code is set to 0 and the function terminates. If both BAPs show busy, the return code is 2, which is the initial return code.

I hope you are wondering why BAP0 is used when BAP1 tests ready first. Recall that I mentioned that one of the Linux 802.11b driver conventions was to use one BAP exclusively for reading and the other BAP exclusively for writing. Also recall that I told you that we won't be doing that in the AirDrop 802.11b driver. The idea in the AirDrop 802.11b driver is to use the alternate BAP if one of the BAPs is busy. The logic behind our *get_free_bap* function says to always use BAP0 if it is available and use BAP1 only when BAP0 is busy. In my experiences with the AirDrop driver, I've never seen BAP1 get used. That's an indication of how fast and efficient the PRISM chipset dynamic memory allocation algorithm is.

Let's assume the *get_free_bap* function returns a 0, which says BAP0 is free. The *bap_write* function is composed of two identical code sets, one for a free BAP0 and one for a free BAP1. A return code of 0 puts the BAP0 code into motion, while a return code of 1 does the same for the BAP1 code segment. Since we are assuming a return code of 0, which sets the *airdrop_cfg* local variable *bapnum* to 0, let's follow the BAP0 code thread.

Please reference Sub Snippet 9.5. From our previous discussion of how to make a BAP work, we know that we will need a FID or RID and possibly an offset value to give to the BAP mechanism before we can ask the BAP to help us access data. In this instance, we have a RID (the PortType RID) value of 0xFC00, which we place in BAP0's Select0_Register. There is no offset value. So, we simply enter 0x0000 into BAP0's Offset0_Register. Writing the Select0_Register and then the Offset0_Register signals to the BAP that we want it to set up a pipeline to the PortType RID. BAP0 will raise the busy bit in the Offset0_Register while it is setting up the virtual data access path. If a BAP setup error occurs, we abort the operation and exit the *bap_write* function. Otherwise, if things are good, we write the data we carefully placed in the *fidrid_buffer* to BAP0.

Sub Snippet 9.5
****
```
 case 0:
  wr_cf_io16(fidrid,Select0_Register);
  wr_cf_io16(offset,Offset0_Register);
  do{
       temp = rd_cf_io16(Offset0_Register);
     }while(temp & BAP0Busy_Bit_Mask);
  temp = rd_cf_io16(Offset0_Register);
  if(temp & BAP0Err_Bit_Mask)
   rc=1;
  else
```

```
   wr_cf_data(BAP0,buffer,count);
*************************************************************
```

The writing of the *fidrid_buffer* buffer contents to the BAP is performed by the *wr_cf_data* function. The original *rid_write* function called for writing the number of bytes specified by *RID_cfgPortType_Length* from the *fidrid_buffer* to address 0xFC00, which is the address of the PortType RID. So far, BAP0 has been selected and initialized with respect to the Port-Type RID. According to our definitions in Code Snippet 9.3, the Data0_Register, which is our data portal to and from the PortType RID, is associated with BAP0. The code that determines the BAP and Data Register relationship is shown in Sub Snippet 9.6.

Sub Snippet 9.6

```
*************************************************************
 if(bapnum)
  addr = Data1_Register;
 else
  addr = Data0_Register;
*************************************************************
```

All we have to do is write the specified number of words from the *fidrid_buffer*, which is specified by *RID_cfgPortType_Length,* to the Data0_Register and BAP0 will take care of the delivery of the data payload to the PortType RID. The *wr_cf_data* function is designed to write an even number of words to either BAP0 or BAP1 and includes a modulus check (*if(count % 2)*) to handle writing any odd character that may exist. The odd character situation should never happen as each of the RID lengths is specified as 2-byte words. Note in Sub Snippet 9.7 that we are writing words and counting them in bytes.

Sub Snippet 9.7

```
*************************************************************
byteptr = (int*)buffer;
 for(i=0;i<(count&0xFFFE);)
  {
   wr_cf_io16(*byteptr++,addr);    //write words
   i+=2;                           //count bytes written
  }
 if(count % 2)                     //write the odd byte if necessary
  {
   data_lo = buffer[i++];
   wr_cf_io16(data_lo,addr);
  }
*************************************************************
```

Our *rid_write* function needs to perform one more task to finish the BSS mode configuration of the 802.11b CompactFlash NIC to BSS mode. An Access Command, such as the one seen in Sub Snippet 9.8, must be issued against the PortType RID with the Access Write bit set.

Sub Snippet 9.8

```
*************************************************************
rc = send_command(Access_Cmd | AccessWrite_Mask, fidrid);
*************************************************************
```

Reading a RID

Suppose that we wanted to know what was data was contained within the PortType RID. That's where the *rid_read* function shown in Code Snippet 9.4 comes into play.

Code Snippet 9.4

```
*****************************
//* RID CONFIGURATION RECORDS
#define  RID_cfgPortType               0xFC00
#define  RID_cfgPortType_Length        2
#define  RID_cfgPortType_IBSS          0        //sets ad-hoc network operation
#define  RID_cfgPortType_Infrastructure 1       //sets BSS Network operation


//* COMMANDS
#define  Access_Cmd                    0x0021

#define  BAP0Busy_Bit_Mask             0x8000 //busy bit in Offset0 register
#define  BAP0Err_Bit_Mask              0x4000 //Err bit in Offset0 register
#define  BAP1Busy_Bit_Mask             0x8000 //busy bit in Offset1 register
#define  BAP1Err_Bit_Mask              0x4000 //Err bit in Offset1 register
#define  AccessWrite_Mask              0x0100 //Write RID to MAC

#define fidrid_buffer_size             256
unsigned int fidrid_buffer[fidrid_buffer_size];
#define  BSS                           0x01   //infrastructure network


//************************************************************
//*    PORT Definitions
//************************************************************
#define data_in      PORTD
#define data_out     LATD
#define addr_hi      LATG
#define addr_lo      LATH
//************************************************************
//*    PORT TRIS Definitions
//************************************************************
#define TO_NIC        TRISD = 0x00
#define FROM_NIC      TRISD = 0xFF
//************************************************************
//*    CF I/O Definitions
//************************************************************
#define OE                  0x01
#define IORD                0x02
#define WE                  0x40
#define IOWR                0x80
#define set_OE              PCTL |= OE
#define clr_OE              PCTL &= ~OE
#define set_IORD            PCTL |= IORD
#define clr_IORD            PCTL &= ~IORD
#define set_IOWR            PCTL |= IOWR
#define clr_IOWR            PCTL &= ~IOWR
```

```
void wr_cf_data(char bapnum,char* buffer,unsigned int count)
{
 unsigned int addr,i;
 int *byteptr;
 char data_lo;

 if(bapnum)
  addr = Data1_Register;
 else
  addr = Data0_Register;

 byteptr = (int*)buffer;
 for(i=0;i<(count&0xFFFE);)
  {
   wr_cf_io16(*byteptr++,addr);
   i+=2;
  }
 if(count % 2)
  {
   data_lo = buffer[i++];
   wr_cf_io16(data_lo,addr);
  }
 }

void wr_cf_addr(unsigned int addr)
{
 addr_hi = (make8(addr,1));
 addr_lo = addr & 0x00FF;
}

unsigned int rd_cf_io16(unsigned int addr)
{
 char data_lo,data_hi,i;
 unsigned int data16;

 wr_cf_addr(addr);
 clr_IORD;
 i=3;
 while(--i);
 data_lo=data_in;
 set_IORD;
 NOP();

 wr_cf_addr(addr+1);
 clr_IORD;
 i=3;
 while(--i);
 data_hi=data_in;
 set_IORD;
 NOP();
 data16 = make16(data_hi,data_lo);
 return(data16);
```

```
}

void wr_cf_io16(unsigned int data16,unsigned int addr)
{
 char i;

 wr_cf_addr(addr);
 data_out = LOW_BYTE(data16);
 TO_NIC;
 clr_IOWR;
 i=1;
 while(--i);
 set_IOWR;
 NOP();
 data_out = 0xFF;

 wr_cf_addr(addr+1);
 data_out = HIGH_BYTE(data16);
 clr_IOWR;
 i=1;
 while(--i);
 set_IOWR;
 NOP();
 data_out = 0xFF;
 FROM_NIC;
}

char  send_command(unsigned int cmd, unsigned int parm0)
{
 char cmd_code,rc;
 unsigned int cmd_status,evstat_data;

 do{
     cmd_data = rd_cf_io16(Command_Register);
    }while(cmd_data &  CmdBusy_Bit_Mask);

 wr_cf_io16(parm0, Param0_Register);
 wr_cf_io16(cmd, Command_Register);

 do{
     cmd_data = rd_cf_io16(Command_Register);
    }while(cmd_data &  CmdBusy_Bit_Mask);

 do{
     evstat_data = rd_cf_io16(EvStat_Register);
    }while(!(evstat_data & EvStat_Cmd_Bit_Mask));

 cmd_status = rd_cf_io16( Status_Register);
 rc = cmd_status &  Status_Result_Mask;
 wr_cf_io16( EvStat_Cmd_Bit_Mask, EvAck_Register);

 switch(rc)
```

```c
  {
   case 0x00:
    rc = 0;           //command completed OK
    break;
   case 0x01:
    rc = 1;           //Card failure
    break;
  case 0x05:
    rc = 2;           //no buffer space
    break;
  case 0x7F:
    rc = 3;           //command error
    break;
  }
 return(rc);
}


//************************************************************
//* get a free bap
//************************************************************
char get_free_bap(void)
{
 unsigned int temp;
 char rc;

 rc = 2;
 temp = rd_cf_io16(Offset1_Register);
 if(!(temp & BAP1Busy_Bit_Mask))
  rc = 1;
 temp = rd_cf_io16(Offset0_Register);
 if(!(temp & BAP0Busy_Bit_Mask))
  rc = 0;
 return(rc);
}


//************************************************************
//* FID read
//************************************************************
char fid_read(unsigned int fid)
{
 unsigned int temp;
 unsigned int fidlen;
 unsigned int fidpointer;
 char rc,bapnum;

 bapnum = get_free_bap();
 rc = 0;
 switch(bapnum)
  {
    case 0:
     wr_cf_io16(fid,Select0_Register);
     wr_cf_io16(0x0000,Offset0_Register);
```

```
     do{
            temp = rd_cf_io16(Offset0_Register);
         }while(temp & BAP0Busy_Bit_Mask);
 temp = rd_cf_io16(Offset0_Register);
 if(temp & BAP0Err_Bit_Mask)
  rc = 1;
 else
 {
  fidlen = rd_cf_io16(Data0_Register);
  fidrid_buffer[0] = fidlen;
  If(fidlen)
  {
   for(fidpointer=0;fidpointer<fidlen;fidpointer++)
   fidrid_buffer[fidpointer+1] = rd_cf_io16(Data0_Register);
   rc = 0;
  }
 }
 break;

   case 1:
    wr_cf_io16(fid,Select1_Register);
    wr_cf_io16(0x0000,Offset1_Register);
    do{
            temp = rd_cf_io16(Offset1_Register);
        }while(temp & BAP1Busy_Bit_Mask);
 temp = rd_cf_io16(Offset1_Register);
 if(temp & BAP1Err_Bit_Mask)
  rc = 1;
 else
 {
  fidlen = rd_cf_io16(Data1_Register);
  fidrid_buffer[0] = fidlen;
  if(fidlen)
  {
   for(fidpointer=0;fidpointer<fidlen;fidpointer++)
   {
    fidrid_buffer[fidpointer+1] = rd_cf_io16(Data1_Register);
   }
  }
 }
 break;
default:
  rc =1 ;
 break;
 }
  return(rc);
 }

//*********************************************************
//*    RID read
//*********************************************************
char rid_read(unsigned int rid)
```

```
{
 unsigned int temp;
 char rc;

 rc = 0;
 temp = send_command(Access_Cmd,rid);
 temp = fid_read(rid);
 if (temp)
  rc=1;

 return(rc);
}
*****************************
void main(void)
{
        unsigned int i,temp,evstat_data;
        char rc;

        init_USART1();
        init_cf_card();
        airdrop_cfg(BSS);
temp = rid_read(RID_cfgPortType);

*****************************
```

Code Snippet 9.3: Just for grins let's pretend we added the source code statements to request a read of the PortType RID.

Instead of sending an Access Command with the Access Write bit set, this time, as seen in Sub Snippet 9.9, we will send the Access Command with the Access Write bit clear indicating that we want to read the RID. We'll assume the *send_command* function returned an OK to our RID access request.

Sub Snippet 9.9

```
***********************************************************
temp = send_command(Access_Cmd,rid);
***********************************************************
```

Reading a RID is just like reading a FID once the Access Command is sent and acknowledged. So, only one FID/RID read function needs to be coded. In the AirDrop 802.11b driver, that multipurpose function is called *fid_read*.

As you already know, BAPs are used for both reading and writing FIDs and RIDs. So, the first logical thing we do in the *fid_read* function is to request a free BAP with the code in Sub Snippet 9.10.

Sub Snippet 9.10

```
***********************************************************
bapnum = get_free_bap();
***********************************************************
```

Like the *bap_write* function, the *fid_read* function contains identical code for BAP0 and BAP1. Let's assume BAP0 was selected for our read of the PortType RID and the code in Sub Snippet 9.11 is executed. To keep things straight in your mind, substitute *rid* for *fid* in the *fid_read* local variable and argument definitions.

To perform a read we must provide an address and offset to the BAP0 mechanism. There is absolutely no difference in the *fid_read* and *bap_write* BAP setup code.

Sub Snippet 9.10
```
    case 0:
      wr_cf_io16(fid,Select0_Register);
      wr_cf_io16(0x0000,Offset0_Register);
      do{
            temp = rd_cf_io16(Offset0_Register);
          }while(temp & BAP0Busy_Bit_Mask);
  temp = rd_cf_io16(Offset0_Register);
  if(temp & BAP0Err_Bit_Mask)
   rc = 1;
```

Once the selected BAP acknowledges the address and offset information, we can proceed with reading from the selected BAP's DataX_Register. Since we're assuming BAP0 was free, the code in Sub Snippet 9.11 shows that our RID data portal is BAP0's Data0_Register. Although the definitions of RIDs in Code Snippet 9.2 don't imply it, the first word we read from a FID or RID is the length of the FID or RID object.

Sub Snippet 9.11
```
else
 {
  fidlen = rd_cf_io16(Data0_Register);
  fidrid_buffer[0] = fidlen;
```

After reading the initial length byte, the rest of the RID read process is just as you would imagine it would be in Sub Snippet 9.12. Simple reads using the *rd_cf_io16* function from BAP0's Data0_Register that store words from the RID in the *fid_read* buffer.

Sub Snippet 9.12
```
  If(fidlen)
  {
   for(fidpointer=0;fidpointer<fidlen;fidpointer++)
   fidrid_buffer[fidpointer+1] = rd_cf_io16(Data0_Register);
   rc = 0;
  }
 }
```

I've preserved a shot of the MPLAB IDE debug windows that show the result of the read operation I posted against the PortType RID.

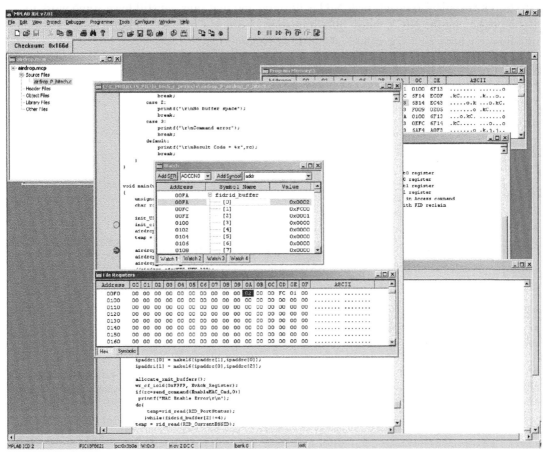

*Screen Capture 9.1: The **fid_read** buffer serves double duty as a holding buffer for FID and RID read operations and a staging area for writing to FIDs and RIDs. Everything data from the PortType RID is in the right place in this shot.*

Thus far, I've showed you how to run bits and bytes between the AirDrop microcontroller and the 802.11b CompactFlash NIC through a BAP portal. There is also a need for pumping string data through the BAP pipe. The SSID is a collection of ASCII characters and the desired SSID, AIRDROP_NETWORK in this case, must be communicated to the PRISM chipset if we want our AirDrop module to search out and join the AIRDROP_NETWORK BSS.

Stringing Up the SSID

We're making good progress. So far we've cranked up the AirDrop microcontroller's USART, initialized the AirDrop's CompactFlash NIC and configured the 802.11b CompactFlash NIC for BSS operation. Although our AirDrop NIC is yet to speak its first RF word, our basic AIRDROP_NETWORK network hardware is working as designed. The Link LED on the AirDrop's 802.11b CompactFlash NIC is flashing and that says we still have some work to do before keying the NIC's radio. Sub Snippet 9.13 is a AirDrop 802.11b driver source code timeline of where we've been and where we are about to go.

Sub Snippet 9.13

```
******************************************************************
void main(void)
{
        unsigned int i,temp,evstat_data;
        char rc;

        init_USART1();
        init_cf_card();
        airdrop_cfg(BSS);
        airdrop_cfg(SSID);
******************************************************************
```

The only source code we haven't examined in the *main* function timeline code is the *airdrop_cfg* call to establish a desired SSID. All of the code that is touched by the *airdrop_cfg(SSID)* call is listed in Code Snippet 9.4.

Code Snippet 9.4

```
****************************
const char ssid[] = "AIRDROP_NETWORK";

//* RID CONFIGURATION RECORDS
#define  RID_cfgDesiredSSID 0xFC02       //the SSID of the BSS we want to join
#define  RID_DesiredSSID_Length   18     //word count of the DesiredSSID RID

//***************************************************************
//*    AIRDROP CONFIGURATION FUNCTION
//***************************************************************
char airdrop_cfg(unsigned int cmd)
{

case SSID:
 fidrid_buffer[0] = RID_DesiredSSID_Length;
 fidrid_buffer[1] = RID_cfgDesiredSSID;
 fidrid_buffer[2] = strlen(ssid);
 j = strlen(ssid);
 i=0;
 k=3;
 while(j > 1)
 {
  fidrid_buffer[k] = (ssid[i+1] << 8) | ssid[i];
  i+=2;
  ++k;
  j-=2;
 }
```

```
while(j > 0)
{
 fidrid_buffer[k] = ssid[i];
 --j;
}
rc = rid_write(RID_cfgDesiredSSID,fidrid_buffer,RID_DesiredSSID_Length);
```
****
Code Snippet 9.4: This snippet shows all of the code associated with pushing an SSID string into the AirDrop's 802.11b CompactFlash NIC. I didn't list all of the supporting functions as I think that by now you are very familiar with the basic AirDrop 802.11b driver building blocks.

In Sub Snippet 9.14 the *fid_read* buffer in employed just as it is with all of the RID operations we have experienced thus far. The actual length of our SSID string, "AIRDROP_ NETWORK" ,is stored at position 2 of the *fidrid_buffer*. If we put a zero-length string into the *fidrid_buffer[2]* slot, the AirDrop's 802.11b CompactFlash NIC will attempt to join the first BSS that it can find that will accept it.

Sub Snippet 9.14
****
```
case SSID:
fidrid_buffer[0] = RID_DesiredSSID_Length;
fidrid_buffer[1] = RID_cfgDesiredSSID;
fidrid_buffer[2] = strlen(ssid);
```
****

The code to prepare the SSID string for presentation to the 802.11b CompactFlash NIC is not complicated at all. All we're doing here is assembling the characters in the *ssid* array into little-endian words. The code between the *while(j>0)* braces in Sub Snippet 9.15 takes care of the odd byte, which we have one of in our 15-byte *ssid* array.

Sub Snippet 9.15
****
```
j = strlen(ssid);
i=0;
k=3;
while(j > 1)
{
 fidrid_buffer[k] = (ssid[i+1] << 8) | ssid[i];
 i+=2;
 ++k;
 j-=2;
}
while(j > 0)
{
 fidrid_buffer[k] = ssid[i];
 --j;
}
```
****

127

Screen Capture 9.2's contents are the results of the work done by the SSID character-to-word translation code behind the *airdrop_cfg(SSID)* function call. We can pull the *rid_write* function trigger in Sub Snippet 9.16 on the *fidrid_buffer* now that all of the SSID bullets are loaded.

Sub Snippet 9.16

```
****************************************************************
rc = rid_write(RID_cfgDesiredSSID,fidrid_buffer,RID_DesiredSSID_Length);
****************************************************************
```

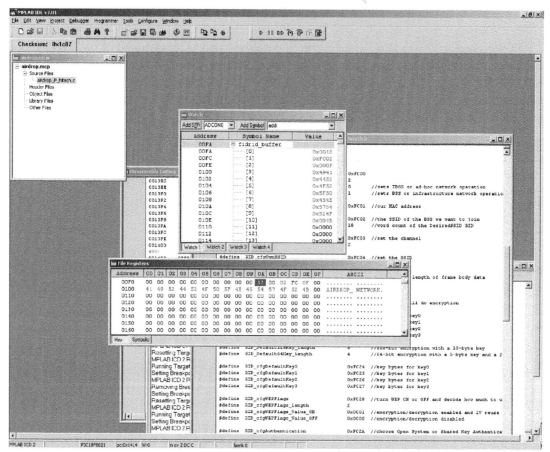

*Screen Capture 9.2: Once the bytes are all lined up in the **fidrid_buffer**, all we have to do is push them out in order through the BAP to the **RID_cfgDesiredSSID** RID.*

See, I told you that this would be easy. Let's get with finishing off the work we have left on the RIDs that involves the *airdrop_cfg* call.

Good RIDdance

There are only two more *airdrop_cfg* call-related configuration tasks we need to perform on our way to finally turning on the RF section of the PRISM chipset. Code Snippet 9.5 uses the *airdrop_cfg* functionality to set the maximum length of the 802.11b frame body data while Code Snippet 9.6 configures the NIC to run at all of the supported 802.11b speeds. These two RID-based functions are the additional *airdrop_cfg* function calls you see in the revised version of our AirDrop 802.11b driver *main* function in Sub Snippet 9.17.

Sub Snippet 9.17
```
************************************************************
void main(void)
{
        unsigned int i,temp,evstat_data;
        char rc;

        init_USART1();
        init_cf_card();
        airdrop_cfg(BSS);
        airdrop_cfg(SSID);
        airdrop_cfg(MAX_DATALEN);
        airdrop_cfg(NIC_RATE);
************************************************************
```

All of the AirDrop 802.11b driver building block code is invoked when necessary to set the maximum data length via the *RID_cfgMaxDataLength* RID. The number 1500 should sound familiar as it is the maximum length of the data area of an Ethernet packet.

Code Snippet 9.5
```
****************************
//* RID CONFIGURATION RECORDS
#define   RID_cfgMaxDataLength            0xFC07
#define   RID_cfgMaxDataLength_Length     2
#define   RID_cfgMaxDataLength_Value      1500   //max length of frame body
data

//************************************************************
//*    AIRDROP CONFIGURATION FUNCTION
//************************************************************
char airdrop_cfg(unsigned int cmd)
{

case MAX_DATALEN:
 fidrid_buffer[0] = RID_cfgMaxDataLength_Length;
 fidrid_buffer[1] = RID_cfgMaxDataLength;
 fidrid_buffer[2] = RID_cfgMaxDataLength_Value;   //0x05DC - 1500 decimal
 rc = rid_write(RID_cfgMaxDataLength, fidrid_buffer,  ID_cfgMaxDataLength_
Length);
****************************
```
Code Snippet 9.5: The beauty of the AirDrop 802.11b driver building block routines lies in their reusability. In most cases, the only variables involved with writing to a RID are the values you place in the first three locations of the **fidrid_buffer**.

The 802.11b CompactFlash NIC mounted in the AirDrop CompactFlash card won't automatically transmit at the full 802.11b speed of 11 Mbps without some help from yet another *airdrop_cfg* call. To allow the PRISM chipset to automatically fall back to a slower speed when the RF signal is suffering from interference, and to allow for full speed operation, we must allow the 802.11b CompactFlash NIC to be able to navigate throughout the entire 802.11b speed range. Using the AirDrop 802.11b driver building block code in conjunction with the *airdrop_cfg(NIC_RATE)* call makes this an easy task.

Code Snippet 9.6

```
*****************************
//* RID CONFIGURATION RECORDS
#define   RID_TxRateControl                 0xFC84 //set transmission bit rates
#define   RID_TxRateControl_Length 2
#define   RID_TxRateControl_Value_10x0001 //1 Mbps
#define   RID_TxRateControl_Value_20x0002 //2 Mbps
#define   RID_TxRateControl_Value_55        0x0004 //5.5 Mbps
#define   RID_TxRateControl_Value_11        0x0008 //11 Mbps
#define   RID_TxRateControl_Value_ALL       0x000F //all 802.11b speeds


//*************************************************************
//*    AIRDROP CONFIGURATION FUNCTION
//*************************************************************
char airdrop_cfg(unsigned int cmd)
{

case NIC_RATE:
fidrid_buffer[0] = RID_TxRateControl_Length;
fidrid_buffer[1] = RID_TxRateControl;
fidrid_buffer[2] = RID_TxRateControl_Value_ALL;
rc = rid_write(RID_TxRateControl, fidrid_buffer, RID_TxRateControl_Length);
*****************************
```

*Code Snippet 9.6: A read of the **RID_TxRateControl** RID revealed that the default maximum 802.11b CompactFlash NIC rate is 2 Mbps.*

Finding the default maximum transmit speed was easy. All I did was insert a *rid_read(RID_TxRateControl)* call into the AirDrop 802.11b driver *main* function before the *RID_TxRateControl* value was set by the *airdrop_cfg* call. The default transmission rate data is shown in Screen Capture 9.3.

Retrieving the MAC Address

This is just as good a time as any to pluck the 802.11b CompactFlash NIC's MAC address from the *RID_cfgOwnMACAddress* RID. All it takes is a simple *rid_read* call like the on in Code Snippet 9.7.

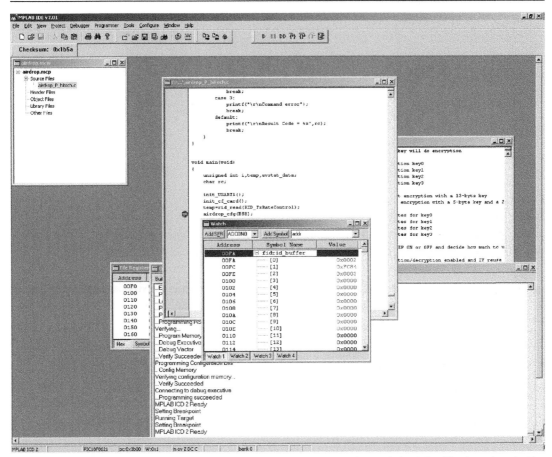

*Screen Capture 9.3: The word of data we are interested in resides at **fidrid_buffer** location 2. The 0x0003 present at **fidrid_buffer** location 2 covers 1 Mbps and 2 Mbps transmit speeds. Writing 0x000F to the **RID_TxRateControl** RID enables all of the 802.11b speed selections.*

Code Snippet 9.7

```
*****************************
//* RID CONFIGURATION RECORDS
#define  RID_cfgOwnMACAddress       0xFC01 //our MAC address

char ipaddrc[4] = { 192,168,0,151};

#define make16(varhigh,varlow) (((unsigned int)varhigh & 0xFF)* 0x100) +
((unsigned int)varlow & 0x00FF)

#define fidrid_buffer_size        256
unsigned int fidrid_buffer[fidrid_buffer_size];

unsigned int macaddri[3],ipaddri[2];
*****************************
```

```
temp = rid_read(RID_cfgOwnMACAddress);
for(i=0;i<6;i++)
{
  if(i%2)
  macaddrc[i] = fidrid_buffer[2+i/2]>>8;
  else
  macaddrc[i] = fidrid_buffer[2+i/2]& 0xFF;
}
for(i=0;i<3;++i)
  macaddri[i] = fidrid_buffer[i+2];
ipaddri[0] = make16(ipaddrc[1],ipaddrc[0]);
ipaddri[1] = make16(ipaddrc[3],ipaddrc[2]);
```

****

Code Snippet 9.7: Now that you know how to use the AirDrop 802.11b driver "wood," reading and writing a RID is just what I say it would be.. child's play. This code snippet shows the reliable way of programmatically obtaining the 802.11b CompactFlash NIC's MAC address.

The results of our fishing trip for the 802.11b CompactFlash NIC MAC address are shown in Screen Capture 9.4 with a supplemental view of the screen capture's hex dump included in Hex Dump 9.2.

The 802.11b CompactFlash NIC MAC address held in the *fidrid_buffer* is first converted to character format and stored in the *macaddrc* array by the code in Sub Snippet 9.18.

Sub Snippet 9.18
****

```
for(i=0;i<6;i++)
 {
  if(i%2)
  macaddrc[i] = fidrid_buffer[2+i/2]>>8;
  else
  macaddrc[i] = fidrid_buffer[2+i/2]& 0xFF;
 }
```
****

Then the same *fidrid_buffer* MAC address contents are formed into an integer format in the array *macaddri* with the *for* loop in Sub Snippet 9.19.

Sub Snippet 9.19
****

```
for(i=0;i<3;++i)
 macaddri[i] = fidrid_buffer[i+2];
```
****

The reason behind translating the 802.11b CompactFlash NIC MAC address into characters and integers hinges on how each AirDrop 802.11b function that has to have access to the MAC address uses the MAC address. For instance, the ARP function wants to see the MAC address as an array of integers, while the AirDrop 802.11b driver *create_comframestructure* function uses the MAC address in character form.

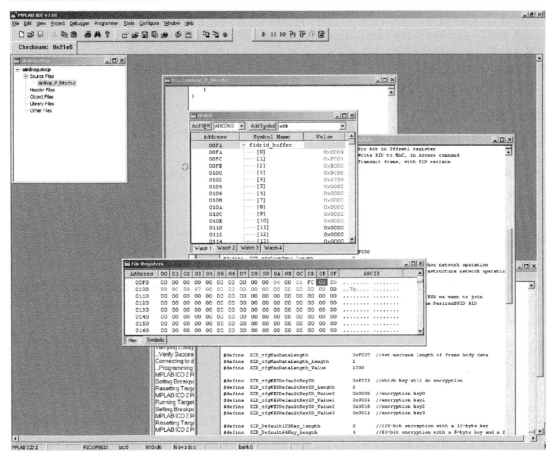

*Screen Capture 9.4: In the formatted dump (Watch Window) you can easily pick out all of the RID components. The 802.11b CompactFlash NIC's MAC address begins at **fidrid_buffer[2]**.*

Hex Dump 9.2

```
**************************************************************
Address   00 01 02 03 04 05 06 07 08 09 0A 0B 0C 0D 0E 0F      ASCII

00F0      00 00 00 00 00 00 00 00 00 00 04 00 01 FC 00 E0  ........ ........
0100      98 BC 59 67 00 00 00 00 00 00 00 00 00 00 00 00  ..Yg.... ........
**************************************************************
```

Hex Dump 9.2: You can also get this information from the back of the 802.11b CompactFlash NIC as shown in Photo 9.1.

Photo 9.1: The 802.11b CompactFlash NIC's MAC address can be seen here on the back of the card written on a barcoded label.

The IP address is not something that the 802.11b CompactFlash NIC cares about. As a result, the IP address is not part of the PRISM chipset or a physical and permanent part of the 802.11b CompactFlash NIC. Basically, we make up an IP address to suit our network needs. I've chosen an IP address of 192.168.0.151 you see in Sub Snippet 9.20 for the AirDrop modules you'll read about in this book.

Sub Snippet 9.20
```
************************************************************
ipaddri[0] = make16(ipaddrc[1],ipaddrc[0]);
ipaddri[1] = make16(ipaddrc[3],ipaddrc[2]);
************************************************************
```

The AirDrop module's IP address is only used in integer format in the AirDrop 802.11b driver.

Status Check

We're gaining altitude and speed really quickly. Talking to the 802.11b CompactFlash NIC has become second nature and we're fluently slinging commands and configuration information at the PRISM chipset. In the process, we've added some additional completed tasks to our *main* function timeline source code in Sub Snippet 9.21.

Sub Snippet 9.21
```
************************************************************
void main(void)
{
 unsigned int i,temp,evstat_data;
 char rc;

 init_USART1();
 init_cf_card();
 airdrop_cfg(BSS);
 airdrop_cfg(SSID);
 airdrop_cfg(MAX_DATALEN);
 airdrop_cfg(NIC_RATE);
 temp = rid_read(RID_cfgOwnMACAddress);
 for(i=0;i<6;i++)
 {
  if(i%2)
   macaddrc[i] = fidrid_buffer[2+i/2]>>8;
 else
  macaddrc[i] = fidrid_buffer[2+i/2]& 0xFF;
 }
 for(i=0;i<3;++i)
  macaddri[i] = fidrid_buffer[i+2];

 ipaddri[0] = make16(ipaddrc[1],ipaddrc[0]);
 ipaddri[1] = make16(ipaddrc[3],ipaddrc[2]);
************************************************************
```

Elvis and a single function are all that stand between us and enabling the 802.11b Compact-Flash NIC RF circuitry. Elvis's first guitar player was Scotty Moore. In July of 1954, Scotty "auditioned" Elvis at Sam Phillips' Sun Records and the result was the song "That's Alright Mama." Do you know what brand of guitar Scotty used for that session?

Putting an AirDrop on a Wireless LAN

Together we've covered quite a bit of ground to get to this point. However, that 802.11b CompactFlash NIC's Link LED is still flashing telling us that we're still shy of our goal of actively participating in a wireless LAN. Here's a list of what we have accomplished thus far:

1. Power up and initialize the AirDrop USART.
2. Initialize the CompactFlash NIC.
3. Configure the CompactFlash NIC for BSS operation.
4. Configure the CompactFlash NIC's desired SSID.
5. Configure the CompactFlash NIC's maximum frame body data length.
6. Configure the CompactFlash NIC's transmission rates.
7. Retrieve and format the CompactFlash NIC's MAC address.
8. Format the AirDrop IP address.

These are the tasks we will accomplish in this chapter:

1. Allocate 802.11b NIC transmit buffers.
2. Enable the 802.11b CompactFlash NIC's MAC.
3. Probe requests will be transmitted by the 802.11b NIC.
4. The AP will acknowledge the 802.11b NIC's Probe request.
5. The 802.11b NIC will request Authentication.
6. The AP will Authenticate the 802.11b NIC.
7. The 802.11b NIC will request Association.
8. The AP will grant the 802.11b NIC's association request.
9. The AirDrop will join the AIRDROP_NETWORK wireless LAN.

The number 9 is 9, whether it's represented in binary (0b00001001) or hexadecimal (0x09). It looks like we have 9 more combat missions to fly before we can get that solid Link LED on our 802.11b CompactFlash NIC, which signifies that we're on the air and actively participating in a wireless LAN.

Bogie Number 1 – Allocating Transmit Buffers

To allocate transmit buffers within out AirDrop 802.11b CompactFlash NIC, we must add yet another line to our AirDrop 802.11b driver *main* function in Sub Snippet 10.1, *allocate_xmit_buffers()*.

Sub Snippet 10.1

```
**************************************************************
void main(void)
{
 unsigned int i,temp,evstat_data;
 char rc;

 init_USART1();
 init_cf_card();
 airdrop_cfg(BSS);
 airdrop_cfg(SSID);
 airdrop_cfg(MAX_DATALEN);
 airdrop_cfg(NIC_RATE);
 temp = rid_read(RID_cfgOwnMACAddress);
 for(i=0;i<6;i++)
 {
  if(i%2)
   macaddrc[i] = fidrid_buffer[2+i/2]>>8;
 else
  macaddrc[i] = fidrid_buffer[2+i/2]& 0xFF;
 }
 for(i=0;i<3;++i)
  macaddri[i] = fidrid_buffer[i+2];

 ipaddri[0] = make16(ipaddrc[1],ipaddrc[0]);
 ipaddri[1] = make16(ipaddrc[3],ipaddrc[2]);

 allocate_xmit_buffers();
**************************************************************
```

The entire *allocate_xmit_buffers* function and the functions and variables that support it are given in Code Snippet 10.1. Let's break it all down line by line and figure out what's going on.

Code Snippet 10.1

```
****************************
//* COMMANDS
#define  Allocate_Cmd             0x000A

//* BIT MASKS
#define  CmdBusy_Bit_Mask         0x8000
#define  EvStat_Alloc_Bit_Mask    0x0008
#define  EvStat_Cmd_Bit_Mask      0x0010

#define ethheader_size            14
#define framestructure_size 46
#define eth_mtu_buffer_size 1500
#define txframe_buffer_size eth_mtu_buffer_size+framestructure_size+ethheader_
size

char comframe_buffer[60];
```

```
unsigned int TxFID_buffers[3];
char free_TxFIDs;
#define txdestaddr_offset   30
char macaddrc[6];

void create_comframestructure(unsigned int len)
{
 unsigned int i;

 for(i=0;i<len;++i)
   comframe_buffer[i] = 0x00;

 comframe_buffer[12] = 0x04;
 comframe_buffer[13] = 0x00;
 if(ad802_11Header)
 {
  comframe_buffer[14] = 0x08;
  comframe_buffer[15] = 0x01;
 }
 else
 {
  for(i=0;i<6;++i)
   comframe_buffer[txsourceaddr_offset+i] = macaddrc[i];
 }
}

char allocate_buffer(unsigned int bufferlen)
{
 char rc;

 rc = send_command(Allocate_Cmd, bufferlen);
 return rc;
}

char  send_command(unsigned int cmd, unsigned int parm0)
{
 char cmd_code,rc;
 unsigned int cmd_status,evstat_data;

 do{
      cmd_data = rd_cf_io16(Command_Register);
    }while(cmd_data &  CmdBusy_Bit_Mask);

 wr_cf_io16(parm0, Param0_Register);
 wr_cf_io16(cmd, Command_Register);

 do{
      cmd_data = rd_cf_io16(Command_Register);
    }while(cmd_data &  CmdBusy_Bit_Mask);

 do{
      evstat_data = rd_cf_io16(EvStat_Register);
```

```
    }while(!(evstat_data & EvStat_Cmd_Bit_Mask));

  cmd_status = rd_cf_io16( Status_Register);
  rc = cmd_status &  Status_Result_Mask;
  wr_cf_io16( EvStat_Cmd_Bit_Mask, EvAck_Register);

  switch(rc)
    {
     case 0x00:
      rc = 0;          //command completed OK
     break;
    case 0x01:
      rc = 1;          //Card failure
     break;
   case 0x05:
     rc = 2;           //no buffer space
     break;
   case 0x7F:
     rc = 3;           //command error
     break;
  }
  return(rc);
}


*****************************

char allocate_xmit_buffers(void)
{
 unsigned int TxFID,buffstoalloc,evstat_data;
 char i,rc;

 buffstoalloc = 3;
 for(i=0;i<3;++i)
  TxFID_buffers[i] = 0;
 free_TxFIDs = 0;
 create_comframestructure(framestructure_size+ethheader_size);
 for (i=0; i<3; i++)
   {
    rc = allocate_buffer(txframe_buffer_size);  // Ask for a buffer
    if (rc)
      {
       printf("\r\nError (%02X) - could not allocate %u-byte buffer", rc,
txframe_buffer_size);
       return 1;
     }
  do{
        evstat_data = rd_cf_io16(EvStat_Register);
      }while(!(evstat_data & EvStat_Alloc_Bit_Mask));

 TxFID = rd_cf_io16(AllocFID_Register); // Get the FID of the allocated buf-
fer
 wr_cf_io16(EvStat_Alloc_Bit_Mask,EvAck_Register );   // Acknowledge we own
```

the FID

```
 rc = bap_write(TxFID,0x0000,comframe_buffer, framestructure_size+ethheader_
size);

 if (rc)
  printf("\r\nCreate TxFrameStructure WriteBAP failed with return code:
%02X", rc);

  TxFID_buffers[i] = TxFID;                    // Remember the FID
  bitset(free_TxFIDs, i);                      // Mark this FID available to us
 }
 return 0;
}
```

★★★★★★★★★★★★★★★★★★★★★★★★★★★★★
Code Snippet 10.1: I liken this code to pizza boxes in a take-out pizza joint. Everybody in the pizza shop knows how big the pizzas are. So, the pizza boxes are ordered to exactly fit the pizzas, and to save time, the boxes are preassembled and stacked up in a corner for quick access. In this code snippet, we make three pizza boxes (transmit buffers), mark them and put them aside for later use. In addition to preassembling the boxes (transmit buffers), there's also some code to help get a head start on putting the ingredients on the pizzas (frames to be transmitted).

In the AirDrop 802.11b driver, we will arbitrarily request the allocation of three transmit buffers. The number of allocated transmit buffers is up to you. I can tell you that during testing I never saw more than one allocated buffer in use at any time. The PRISM chipset is speedy and usually can use and reclaim the same buffer before you need it again.

The *TxFID_buffers* array holds the three FIDs that we will receive after the successful allocation of the three transmit buffers. We will keep up with the number of free transmit buffers with the *free_TxFIDs* variable depicted in Sub Snippet 10.2.

Sub Snippet 10.2
★★★
```
buffstoalloc = 3;
 for(i=0;i<3;++i)
  TxFID_buffers[i] = 0;
 free_TxFIDs = 0;
```
★★★

If your mind is still on those pizzas and pizza boxes, the *create_comframestructure* function call in Sub Snippet 10.3 is responsible for putting an essential ingredient or two to the pizzas (frames to be transmitted) that will go in the preassembled boxes (transmit buffers).

Sub Snippet 10.3
★★
```
create_comframestructure(framestructure_size+ethheader_size);
```
★★

The *framestructure_size* represents the number of bytes that when combined make up the NIC's common and 802.11 header areas plus a data length word. With the exception of the byte at *comframe_buffer[12]*, the NIC's common area bytes are not under the control of the AirDrop 802.11b driver. The *ethheader_size* is determined by the number of bytes contained within an 802.3 header. Thus, the *comframestruture* is made up of a common area (14 bytes), an 802.11 header area (30 bytes), an an 802.3 header area (14 bytes) and a data length word (2 bytes). The header layouts in Figure 10.1 demonstrate the differences in the 802.3 and 802.11 headers.

802.3 HEADER			
# BYTES	6	6	2
DESC	DADDR	SRCADDR	LEN/TYPE

802.11 HEADER							
# BYTES	2	2	6	6	6	2	6
DESC	FC	DU/ID	ADDR1	ADDR2	ADDR3	SEQ-CTL	ADDR4

DADDR	= DESTINATION ADDRESS
SRCADDR	= SOURCE ADDRESS
LEN/TYPE	= LENGTH/TYPE
FC	= FRAME CONTROL
DU/ID	= DURATION/ID
ADDR1	= ADDRESS 1
ADDR2	= ADDRESS 2
ADDR3	= ADDRESS 3
SEQ-CTL	= SEQUENCE-CONTROL
ADDR4	= ADDRESS 4

Figure 10.1: The extra addresses in the 802.11 header are there to satisfy the needs of the 802.11 protocol. These are all MAC addresses.

Don't let the extra 802.11 addresses get to you. Unless you wish to run your AirDrop in 802.11 header mode, you'll never have to worry about them. Let me start by telling you that Address 4 is optional and is only used in certain situations. Address 1 is always the address of the receiver. Thus, it is also the destination address. As you've probably guessed, Address 2 is the sender's address, or the source address. Address 2 is always the transmitter's address. In a BSS, Address 3 can be either a source or destination address depending on which way the frame is flowing (to or away from the AP). With that thought, Address 3 could be the destination address if the frame is coming into the AP and destined for another station on the BSS. On the flip side of that, Address 3 would represent the source address for frames leaving the AP that originate from a station on the BSS. If this seems a bit confusing right now, don't worry about it. I'll be sure to point out the addressing schemes when we start looking at 802.11b frames with the Netasyst Sniffer. The good news is that if we choose to use 802.3 header mode, the NIC takes care of the 802.11 addressing for us. I've supplied a logic table for 802.11b addressing for you in Table 10.1.

802.11B ADDRESS FIELD LOGIC						
DEST	ADDR1	ADDR2	ADDR3	ADDR4	To DS	From DS
TO AP	BSSID	SA	DA	NA	1	0
FROM AP	DA	BSSID	SA	NA	0	1
IBSS	DA	SA	BSSID	NA	0	0

DEST   = DESTINATION
To DS   = TO DISTRIBUTION SYSTEM BIT
From DS   = FROM DISTRIBUTION SYSTEM BIT
NA   = NOT USED

Table 10.1: The Distribution System bits are found in the Frame Control #2 field of the Sniffer captures. Address 4 is only used in bridge mode and we aren't going there anywhere in this book.

The call to the *comframestruture* function is only made once. And, since the contents of the the *comframe_buffer* in Sub Snippet 10.4 will ultimately end up in all of our preallocated transmit buffers, we cannot assume that the *comframe_buffer* array is clear. So, to be on the safe side, we initialize all of the *comframe_buffer* array elements to 0x00 before continuing.

Sub Snippet 10.4
```
************************************************************
for(i=0;i<len;++i)
   comframe_buffer[i] = 0x00;
************************************************************
```

Trust me on this one. Writing a 0x04 to *comframe_buffer[12]* to use 802.3 headers in Sub Snippet 10.5 is all you really need to know about 802.11b addressing in the AirDrop 802.11b driver. There is no direct 802.11 header support in the AirDrop 802.11b driver code.

Sub Snippet 10.5
```
************************************************************
comframe_buffer[12] = 0x04;
comframe_buffer[13] = 0x00;
************************************************************
```

With the 802.3 header mode selected, our code bypasses the 802.11 header option in Sub Snippet 10.6 and falls through to insert our current NIC's MAC address into its reserved position inside the *comframe_buffer* array. Since the *comframe_buffer* array will only be used by transmit FIDs, the current NIC's MAC address is always considered the source address and is placed at the *txsourceaddr_offset* in the *comframestructure*, which is really the same thing as the *comframe_buffer* array. Note that I used the term "current NIC". That's because every time we power up the AirDrop module it will read, translate and store the MAC address of the 802.11b CompactFlash NIC that is mounted in its CompactFlash card connector. Now you know why earlier on we took the time to retrieve the NIC's MAC programmatically.

Sub Snippet 10.6
```
************************************************************
if(ad802_11Header)
```

```
 {
  comframe_buffer[14]  = 0x08;
  comframe_buffer[15]  = 0x01;
 }
else
 {
  for(i=0;i<6;++i)
   comframe_buffer[txsourceaddr_offset+i] = macaddrc[i];
 }
}
```
****

OK...Now that the *comframestructure* has been conceived and initialized, let's put it to use. We stated earlier that we wanted to allocate three transmit buffers. So it is written, so shall it be done. The *allocate_buffer* function in Sub Snippet 10.7 simply invokes the *send_command* function, which delivers the Allocate Command and the desired buffer length to the PRISM chipset.

Sub Snippet 10.7
****
```
char allocate_buffer(unsigned int bufferlen)
{
 char rc;

 rc = send_command(Allocate_Cmd, bufferlen);
 return rc;
}
```
****

In this instance, our buffer length is equal to the size of the *txframe_buffer_size* variable in Sub Snippet 10.8, which is 1500 + 46 + 14, or 1560 bytes.

Sub Snippet 10.8
****
```
for  (i=0;  i<3;  i++)
  {
   rc = allocate_buffer(txframe_buffer_size);  // Ask for a buffer
   if (rc)
    {
     printf("\r\nError (%02X) - could not allocate %u-byte buffer", rc,
txframe_buffer_size);
     return 1;
    }
```
****

You guessed it. There's an event bit generated when a buffer is successfully allocated and we must check for it in our AirDrop 802.11b driver code shown in Sub Snippet 10.9.

Sub Snippet 10.9
```
*************************************************************
do{
        evstat_data = rd_cf_io16(EvStat_Register);
      }while(!(evstat_data & EvStat_Alloc_Bit_Mask));
*************************************************************
```

When the *Alloc_Bit* appears, we can pull in the FID, which is the identifier for the buffer that was just allocated. To gain ownership of our new FID we corralled in Sub Snippet 10.10, we use one of the AirDrop 802.11b driver building blocks, *wr_cf_io16*, to acknowledge our receipt of the FID.

Sub Snippet 10.10
```
*************************************************************
 TxFID = rd_cf_io16(AllocFID_Register);  // Get the FID of the allocated buffer
 wr_cf_io16(EvStat_Alloc_Bit_Mask,EvAck_Register); // Acknowledge we own the FID
*************************************************************
```

Now that we are the owners of the buffer space represented by the FID, we can do whatever we want to it until we turn it back over to the MAC subsystem for transmission. Once we give the FID back to the MAC, we can't retrieve it for any further modification. The only thing we want to do to the FID right now is put our initialized *comframe_buffer* contents into it with the code shown in Sub Snippet 10.11. This is a one-for-one transfer as far as addressing is concerned. The first 60 bytes of a FID we wish to eventually transmit must be the contents of our initialized *comframe_buffer*. Recall that the BAP will automatically increment the transmit buffer (TxFID) address as we feed in the bytes. Since we're only writing to the TxFID and not handing it over to the MAC, using the sevices of a BAP, we can fill in the blanks within the *comframe_buffer* later if we need to. For now, we'll just load the initialized *comframe_buffer* contents, store the TxFID identifier in the *TxFID_buffers* array and put a "free" marker on the FID using the character bit field *free_TxFIDs*. We wanted three TxFIDs. So, we do the allocate-write *comframe_buffer*-save TxFID-mark TxFID free process three times.

Sub Snippet 10.11
```
*************************************************************
 rc = bap_write(TxFID,0x0000,comframe_buffer, framestructure_size+ethheader_
size);

 if (rc)
  printf("\r\nCreate TxFrameStructure WriteBAP failed with return code:
%02X", rc);

 TxFID_buffers[i] = TxFID;               // Remember the FID
 bitset(free_TxFIDs, i);                 // Mark this FID available to us
 }

*************************************************************
```

The three FIDs returned the 802.11b CompactFlash NIC that respectively represent our three TxFIDs are shown in the Watch Window of Screen Capture 10.1.

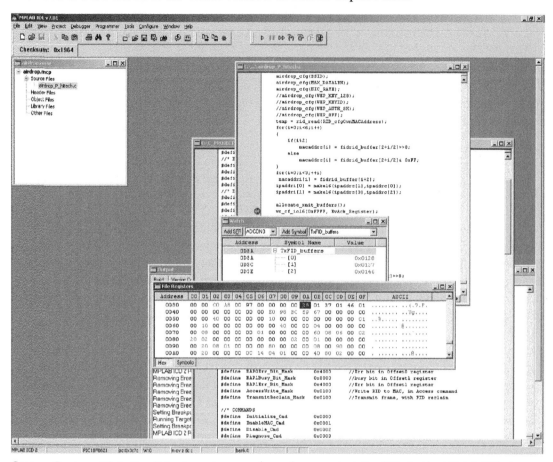

Screen Capture 10.1: Don't try to put any real numbers against the FIDs in the Watch Window as there is no correlation to the size of the FID versus what you get back as an identifier. There are a couple of other identifiers that should stand out to you in the File Registers Window. Our self-proclaimed IP address of 192.168.0.151 (hexadecimal C0 A8 00 97) is present at address 0x0D32 and our current NIC's MAC address (hexadecimal 00 E0 98 BC 59 67) is shown beginning at address 0x0D46.

The transmit buffers have been allocated and we've initialized, categorized and stored them. Since we're flying pretty high right now and you're getting better and better at the Air-Drop 802.11b driver controls, we'll use a little naval aviator term to signal the completion of the first task of nine we specified at the beginning of this chapter. Splash 1.

Bogie Number 2 – Enabling the MAC

802.11b frames are about to start flying all over the place. I've posted our current position and the six lines of additional code out on the horizon in Sub Snippet 10.12. For now, everything "RID" is in place. Once the MAC is enabled things will move forward very quickly. So, I'll go ahead and start the Netasyst Sniffer.

Sub Snippet 10.12

```
*************************************************************
void main(void)
{
 unsigned int i,temp,evstat_data;
 char rc;

 init_USART1();
 init_cf_card();
 airdrop_cfg(BSS);
 airdrop_cfg(SSID);
 airdrop_cfg(MAX_DATALEN);
 airdrop_cfg(NIC_RATE);
 temp = rid_read(RID_cfgOwnMACAddress);
 for(i=0;i<6;i++)
 {
  if(i%2)
   macaddrc[i] = fidrid_buffer[2+i/2]>>8;
  else
  macaddrc[i] = fidrid_buffer[2+i/2]& 0xFF;
 }
 for(i=0;i<3;++i)
  macaddri[i] = fidrid_buffer[i+2];

 ipaddri[0] = make16(ipaddrc[1],ipaddrc[0]);
 ipaddri[1] = make16(ipaddrc[3],ipaddrc[2]);

 allocate_xmit_buffers();

 wr_cf_io16(0xFFFF, EvAck_Register);
 if(rc=send_command(EnableMAC_Cmd,0))
  printf("MAC Enable Error\r\n");
 do{
     temp=rid_read(RID_PortStatus);
   }while(fidrid_buffer[2]!=4);
*************************************************************
```

Just in case an event has slipped by us unnoticed, the first thing we do before sending the EnableMAC command in Sub Snippet 10.13 is acknowledge all pending events by writing all ones to the EvAck Register. The EnableMAC command is then immediately sent. Roger that! Splash 2!

Sub Snippet 10.13

```
***********************************************************
wr_cf_io16(0xFFFF, EvAck_Register);
if(rc=send_command(EnableMAC_Cmd,0))
 printf("MAC Enable Error\r\n");
***********************************************************
```

As soon as the EnableMAC command completes successfully, the 802.11b Compact-Flash NIC immediately starts looking for the AIRDROP_NETWORK AP by issuing Probe Requests. Tally Ho! Splash 3! The Probe Request that caught the AIRDROP_NETWORK AP's eye is displayed in Sniffer Capture 10.1. Since you can't see the whole sniff, the meat of Sniffer Capture 10.1 can be seen in Sniffer Text 10.1.

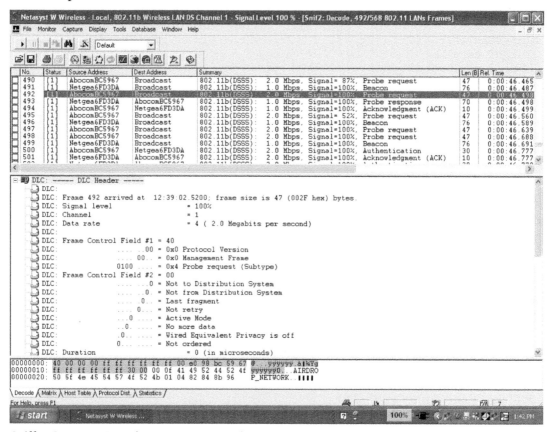

Sniffer Capture 10.1: The Netgea6FD3DA is our AIRDROP_NETWORK AP. We're only interested in exchanges between the AIRDROP_NETWORK AP and our wireless LAN station AbocomBC5967.

Sniffer Text 10.1

PROBE REQUEST							
# BYTES	2	2	6	6	6	2	VARIABLE
DESC	FC	DU/ID	DADDR	SRCADDR	BSSID	SEQ-CTL	FRAME BODY

FC = FRAME CONTROL
DU/ID = DURATION/ID
DADDR = DESTINATION ADDRESS
SRCADDR = SOURCE ADDRESS
BSSID = BASIC SERVICE SET ID
SEQ-CTL = SEQUENCE-CONTROL

```
DLC: ----- DLC Header -----
     DLC:
     DLC: Frame 492 arrived at  12:39:02.5200; frame size is 47 (002F hex)
bytes.
     DLC: Signal level              = 100%
     DLC: Channel                   = 1
     DLC: Data rate                 = 4 ( 2.0 Megabits per second)
     DLC:
     DLC: Frame Control Field #1 = 40
     DLC:                .... ..00 = 0x0 Protocol Version
     DLC:                .... 00.. = 0x0 Management Frame
     DLC:                0100 .... = 0x4 Probe request (Subtype)
     DLC: Frame Control Field #2 = 00
     DLC:                .... ...0 = Not to Distribution System
     DLC:                .... ..0. = Not from Distribution System
     DLC:                .... .0.. = Last fragment
     DLC:                .... 0... = Not retry
     DLC:                ...0 .... = Active Mode
     DLC:                ..0. .... = No more data
     DLC:                .0.. .... = Wired Equivalent Privacy is off
     DLC:                0... .... = Not ordered
     DLC: Duration                  = 0 (in microseconds)
     DLC: Destination Address       = BROADCAST FFFFFFFFFFFF, Broadcast
     DLC: Source Address            = Station AbocomBC5967
     DLC: Basic Service Set ID      = BROADCAST FFFFFFFFFFFF, Broadcast
     DLC: Sequence Control          = 0x0090
     DLC: ...Sequence Number        = 0x009 (9)
     DLC: ...Fragment Number        = 0x0   (0)
     DLC: Element ID                = 0 (Service Set Identifier)
     DLC: ...Length                 = 15 octet(s)
     DLC: ...Service Set Identity   = "AIRDROP_NETWORK"
     DLC:
     DLC: Element ID                = 1 (Supported Rates)
     DLC: ...Length                 = 4 octet(s)
     DLC: ...Supported Rates information field = 82
     DLC:                     1... .... = Basic Service Set Basic Rate
     DLC:                     .000 0010 =  1.0 Megabits per second
     DLC: ...Supported Rates information field = 84
```

```
DLC:                              1... .... = Basic Service Set Basic Rate
DLC:                              .000 0100 = 2.0 Megabits per second
DLC: ...Supported Rates information field = 8B
DLC:                              1... .... = Basic Service Set Basic Rate
DLC:                              .000 1011 = 5.5 Megabits per second
DLC: ...Supported Rates information field = 96
DLC:                              1... .... = Basic Service Set Basic Rate
DLC:                              .001 0110 = 11.0 Megabits per second
DLC:
ADDR  HEX                                              ASCII
0000: 40 00 00 00 ff ff ff ff ff ff 00 e0 98 bc 59 67 | @...ÿÿÿÿÿÿ.à˜¼Yg
0010: ff ff ff ff ff ff 90 00 00 0f 41 49 52 44 52 4f | ÿÿÿÿÿÿ□...AIRDRO
0020: 50 5f 4e 45 54 57 4f 52 4b 01 04 82 84 8b 96     | P_NETWORK..,„<-
```

I've already informed you as to what type of frame we would be analyzing and the hard evidence can be seen in Mini Sniff 10.1.

Mini Sniff 10.1

```
DLC: Frame Control Field #1 = 40
DLC:                    .... ..00 = 0x0 Protocol Version
DLC:                    .... 00.. = 0x0 Management Frame
DLC:                    0100 .... = 0x4 Probe request (Subtype)
ADDR  HEX                                              ASCII
0000: 40 00 00 00 ff ff ff ff ff ff 00 e0 98 bc 59 67 | @...ÿÿÿÿÿÿ.à˜¼Yg
0010: ff ff ff ff ff ff 90 00 00 0f 41 49 52 44 52 4f | ÿÿÿÿÿÿ□...AIRDRO
0020: 50 5f 4e 45 54 57 4f 52 4b 01 04 82 84 8b 96     | P_NETWORK..,„<-
```

Since the Probe Request frame is a Management Frame and not a Data Frame, we can expect to see zeroes in the Frame Control Field #2 of Mini Sniff 10.2.

Mini Sniff 10.2

```
DLC: Frame Control Field #2 = 00
DLC:                    .... ...0 = Not to Distribution System
DLC:                    .... ..0. = Not from Distribution System
DLC:                    .... .0.. = Last fragment
DLC:                    .... 0... = Not retry
DLC:                    ...0 .... = Active Mode
DLC:                    ..0. .... = No more data
DLC:                    .0.. .... = Wired Equivalent Privacy is off
DLC:                    0... .... = Not ordered
ADDR  HEX                                              ASCII
0000: 40 00 00 00 ff ff ff ff ff ff 00 e0 98 bc 59 67 | @...ÿÿÿÿÿÿ.à˜¼Yg
0010: ff ff ff ff ff ff 90 00 00 0f 41 49 52 44 52 4f | ÿÿÿÿÿÿ□...AIRDRO
0020: 50 5f 4e 45 54 57 4f 52 4b 01 04 82 84 8b 96     | P_NETWORK..,„<-
```

Normally, the Duration field in Mini Sniff 10.3 is directly tied to the Network Allocation Vector or NAV. The NAV is actually a count down timer that is used by stations to reserve the ether long enough to allow the completion of all of the operations associated with their current transmission. When a transmitting station deposits a number the Duration field, the other stations on the wireless LAN use that number to count down to zero before assuming that the ether has been freed up by the transmitting station. In the case of the Probe Request, no ACK, or acknowledgment frame, will be forthcoming from the AP. So, there's no reason to hold the ether and wait for it. Thus, the Duration field for a Probe Request is always set to zero.

Mini Sniff 10.3

```
     DLC: Duration                       = 0 (in microseconds)
ADDR HEX                                            ASCII
0000: 40 00 00 00 ff ff ff ff ff ff 00 e0 98 bc 59 67 | @...ÿÿÿÿÿÿ.à˜¼Yg
0010: ff ff ff ff ff ff 90 00 00 0f 41 49 52 44 52 4f | ÿÿÿÿÿÿ□...AIRDRO
0020: 50 5f 4e 45 54 57 4f 52 4b 01 04 82 84 8b 96    | P_NETWORK..,„<-
```

I've stacked the raw 802.11 header layout over the Probe Request layout so you can see the similarities. From left to right it's just like I said it should be earlier. Address 1 of the raw 802.11 header is the destination address in the Probe Request frame. The source address field of the Probe Request frame is analogous to the Address 2 field in the raw 802.11 header. The wireless LAN station would normally know which BSSID it is associated with. In this case, Address 3 is used to carry the BSSID data.

802.11 HEADER						
# BYTES	2	2	6	6	6	2
DESC	FC	DU/ID	ADDR1	ADDR2	ADDR3	SEQ-CTL

PROBE REQUEST							
# BYTES	2	2	6	6	6	2	VARIABLE
DESC	FC	DU/ID	DADDR	SRCADDR	BSSID	SEQ-CTL	FRAME BODY

A Probe Request is intended to seek out any AP in range. That's why the destination address in Mini Sniff 10.4 is a Broadcast Address that contains all hexadecimal 'F's'.

Mini Sniff 10.4

```
     DLC: Destination Address           = BROADCAST FFFFFFFFFFFF, Broadcast
     DLC: Source Address                = Station AbocomBC5967
     DLC: Basic Service Set ID          = BROADCAST FFFFFFFFFFFF, Broadcast
ADDR HEX                                            ASCII
0000: 40 00 00 00 ff ff ff ff ff ff 00 e0 98 bc 59 67 | @...ÿÿÿÿÿÿ.à˜¼Yg
0010: ff ff ff ff ff ff 90 00 00 0f 41 49 52 44 52 4f | ÿÿÿÿÿÿ□...AIRDRO
0020: 50 5f 4e 45 54 57 4f 52 4b 01 04 82 84 8b 96    | P_NETWORK..,„<-
```

Once an AP receives the wireless LAN stations Probe Request, it must have a return address so that is can respond to the wireless station's Probe Request. The AP gets its return address from the source address field of the Probe Request. By the way, notice that in Mini Sniff 10.5 the Netasyst Sniffer substituted "Abocom" for the three hexadecimal characters 00 E0 98.

Mini Sniff 10.5

```
     DLC: Destination Address          = BROADCAST FFFFFFFFFFFF, Broadcast
     DLC: Source Address               = Station AbocomBC5967
     DLC: Basic Service Set ID         = BROADCAST FFFFFFFFFFFF, Broadcast
ADDR HEX                                            ASCII
0000: 40 00 00 00 ff ff ff ff ff ff 00 e0 98 bc 59 67 | @...ÿÿÿÿÿÿ.à˜¼Yg
0010: ff ff ff ff ff ff 90 00 00 0f 41 49 52 44 52 4f | ÿÿÿÿÿÿ□...AIRDRO
0020: 50 5f 4e 45 54 57 4f 52 4b 01 04 82 84 8b 96    | P_NETWORK..,„<-
```

Once the wireless station associates with the AP, the BSSID field will be filled in with the associated AP's BSSID, which is the AP's MAC address. Since the wireless LAN station is not associated with any AP at this point, the BSSID field must contain the Broadcast bit pattern of all ones as shown in Mini Sniff 10.6.

Mini Sniff 10.6

```
     DLC: Destination Address          = BROADCAST FFFFFFFFFFFF, Broadcast
     DLC: Source Address               = Station AbocomBC5967
     DLC: Basic Service Set ID         = BROADCAST FFFFFFFFFFFF, Broadcast
ADDR HEX                                            ASCII
0000: 40 00 00 00 ff ff ff ff ff ff 00 e0 98 bc 59 67 | @...ÿÿÿÿÿÿ.à˜¼Yg
0010: ff ff ff ff ff ff 90 00 00 0f 41 49 52 44 52 4f | ÿÿÿÿÿÿ□...AIRDRO
0020: 50 5f 4e 45 54 57 4f 52 4b 01 04 82 84 8b 96    | P_NETWORK..,„<-
```

No fragmentation has occurred and the fact of the matter is backed up by the zero in the Fragment Number bit field of the Sequence Control word in Mini Sniff 10.7.

Mini Sniff 10.7

```
     DLC: ...Sequence Number           = 0x009 (9)
     DLC: ...Fragment Number           = 0x0   (0)
ADDR HEX                                            ASCII
0000: 40 00 00 00 ff ff ff ff ff ff 00 e0 98 bc 59 67 | @...ÿÿÿÿÿÿ.à˜¼Yg
0010: ff ff ff ff ff ff 90 00 00 0f 41 49 52 44 52 4f | ÿÿÿÿÿÿ□...AIRDRO
0020: 50 5f 4e 45 54 57 4f 52 4b 01 04 82 84 8b 96    | P_NETWORK..,„<-
```

The first of the Element IDs is the Service Set Identifier, which in our case is the BSS that the wireless LAN station desires to associate with. The SSID in Mini Sniff 10.8, AIRDROP_ NETWORK, should look very familiar.

Mini Sniff 10.8

```
     DLC: Element ID                   = 0 (Service Set Identifier)
     DLC: ...Length                    = 15 octet(s)
     DLC: ...Service Set Identity      = "AIRDROP_NETWORK"
ADDR HEX                                            ASCII
0000: 40 00 00 00 ff ff ff ff ff ff 00 e0 98 bc 59 67 | @...ÿÿÿÿÿÿ.à˜¼Yg
0010: ff ff ff ff ff ff 90 00 00 0f 41 49 52 44 52 4f | ÿÿÿÿÿÿ□...AIRDRO
0020: 50 5f 4e 45 54 57 4f 52 4b 01 04 82 84 8b 96    | P_NETWORK..,„<-
```

In Mini Sniff 10.9, the wireless LAN station is telling any AP that can or will listen that it supports all of the 802.11b rates.

Mini Sniff 10.9

```
DLC: Element ID                       = 1 (Supported Rates)
DLC: ...Length                        = 4 octet(s)
DLC: ...Supported Rates information field = 82
DLC:                     1... .... = Basic Service Set Basic Rate
DLC:                     .000 0010 =  1.0 Megabits per second
DLC: ...Supported Rates information field = 84
DLC:                     1... .... = Basic Service Set Basic Rate
DLC:                     .000 0100 =  2.0 Megabits per second
DLC: ...Supported Rates information field = 8B
DLC:                     1... .... = Basic Service Set Basic Rate
DLC:                     .000 1011 =  5.5 Megabits per second
DLC: ...Supported Rates information field = 96
DLC:                     1... .... = Basic Service Set Basic Rate
DLC:                     .001 0110 = 11.0 Megabits per second
ADDR  HEX                                              ASCII
0000: 40 00 00 00 ff ff ff ff ff ff 00 e0 98 bc 59 67 | @...ÿÿÿÿÿÿ.à~¼Yg
0010: ff ff ff ff ff ff 90 00 00 0f 41 49 52 44 52 4f | ÿÿÿÿÿÿ□...AIRDRO
0020: 50 5f 4e 45 54 57 4f 52 4b 01 04 82 84 8b 96     | P_NETWORK..‚„<¬
```

Lots of things can happen in the meantime, but after our wireless LAN station issues its Probe Request, it expects to receive a Probe Response from an AP if there's one in range. According to Sniffer Capture 10.1, we did indeed receive a Probe Response frame from an AP. So, let's examine the Probe Response in Sniffer Text 10.2 that our AirDrop wireless LAN station received.

Sniffer Text 10.2

```
DLC: ----- DLC Header -----
     DLC:
     DLC: Frame 493 arrived at 12:39:02.5209; frame size is 70 (0046 hex) bytes.
     DLC: Signal level              = 100%
     DLC: Channel                   = 1
     DLC: Data rate                 = 2 ( 1.0 Megabits per second)
     DLC:
     DLC: Frame Control Field #1 = 50
     DLC:                 .... ..00 = 0x0 Protocol Version
     DLC:                 .... 00.. = 0x0 Management Frame
     DLC:                 0101 .... = 0x5 Probe response (Subtype)
     DLC: Frame Control Field #2 = 00
     DLC:                 .... ...0 = Not to Distribution System
     DLC:                 .... ..0. = Not from Distribution System
     DLC:                 .... .0.. = Last fragment
     DLC:                 .... 0... = Not retry
     DLC:                 ...0 .... = Active Mode
     DLC:                 ..0. .... = No more data
     DLC:                 .0.. .... = Wired Equivalent Privacy is off
     DLC:                 0... .... = Not ordered
     DLC: Duration                  = 258 (in microseconds)
     DLC: Destination Address       = Station AbocomBC5967
     DLC: Source Address            = Station Netgea6FD3DA
```

```
DLC: Basic Service Set ID            = Netgea6FD3DA
DLC: Sequence Control                = 0x9CD0
DLC: ...Sequence Number              = 0x9CD (2509)
DLC: ...Fragment Number              = 0x0   (0)
DLC: Timestamp                       = 278013452219
DLC: Beacon Interval                 = 100
DLC: Capability information field #1 = 05
DLC:                     .... ...1 = Extended Service Set is on
DLC:                     .... ..0. = Independent Basic Service Set is off
DLC:                     .... 01.. = Point coordinator at Access Point for
delivery and polling
DLC:                   ...0 .... = No privacy
DLC:                   ..0. .... = Short Preamble option is not allowed
DLC:                   .0.. .... = Packet Binary Convolutional Coding
Modulation mode option is not allowed
DLC:                   0... .... = Channel agility is not in use
DLC: Capability information field #2 = 00
DLC:                    0000 0000 = Reserved
DLC:
DLC: Element ID                      = 0 (Service Set Identifier)
DLC: ...Length                       = 15 octet(s)
DLC: ...Service Set Identity         = "AIRDROP_NETWORK"
DLC:
DLC: Element ID                      = 1 (Supported Rates)
DLC: ...Length                       = 4 octet(s)
DLC: ...Supported Rates information field = 82
DLC:              1... .... = Basic Service Set Basic Rate
DLC:              .000 0010 =  1.0 Megabits per second
DLC: ...Supported Rates information field = 84
DLC:              1... .... = Basic Service Set Basic Rate
DLC:              .000 0100 =  2.0 Megabits per second
DLC: ...Supported Rates information field = 8B
DLC:              1... .... = Basic Service Set Basic Rate
DLC:              .000 1011 =  5.5 Megabits per second
DLC: ...Supported Rates information field = 96
DLC:              1... .... = Basic Service Set Basic Rate
DLC:              .001 0110 = 11.0 Megabits per second
DLC:
DLC: Element ID                      = 3 (Direct Sequence Parameter set)
DLC: ...Length                       = 1 octet(s)
DLC: ...dot11CurrentChannelNumber    = 1
DLC:
DLC: Element ID                      = 4 (Contention Free Parameter set)
DLC: ...Length                       = 6 octet(s)
DLC: ...CFP Count                    = 0
DLC: ...CFP Period                   = 2
DLC: ...CFP Maximum Duration         = 0 (in Time Unit)
DLC: ...CFP Duration Remaining       = 0 (dot11CurrentPattern)
DLC:
```

```
ADDR   HEX                                                         ASCII
0000:  50 00 02 01 00 e0 98 bc 59 67 00 09 5b 6f d3 da   | P....à˜¼Yg..[oÓÚ
0010:  00 09 5b 6f d3 da d0 9c bb 9f e4 ba 40 00 00 00   | ..[oÓÚÐœ»Ÿäº@...
0020:  64 00 05 00 00 0f 41 49 52 44 52 4f 50 5f 4e 45   | d.....AIRDROP_NE
0030:  54 57 4f 52 4b 01 04 82 84 8b 96 03 01 01 04 06   | TWORK..,„<-.....
0040:  00 02 00 00 00 00                                 | ......
```

Sniffer Text 10.2

This is sorta like looking at the menu in a fine dining establishment.

In Mini Sniff 10.10, I've included the frame size/signal level/channel/data rate information that the Netasyst Sniffer automatically adds at the beginning of each frame breakdown to point out that even though the AirDrop wireless LAN station is speaking to the AP at 2 Mbps, the AP realizes that the AirDrop wireless LAN station has not yet associated with it and replies to the AirDrop wireless LAN station at 1 Mbps. Frame Control Field #1 tells us that the bit pattern in the Management Frame sent by the AP represents a Probe Response frame. The AP is looking for an ACK from the probing wireless LAN station and thus sets the Duration field.

Mini Sniff 10.10

```
      DLC: Frame 493 arrived at 12:39:02.5209; frame size is 70 (0046 hex) bytes.
      DLC: Signal level              = 100%
      DLC: Channel                   = 1
      DLC: Data rate                 = 2 ( 1.0 Megabits per second)
      DLC:
      DLC: Frame Control Field #1 = 50
      DLC:                 .... ..00 = 0x0 Protocol Version
      DLC:                 .... 00.. = 0x0 Management Frame
      DLC:                 0101 .... = 0x5 Probe response (Subtype)
      DLC: Frame Control Field #2 = 00
      DLC:                 .... ...0 = Not to Distribution System
      DLC:                 .... ..0. = Not from Distribution System
      DLC:                 .... .0.. = Last fragment
      DLC:                 .... 0... = Not retry
      DLC:                 ...0 .... = Active Mode
      DLC:                 ..0. .... = No more data
      DLC:                 .0.. .... = Wired Equivalent Privacy is off
      DLC:                 0... .... = Not ordered
      DLC: Duration                  = 258 (in microseconds)
ADDR   HEX                                                         ASCII
0000:  50 00 02 01 00 e0 98 bc 59 67 00 09 5b 6f d3 da   | P....à˜¼Yg..[oÓÚ
0010:  00 09 5b 6f d3 da d0 9c bb 9f e4 ba 40 00 00 00   | ..[oÓÚÐœ»Ÿäº@...
0020:  64 00 05 00 00 0f 41 49 52 44 52 4f 50 5f 4e 45   | d.....AIRDROP_NE
0030:  54 57 4f 52 4b 01 04 82 84 8b 96 03 01 01 04 06   | TWORK..,„<-.....
0040:  00 02 00 00 00 00                                 | ......
```

In this address exchange, the AP reveals its BSSID to the wireless LAN station in the Address 3 field of Mini Sniff 10.11.

Mini Sniff 10.11

```
        DLC: Destination Address        = Station AbocomBC5967
        DLC: Source Address             = Station Netgea6FD3DA
        DLC: Basic Service Set ID       = Netgea6FD3DA
ADDR   HEX                                                  ASCII
0000:  50 00 02 01 00 e0 98 bc 59 67 00 09 5b 6f d3 da  |  P....à~¼Yg..[oÓÚ
0010:  00 09 5b 6f d3 da d0 9c bb 9f e4 ba 40 00 00 00  |  ..[oÓÚÐœ»Ÿä°@...
0020:  64 00 05 00 00 0f 41 49 52 44 52 4f 50 5f 4e 45  |  d.....AIRDROP_NE
0030:  54 57 4f 52 4b 01 04 82 84 8b 96 03 01 01 04 06  |  TWORK..,„<-.....
0040:  00 02 00 00 00 00                                |  ......
```

The sequence in Mini Sniff 10.12 shows us that the AP continues to inform the probing wireless LAN station about it habits and capabilities. Again, the frame is not fragmented. However, the AP has provided a time stamp and Beacon interval value in its Probe Response. The only bit in the Capability information field #1 we are interested in is the "Extended Service Set is on" bit, which signifies a BSS network.

Mini Sniff 10.11

```
        DLC: Sequence Control           = 0x9CD0
        DLC: ...Sequence Number         = 0x9CD (2509)
        DLC: ...Fragment Number         = 0x0   (0)
        DLC: Timestamp                  = 278013452219
        DLC: Beacon Interval            = 100
        DLC: Capability information field #1 = 05
        DLC:                    .... ...1 = Extended Service Set is on
        DLC:                    .... ..0. = Independent Basic Service Set is off
        DLC:                    .... 01.. = Point coordinator at Access Point
for delivery and polling
        DLC:                    ...0 .... = No privacy
        DLC:                    ..0. .... = Short Preamble option is not allowed
        DLC:                    .0.. .... = Packet Binary Convolutional Coding
Modulation mode option is not allowed
        DLC:                    0... .... = Channel agility is not in use
        DLC: Capability information field #2 = 00
        DLC:                    0000 0000 = Reserved
ADDR   HEX                                                  ASCII
0000:  50 00 02 01 00 e0 98 bc 59 67 00 09 5b 6f d3 da  |  P....à~¼Yg..[oÓÚ
0010:  00 09 5b 6f d3 da d0 9c bb 9f e4 ba 40 00 00 00  |  ..[oÓÚÐœ»Ÿä°@...
0020:  64 00 05 00 00 0f 41 49 52 44 52 4f 50 5f 4e 45  |  d.....AIRDROP_NE
0030:  54 57 4f 52 4b 01 04 82 84 8b 96 03 01 01 04 06  |  TWORK..,„<-.....
0040:  00 02 00 00 00 00                                |  ......
```

There's really nothing new to discuss in Mini Sniff 10.12 except that the AP is revealing its SSID to the AirDrop wireless LAN station.

Mini Sniff 10.12

```
        DLC: Element ID                 = 0 (Service Set Identifier)
        DLC: ...Length                  = 15 octet(s)
        DLC: ...Service Set Identity    = "AIRDROP_NETWORK"
```

```
        DLC:
        DLC: Element ID                      = 1 (Supported Rates)
        DLC: ...Length                       = 4 octet(s)
        DLC: ...Supported Rates information field = 82
        DLC:                      1... .... = Basic Service Set Basic Rate
        DLC:                      .000 0010 =  1.0 Megabits per second
        DLC: ...Supported Rates information field = 84
        DLC:                      1... .... = Basic Service Set Basic Rate
        DLC:                      .000 0100 =  2.0 Megabits per second
        DLC: ...Supported Rates information field = 8B
        DLC:                      1... .... = Basic Service Set Basic Rate
        DLC:                      .000 1011 =  5.5 Megabits per second
        DLC: ...Supported Rates information field = 96
        DLC:                      1... .... = Basic Service Set Basic Rate
        DLC:                      .001 0110 = 11.0 Megabits per second
ADDR  HEX                                              ASCII
0000: 50 00 02 01 00 e0 98 bc 59 67 00 09 5b 6f d3 da | P....à~¼Yg..[oÓÚ
0010: 00 09 5b 6f d3 da d0 9c bb 9f e4 ba 40 00 00 00 | ..[oÓÚÐœ»Ÿäº@...
0020: 64 00 05 00 00 0f 41 49 52 44 52 4f 50 5f 4e 45 | d.....AIRDROP_NE
0030: 54 57 4f 52 4b 01 04 82 84 8b 96 03 01 01 04 06 | TWORK..,„<-.....
0040: 00 02 00 00 00 00                               | ......
```

Channel information is all that is of any importance in Mini Sniff 10.13. The Contention Free Parameter set Element ID is describing a feature supported by the AP. However, the zeros in the CFP (Contention Free Polling) Count, CFP Maximum Duration and CFP Duration Remaining fields render the feature useless in our wireless LAN. The Probe Response task has fallen from our tasklist. Splash 4!

Mini Sniff 10.13

```
        DLC: Element ID                 = 3 (Direct Sequence Parameter set)
        DLC: ...Length                  = 1 octet(s)
        DLC: ...dot11CurrentChannelNumber  = 1
        DLC:
        DLC: Element ID                 = 4 (Contention Free Parameter set)
        DLC: ...Length                  = 6 octet(s)
        DLC: ...CFP Count               = 0
        DLC: ...CFP Period              = 2
        DLC: ...CFP Maximum Duration    = 0 (in Time Unit)
        DLC: ...CFP Duration Remaining  = 0 (dot11CurrentPattern)
ADDR  HEX                                              ASCII
0000: 50 00 02 01 00 e0 98 bc 59 67 00 09 5b 6f d3 da | P....à~¼Yg..[oÓÚ
0010: 00 09 5b 6f d3 da d0 9c bb 9f e4 ba 40 00 00 00 | ..[oÓÚÐœ»Ÿäº@...
0020: 64 00 05 00 00 0f 41 49 52 44 52 4f 50 5f 4e 45 | d.....AIRDROP_NE
0030: 54 57 4f 52 4b 01 04 82 84 8b 96 03 01 01 04 06 | TWORK..,„<-.....
0040: 00 02 00 00 00 00                               | ......
```

Since the Duration field was populated with a nonzero value in the AP's Probe Response frame, the AP is expecting a response from the probing wireless LAN station. That response the AP is waiting for would be an ACK.

Note that the ACK is a Control Frame not a Management Frame. The Implied Transmitter Address in Mini Sniff 10.14 is not part of the actual ACK frame and is filled in by the Netasyst Sniffer. The Receiver Address is pulled from the Address 2 field of the frame that is being acknowledged.

Mini Sniff 10.14

```
DLC: ----- DLC Header -----
     DLC:
     DLC: Frame 494 arrived at 12:39:02.5211; frame size is 10 (000A hex) bytes.
     DLC: Signal level            = 100%
     DLC: Channel                 = 1
     DLC: Data rate               = 2 ( 1.0 Megabits per second)
     DLC:
     DLC: Frame Control Field #1 = D4
     DLC:              .... ..00 = 0x0 Protocol Version
     DLC:              .... 01.. = 0x1 Control Frame
     DLC:              1101 .... = 0xD Acknowledgment (ACK) (Subtype)
     DLC: Frame Control Field #2 = 00
     DLC:              .... ...0 = Not to Distribution System
     DLC:              .... ..0. = Not from Distribution System
     DLC:              .... .0.. = Last fragment
     DLC:              .... 0... = Not retry
     DLC:              ...0 .... = Active Mode
     DLC:              ..0. .... = No more data
     DLC:              .0.. .... = Wired Equivalent Privacy is off
     DLC:              0... .... = Not ordered
     DLC: Duration                = 0 (in microseconds)
     DLC: Receiver Address             = Station Netgea6FD3DA
     DLC: Implied Transmitter Address  = Station AbocomBC5967
ADDR HEX                                            ASCII
0000: d4 00 00 00 00 09 5b 6f d3 da                 | Ô....[oóÚ
```

Authenticating the AirDrop Wireless LAN Station

We got jumped by some pretty fast Bogies when we unleashed the AirDrop's 802.11b CompactFlash NIC radio. In the blink of an eye the AirDrop 802.11b CompactFlash NIC probed the surrounding ether and found its target AP. In the next step towards joining the AIRDROP_ NETWORK BSS, the AirDrop wireless LAN station must be identified as friend or foe. This step in the process is called authentication and begins with an Authentication Frame being transmitted by the AirDrop wireless LAN station.

There's not much in the Authentication Frame in Mini Sniff 10.15 you haven't already seen. The AirDrop wireless LAN station is asking to be authenticated using the Open System method. Open System authentication pretty much says "come on in" to any wireless LAN station. No shirt, no shoes, no problem.

Mini Sniff 10.15

```
DLC: ----- DLC Header -----
     DLC:
     DLC: Frame 500 arrived at 12:39:02.7998; frame size is 30 (001E hex) bytes.
     DLC: Signal level                  = 100%
     DLC: Channel                       = 1
     DLC: Data rate                     = 4 ( 2.0 Megabits per second)
     DLC:
     DLC: Frame Control Field #1 = B0
     DLC:              .... ..00 = 0x0 Protocol Version
     DLC:              .... 00.. = 0x0 Management Frame
     DLC:              1011 .... = 0xB Authentication (Subtype)
     DLC: Frame Control Field #2 = 00
     DLC:              .... ...0 = Not to Distribution System
     DLC:              .... ..0. = Not from Distribution System
     DLC:              .... .0.. = Last fragment
     DLC:              .... 0... = Not retry
     DLC:              ...0 .... = Active Mode
     DLC:              ..0. .... = No more data
     DLC:              .0.. .... = Wired Equivalent Privacy is off
     DLC:              0... .... = Not ordered
     DLC: Duration                      = 43564 (in microseconds)
     DLC: Destination Address           = Station Netgea6FD3DA
     DLC: Source Address                = Station AbocomBC5967
     DLC: Basic Service Set ID          = Netgea6FD3DA
     DLC: Sequence Control              = 0x00B0
     DLC: ...Sequence Number            = 0x00B (11)
     DLC: ...Fragment Number            = 0x0    (0)
     DLC: Authentication algorithm number = 0 (Open System)
     DLC: Authentication transaction sequence number = 1
     DLC: Status code                   = 0 (Reserved)
ADDR  HEX                                                    ASCII
0000: b0 00 2c aa 00 09 5b 6f d3 da 00 e0 98 bc 59 67 | °.,ª..[oóÚ.à˜¼Yg
0010: 00 09 5b 6f d3 da b0 00 00 00 01 00 00 00        | ..[oóÚ°.....
```

A look the big picture in Sniffer Capture 10.1 shows that the AP acknowledged the AirDrop's authentication request. Task number 5 has fallen from the sky. Splash 5!

As we would expect with Open System Authentication, in Mini Sniff 10.16 permission has been granted for our AirDrop wireless LAN station to attempt to associate with the AIRDROP_NETWORK AP. Here's what the Authentication Response Frame from the AIR-DROP_NETWORK AP looks like.

Mini Sniff 10.16

```
DLC: ----- DLC Header -----
     DLC:
     DLC: Frame 502 arrived at 12:39:02.8005; frame size is 30 (001E hex) bytes.
     DLC: Signal level                  = 100%
     DLC: Channel                       = 1
     DLC: Data rate                     = 4 ( 2.0 Megabits per second)
     DLC:
```

```
       DLC: Frame Control Field #1 = B0
       DLC:                .... ..00 = 0x0 Protocol Version
       DLC:                .... 00.. = 0x0 Management Frame
       DLC:                1011 .... = 0xB Authentication (Subtype)
       DLC: Frame Control Field #2 = 00
       DLC:                .... ...0 = Not to Distribution System
       DLC:                .... ..0. = Not from Distribution System
       DLC:                .... .0.. = Last fragment
       DLC:                .... 0... = Not retry
       DLC:                ...0 .... = Active Mode
       DLC:                ..0. .... = No more data
       DLC:                .0.. .... = Wired Equivalent Privacy is off
       DLC:                0... .... = Not ordered
       DLC: Duration                     = 300 (in microseconds)
       DLC: Destination Address          = Station AbocomBC5967
       DLC: Source Address               = Station Netgea6FD3DA
       DLC: Basic Service Set ID         = Netgea6FD3DA
       DLC: Sequence Control             = 0x9D00
       DLC: ...Sequence Number           = 0x9D0 (2512)
       DLC: ...Fragment Number           = 0x0   (0)
       DLC: Authentication algorithm number = 0 (Open System)
       DLC: Authentication transaction sequence number = 2
       DLC: Status code                  = 0 (Successful)
ADDR   HEX                                              ASCII
0000:  b0 00 2c 01 00 e0 98 bc 59 67 00 09 5b 6f d3 da | °.,...à˜¼Yg..[oóÚ
0010:  00 09 5b 6f d3 da 00 9d 00 00 02 00 00 00       | ..[oóÚ.□.....
```

A positive acknowledgment from the AirDrop wireless LAN station completes the authentication process. Splash 6!

Associating with the AIRDROP_NETWORK AP

We're running low on ammo but we've almost accomplished our mission. The sniffer capture text in Mini Sniff 10.17 is identified in Frame Control Field #1 as an Association Request frame. You've seen just about everything contained within this frame before. Do you recall earlier that I told you that the CFP fields being equal to zero in the AP's Probe Request response frame effectively disabled the AP's Contention Free Polling feature? Well, the Air-Drop wireless LAN station is telling the AP that it can't do CFP anyway.

The Listen interval value of 1 tells the AP that the AirDrop wireless LAN station will check for buffered messages every Beacon interval if the AirDrop wireless LAN station falls into sleep mode. The Listen interval is defined as the number of Beacon intervals a station will wait before waking up to check for buffered frames. A value of 2 in the Listen interval field would tell the AP that the AirDrop wireless LAN station would check every 2 Beacon intervals, or every 200 mS (recall from Chapter 8, Table 8.1, that our AIRDROP_NETWORK AP's Beacon interval is 100 mS).

Mini Sniff 10.17

```
DLC: ----- DLC Header -----
      DLC:
      DLC: Frame 504 arrived at 12:39:02.8014; frame size is 51 (0033 hex) bytes.
      DLC: Signal level              = 100%
      DLC: Channel                   = 1
      DLC: Data rate                 = 4 ( 2.0 Megabits per second)
      DLC:
      DLC: Frame Control Field #1 = 00
      DLC:                  .... ..00 = 0x0 Protocol Version
      DLC:                  .... 00.. = 0x0 Management Frame
      DLC:                  0000 .... = 0x0 Association request (Subtype)
      DLC: Frame Control Field #2 = 00
      DLC:                  .... ...0 = Not to Distribution System
      DLC:                  .... ..0. = Not from Distribution System
      DLC:                  .... .0.. = Last fragment
      DLC:                  .... 0... = Not retry
      DLC:                  ...0 .... = Active Mode
      DLC:                  ..0. .... = No more data
      DLC:                  .0.. .... = Wired Equivalent Privacy is off
      DLC:                  0... .... = Not ordered
      DLC: Duration                  = 43266 (in microseconds)
      DLC: Destination Address       = Station Netgea6FD3DA
      DLC: Source Address            = Station AbocomBC5967
      DLC: Basic Service Set ID      = Netgea6FD3DA
      DLC: Sequence Control          = 0x00C0
      DLC: ...Sequence Number        = 0x00C (12)
      DLC: ...Fragment Number        = 0x0   (0)
      DLC: Capability information field #1 = 01
      DLC:                  .... ...1 = Extended Service Set is on
      DLC:                  .... ..0. = Independent Basic Service Set is off
      DLC:                  .... 00.. = STA is not Contention Free-Pollable
      DLC:                  ...0 .... = No privacy
      DLC:                  ..0. .... = Short Preamble option is not
implemented
      DLC:                  .0.. .... = Packet Binary Convolutional Coding
Modulation mode option is not implemented
      DLC:                  0... .... = Channel agility is not in use
      DLC: Capability information field #2 = 00
      DLC:                  0000 0000 = Reserved
      DLC: Listen interval           = 1
      DLC:
      DLC: Element ID                = 0 (Service Set Identifier)
      DLC: ...Length                 = 15 octet(s)
      DLC: ...Service Set Identity   = "AIRDROP_NETWORK"
      DLC:
      DLC: Element ID                = 1 (Supported Rates)
      DLC: ...Length                 = 4 octet(s)
      DLC: ...Supported Rates information field = 82
      DLC:                  1... .... = Basic Service Set Basic Rate
```

```
DLC:                                      .000 0010 =  1.0 Megabits per second
DLC: ...Supported Rates information field = 84
DLC:                              1... .... = Basic Service Set Basic Rate
DLC:                              .000 0100 =  2.0 Megabits per second
DLC: ...Supported Rates information field = 8B
DLC:                              1... .... = Basic Service Set Basic Rate
DLC:                              .000 1011 =  5.5 Megabits per second
DLC: ...Supported Rates information field = 96
DLC:                              1... .... = Basic Service Set Basic Rate
DLC:                              .001 0110 = 11.0 Megabits per second
DLC:
ADDR  HEX                                                        ASCII
0000: 00 00 02 a9 00 09 5b 6f d3 da 00 e0 98 bc 59 67 | ...©..[oÓÚ.à~¼Yg
0010: 00 09 5b 6f d3 da c0 00 01 00 01 00 00 0f 41 49 | ..[oÓÚÀ.▪.....AI
0020: 52 44 52 4f 50 5f 4e 45 54 57 4f 52 4b 01 04 82 | RDROP_NETWORK..,
0030: 84 8b 96                                         | „<-
```

An acknowledgment from the AIRDROP_NETWORK AP knocks another Bogie out fo the sky. Splash 8!

There's only one field we really care about as humans in the Association Response frame in Mini Sniff 10.18, the Status code.

Mini Sniff 10.18

```
DLC: ----- DLC Header -----
     DLC:
     DLC: Frame 506 arrived at 12:39:02.8024; frame size is 36 (0024 hex) bytes.
     DLC: Signal level            = 100%
     DLC: Channel                 = 1
     DLC: Data rate               = 4 ( 2.0 Megabits per second)
     DLC:
     DLC: Frame Control Field #1 = 10
     DLC:              .... ..00 = 0x0 Protocol Version
     DLC:              .... 00.. = 0x0 Management Frame
     DLC:              0001 .... = 0x1 Association response (Subtype)
     DLC: Frame Control Field #2 = 00
     DLC:              .... ...0 = Not to Distribution System
     DLC:              .... ..0. = Not from Distribution System
     DLC:              .... .0.. = Last fragment
     DLC:              .... 0... = Not retry
     DLC:              ...0 .... = Active Mode
     DLC:              ..0. .... = No more data
     DLC:              .0.. .... = Wired Equivalent Privacy is off
     DLC:              0... .... = Not ordered
     DLC: Duration                = 300 (in microseconds)
     DLC: Destination Address     = Station AbocomBC5967
     DLC: Source Address          = Station Netgea6FD3DA
     DLC: Basic Service Set ID    = Netgea6FD3DA
     DLC: Sequence Control        = 0x9D10
     DLC: ...Sequence Number      = 0x9D1 (2513)
```

```
         DLC: ...Fragment Number            = 0x0   (0)
         DLC: Capability information field #1 = 05
         DLC:                   .... ...1 = Extended Service Set is on
         DLC:                   .... ..0. = Independent Basic Service Set is off
         DLC:                   .... 01.. = Point coordinator at Access Point for
delivery and polling
         DLC:                   ...0 .... = No privacy
         DLC:                   ..0. .... = Short Preamble option is not allowed
         DLC:                   .0.. .... = Packet Binary Convolutional Coding
Modulation mode option is not allowed
         DLC:                   0... .... = Channel agility is not in use
         DLC: Capability information field #2 = 00
         DLC:                   0000 0000 = Reserved
         DLC: Status code                   = 0 (Successful)
         DLC: Association ID                = 2
         DLC:
         DLC: Element ID                    = 1 (Supported Rates)
         DLC: ...Length                     = 4 octet(s)
         DLC: ...Supported Rates information field = 82
         DLC:                   1... .... = Basic Service Set Basic Rate
         DLC:                   .000 0010 =   1.0 Megabits per second
         DLC: ...Supported Rates information field = 84
         DLC:                   1... .... = Basic Service Set Basic Rate
         DLC:                   .000 0100 =   2.0 Megabits per second
         DLC: ...Supported Rates information field = 8B
         DLC:                   1... .... = Basic Service Set Basic Rate
         DLC:                   .000 1011 =   5.5 Megabits per second
         DLC: ...Supported Rates information field = 96
         DLC:                   1... .... = Basic Service Set Basic Rate
         DLC:                   .001 0110 =  11.0 Megabits per second
         DLC:
ADDR  HEX                                             ASCII
0000: 10 00 2c 01 00 e0 98 bc 59 67 00 09 5b 6f d3 da | ..,..à˜¼Yg..[oÓÚ
0010: 00 09 5b 6f d3 da 10 9d 05 00 00 00 02 c0 01 04 | ..[oÓÚ.☐..␢..À..
0020: 82 84 8b 96                                     | ‚„‹<–
```

An acknowledgment frame from the AirDrop wireless LAN station puts the last Bogie into the dirt. Our AirDrop is associated with the AIRDROP_NETWORK AP and is actively participating in the AIRDROP_NETWORK BSS. Splash 9! Returning to the carrier!

While we were flying around probing, acknowledging, authenticating and finally associating, the AirDrop 802.11b CompactFlash NIC's Link LED was flashing and the 802.11b CompactFlash NIC's Activity LED was chattering as well. While the NIC LEDs were giving a visual indication that things were happening out on the wireless LAN ether, the AirDrop 802.11b driver was spinning around in a *rid_read* loop checking the 802.11b CompactFlash NIC's port status. I'll bet you can't guess which one of the RID_PortStatus values the Air-Drop 802.11b driver was looking for in Sub Snippet 10.14?

Sub Snippet 10.14

```
//* RID INFORMATION RECORDS
#define  RID_PortStatus                 0xFD40
#define  RID_PortStatus_Disabled        0x0001
#define  RID_PortStatus_Searching 0x0002
#define  RID_PortStatus_IBSS            0x0003
#define  RID_PortStatus_BSS       0x0004
#define  RID_PortStatus_OutofRange 0x0005
****************************
do{
     temp=rid_read(RID_PortStatus);
   }while(fidrid_buffer[2]!=4);
**********************************************************
```

Now that the AirDrop wireless LAN station is associated with the AIRDROP_NET-WORK AP, we want to save the BSSID information in an array just in case we may need to stuff it into a frame we will need to transmit later. I've added the "Save-The-BSSID" code at the end of our ever-growing AirDrop 802.11b driver *main* function in Sub Snippet 10.15.

Sub Snippet 10.15

```
**********************************************************
void main(void)
{
 unsigned int i,temp,evstat_data;
 char rc;

 init_USART1();
 init_cf_card();
 airdrop_cfg(BSS);
 airdrop_cfg(SSID);
 airdrop_cfg(MAX_DATALEN);
 airdrop_cfg(NIC_RATE);
 temp = rid_read(RID_cfgOwnMACAddress);
 for(i=0;i<6;i++)
 {
  if(i%2)
   macaddrc[i] = fidrid_buffer[2+i/2]>>8;
 else
  macaddrc[i] = fidrid_buffer[2+i/2]& 0xFF;
 }
 for(i=0;i<3;++i)
  macaddri[i] = fidrid_buffer[i+2];

 ipaddri[0] = make16(ipaddrc[1],ipaddrc[0]);
 ipaddri[1] = make16(ipaddrc[3],ipaddrc[2]);

 allocate_xmit_buffers();

 wr_cf_io16(0xFFFF, EvAck_Register);
 if(rc=send_command(EnableMAC_Cmd,0))
```

```
 printf("MAC Enable Error\r\n");
do{
      temp=rid_read(RID_PortStatus);
    }while(fidrid_buffer[2]!=4);


temp = rid_read(RID_CurrentBSSID);
for(i=0;i<6;i++)
{
 if(i%2)
  bssidc[i] = fidrid_buffer[2+i/2]>>8;
 else
  bssidc[i] = fidrid_buffer[2+i/2]& 0xFF;
}
for(i=0;i<3;++i)
 bssidi[i] = fidrid_buffer[i+2];
```
****

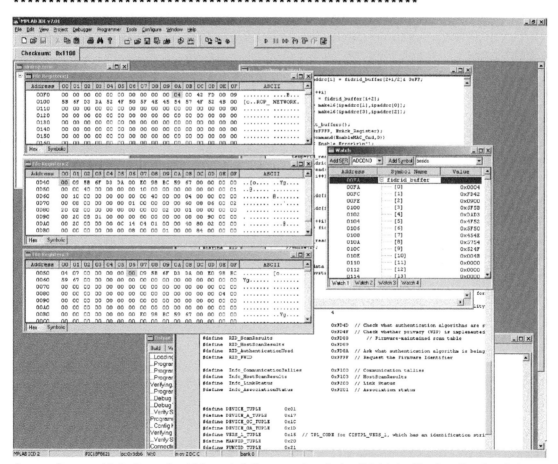

*Screen Capture 10.2: Once the AirDrop is fully operational, the **fidrid_buffer** is utilized and always is holding data garnered from one place or another. The extra "dirty" bytes following the BSSID in the **fidrid_buffer** are ignored as the first word of the **fidrid_buffer** holds the number of words in the RID that follows.*

Note that the BSSID data is stored in an identical manner to the MAC data using character and integer arrays. In Screen Capture 10.2, the File Registers:1 window shows the retrieved BSSID beginning at address 0x00FA, which also happens to be the beginning of the *fidrid_buffer*. The File Registers:2 window shows the data inside the *bssidi* array beginning at address 0x0D40. The *bssidc* array and its BSSID contents are found in the File Registers:3 window beginning at address 0x0056.

Getting the AirDrop online is quite a bit different from getting a wired LAN device up on a network but the results are the same. It doesn't matter if the LAN station is wired or wireless, the same internet protocols apply to either. In the next chapter, we'll actively seek out and process frames from other stations on the AIRDROP_NETWORK BSS.

Scotty's first guitar was a Fender Esquire. He didn't like it and called it a "girly thing" to hold. All of Scotty's early work was done with a full-bodied Gibson guitar with heavy weight strings.

Processing 802.11b
Frames with the AirDrop

Our AirDrop wireless LAN module has successfully negotiated its way onto our little AIR-DROP_NETWORK wireless LAN. The addition of the last five lines of AirDrop 802.11b driver source code to the AirDrop 802.11b driver *main* function in Sub Snippet 11.1 gets the AirDrop wireless LAN station involved with the other members of the AIRDROP_NETWORK BSS.

Sub Snippet 11.1

```
**********************************************************
void main(void)
{
 unsigned int i,temp,evstat_data;
 char rc;

 init_USART1();
 init_cf_card();
 airdrop_cfg(BSS);
 airdrop_cfg(SSID);
 airdrop_cfg(MAX_DATALEN);
 airdrop_cfg(NIC_RATE);
 temp = rid_read(RID_cfgOwnMACAddress);
 for(i=0;i<6;i++)
 {
  if(i%2)
   macaddrc[i] = fidrid_buffer[2+i/2]>>8;
 else
  macaddrc[i] = fidrid_buffer[2+i/2]& 0xFF;
 }
 for(i=0;i<3;++i)
  macaddri[i] = fidrid_buffer[i+2];

 ipaddri[0] = make16(ipaddrc[1],ipaddrc[0]);
 ipaddri[1] = make16(ipaddrc[3],ipaddrc[2]);

 allocate_xmit_buffers();

 wr_cf_io16(0xFFFF, EvAck_Register);
 if(rc=send_command(EnableMAC_Cmd,0))
  printf("MAC Enable Error\r\n");
 do{
     temp=rid_read(RID_PortStatus);
   }while(fidrid_buffer[2]!=4);
```

```
temp = rid_read(RID_CurrentBSSID);
for(i=0;i<6;i++)
{
 if(i%2)
  bssidc[i] = fidrid_buffer[2+i/2]>>8;
 else
  bssidc[i] = fidrid_buffer[2+i/2]& 0xFF;
}
for(i=0;i<3;++i)
 bssidi[i] = fidrid_buffer[i+2];


while(1){
   do{
             evstat_data = rd_cf_io16(EvStat_Register);
             }while(!(evstat_data & EvStat_Rx_Bit_Mask));
         get_frame();
}
}
```
****

AirDrop Frame Structure

The easiest and most logical way to look at an 802.11b frame as it relates to an AirDrop is to send a frame from the SNOOPER laptop and capture that frame with both the Netasyst Sniffer and the MPLAB ICD 2.

The AirDrop-P will remain in the do-while loop shown in Sub Snippet 11.2 until a receive event is logged in the Event Status Register. Once a valid frame is process by the AirDrop-P MAC, the fun begins with a call to the *get_frame* function.

Sub Snippet 11.2
****
```
while(1){
    do{
             evstat_data = rd_cf_io16(EvStat_Register);
             }while(!(evstat_data & EvStat_Rx_Bit_Mask));
         get_frame();
   }
```
****

I've setup an MPLAB ICD 2 and the Netasyst Sniffer to show you what goes on inside the AirDrop-P's *get_frame* function. I'll use the SNOOPER laptop to issue a PING command. That will kick off the ARP frame capture process.

First, I'll make sure that SNOOPER doesn't already have the AirDrop-P's socket information (IP address and MAC address) in its ARP cache by issuing the "ARP -a" command. If the AirDrop-P address information is present, I'll use the "ARP – d" command to clear SNOOPER's ARP cache. Once I've determined that the AirDrop-P module information isn't in SNOOPER's ARP cache, I'll issue "ping 192.168.0.151" from SNOOPER. As you already know, IP address 192.168.0.151 belongs to the AirDrop-P module.

OK...The Netasyst Sniffer is started and I've just issued the PING request from SNOOP-ER. The resulting ARP-attempt Netasyst Sniffer capture can be seen in Sniffer Capture 11.1. The text detail inside Sniffer Capture 11.1 is presented as Sniffer Text 11.1.

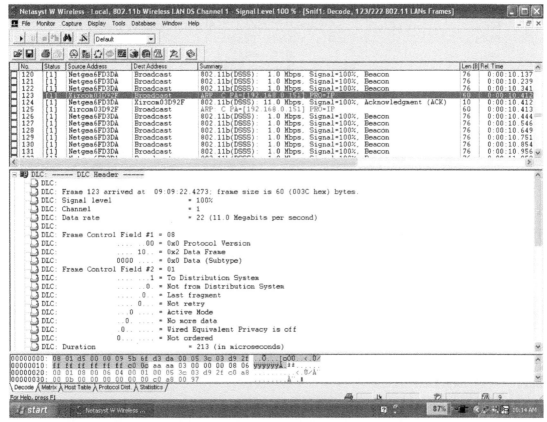

Sniffer Capture 11.1: The SNOOPER laptop is identified as Xircom03D92F in the sniffer capture. Now that real data is starting to fly around, the speeds have gotten serious. SNOOPER is not fooling around and is transmitting at the maximum 802.11b rate of 11 Mbps.

The ARP frame data shown in Sniffer Text 11.1 is basically a standard ARP frame that you would expect from a wired Ethernet LAN station. However, there's some 802.11b spice mixed in.

Sniffer Text 11.1

```
DLC: ----- DLC Header -----
    DLC:
    DLC: Frame 123 arrived at 09:09:22.4273; frame size is 60 (003C hex) bytes.
    DLC: Signal level                = 100%
    DLC: Channel                     = 1
    DLC: Data rate                   = 22 (11.0 Megabits per second)
    DLC:
    DLC: Frame Control Field #1 = 08
```

```
         DLC:                     .... ..00 = 0x0 Protocol Version
         DLC:                     .... 10.. = 0x2 Data Frame
         DLC:                     0000 .... = 0x0 Data (Subtype)
         DLC: Frame Control Field #2 = 01
         DLC:                     .... ...1 = To Distribution System
         DLC:                     .... ..0. = Not from Distribution System
         DLC:                     .... .0.. = Last fragment
         DLC:                     .... 0... = Not retry
         DLC:                     ...0 .... = Active Mode
         DLC:                     ..0. .... = No more data
         DLC:                     .0.. .... = Wired Equivalent Privacy is off
         DLC:                     0... .... = Not ordered
         DLC: Duration                   = 213 (in microseconds)
         DLC: Basic Service Set ID       = Station Netgea6FD3DA
         DLC: Source Address             = Station Xircom03D92F
         DLC: Destination Address        = BROADCAST FFFFFFFFFFFF, Broadcast
         DLC: Sequence Control           = 0x0CC0
         DLC: ...Sequence Number         = 0x0CC (204)
         DLC: ...Fragment Number         = 0x0   (0)
         DLC:
LLC:     ----- LLC Header -----
         LLC:
         LLC:  DSAP Address = AA, DSAP IG Bit = 00 (Individual Address)
         LLC:  SSAP Address = AA, SSAP CR Bit = 00 (Command)
         LLC:  Unnumbered frame: UI
         LLC:
SNAP:    ----- SNAP Header -----
         SNAP:
         SNAP: Type       = 0806 (ARP)
         SNAP:
ARP:     ----- ARP/RARP frame -----
         ARP:
         ARP: Hardware type = 1 (10Mb Ethernet)
         ARP: Protocol type = 0800 (IP)
         ARP: Length of hardware address = 6 bytes
         ARP: Length of protocol address = 4 bytes
         ARP: Opcode 1 (ARP request)
         ARP: Sender's hardware address = 00053C03D92F
         ARP: Sender's protocol address = [192.168.0.11]
         ARP: Target hardware address  = 000000000000
         ARP: Target protocol address  = [192.168.0.151]
         ARP:
         ARP:
ADDR  HEX                                                    ASCII
0000: 08 01 d5 00 00 09 5b 6f d3 da 00 05 3c 03 d9 2f | ..õ...[oÓÚ..<.Ù/
0010: ff ff ff ff ff ff c0 0c aa aa 03 00 00 00 08 06 | ÿÿÿÿÿÿÀ.ªª......
0020: 00 01 08 00 06 04 00 01 00 05 3c 03 d9 2f c0 a8 | .........<.Ù/À¨
0030: 00 0b 00 00 00 00 00 00 c0 a8 00 97             | .......À¨.—
```

Sniffer Text 11.1: The only difference in this frame and a wired Ethernet frame is the addition of specific 802.11b fields. This 802.11b ARP request will never be answered as I stopped the AirDrop-P module with a breakpoint right after the packet was received into the AirDrop-P packet buffer.

The size of the ARP frame is identical to that of a standard wired Ethernet LAN frame. That's because 802.11b frames use protocol encapsulation just like the wired Ethernet frames do. Note in Mini Sniff 11.1 that the SNOOPER laptop is serious about getting data through the wireless medium as quickly as possible.

Mini Sniff 11.1

```
DLC: Frame 123 arrived at 09:09:22.4273; frame size is 60 (003C hex) bytes.
DLC: Signal level                      = 100%
DLC: Channel                           = 1
DLC: Data rate                         = 22 (11.0 Megabits per second)
```

Mini Sniff 11.2 is the first data frame we've seen thus far. In a BSS, data frames must use the 802.11b Distribution System to get from Point A to Point B. In this case, the SNOOPER laptop fired off an ARP request frame with intentions of finding and contacting a station on the AIRDROP_NETWORK. Thus, the ARP frame is directed to the 802.11b Distribution System. This is reflected in the Frame Control Field #2.

Mini Sniff 11.2

```
DLC: Frame Control Field #1 = 08
DLC:                  .... ..00 = 0x0 Protocol Version
DLC:                  .... 10.. = 0x2 Data Frame
DLC:                  0000 .... = 0x0 Data (Subtype)
DLC: Frame Control Field #2 = 01
DLC:                  .... ...1 = To Distribution System
DLC:                  .... ..0. = Not from Distribution System
DLC:                  .... .0.. = Last fragment
DLC:                  .... 0... = Not retry
DLC:                  ...0 .... = Active Mode
DLC:                  ..0. .... = No more data
DLC:                  .0.. .... = Wired Equivalent Privacy is off
DLC:                  0... .... = Not ordered
ADDR  HEX                                              ASCII
0000: 08 01 d5 00 00 09 5b 6f d3 da 00 05 3c 03 d9 2f | ..õ...[oóú..<.Ù/
0010: ff ff ff ff ff ff c0 0c aa aa 03 00 00 00 08 06 | ÿÿÿÿÿÿÀ.ªª......
0020: 00 01 08 00 06 04 00 01 00 05 3c 03 d9 2f c0 a8 | .........<.Ù/À¨
0030: 00 0b 00 00 00 00 00 00 c0 a8 00 97             | .......À¨.─
```

The presence of a Duration value in Mini Sniff 11.3 tells us that the sending 802.11b station is expecting to hear from the target 802.11b station before giving up the wireless ether to another station on the wireless LAN.

Mini Sniff 11.3

```
DLC: Duration                          = 213 (in microseconds)
ADDR  HEX                                              ASCII
0000: 08 01 d5 00 00 09 5b 6f d3 da 00 05 3c 03 d9 2f | ..õ...[oóú..<.Ù/
0010: ff ff ff ff ff ff c0 0c aa aa 03 00 00 00 08 06 | ÿÿÿÿÿÿÀ.ªª......
0020: 00 01 08 00 06 04 00 01 00 05 3c 03 d9 2f c0 a8 | .........<.Ù/À¨
0030: 00 0b 00 00 00 00 00 00 c0 a8 00 97             | .......À¨.─
```

Two 802.11b devices are involved in getting this ARP request to its destination, the station issuing the ARP request and the AP. The destination address should be our first address entry. Since the AIRDROP_NETWORK is an infrastructure network and the To Distribution System bit is set, the AIRDROP_NETWORK AP should be the destination. For your convenience, I've reprised Table 10.1 in Mini Sniff 11.4.

Mini Sniff 11.4

802.11B ADDRESS FIELD LOGIC						
DEST	ADDR1	ADDR2	ADDR3	ADDR4	To DS	From DS
TO AP	BSSID	SA	DA	NA	1	0
FROM AP	DA	BSSID	SA	NA	0	1
IBSS	DA	SA	BSSID	NA	0	0

```
        DEST        = DESTINATION
        To DS       = TO DISTRIBUTION SYSTEM BIT
        From DS     = FROM DISTRIBUTION SYSTEM BIT
        NA          = NOT USED
```

```
      DLC: Basic Service Set ID       = Station Netgea6FD3DA
      DLC: Source Address             = Station Xircom03D92F
      DLC: Destination Address        = BROADCAST FFFFFFFFFFFF, Broadcast
      DLC: Sequence Control           = 0x0CC0
      DLC: ...Sequence Number         = 0x0CC (204)
      DLC: ...Fragment Number         = 0x0    (0)
ADDR  HEX                                       ASCII
0000: 08 01 d5 00 00 09 5b 6f d3 da 00 05 3c 03 d9 2f | ..õ...[oóÚ..<.Ù/
0010: ff ff ff ff ff ff c0 0c aa aa 03 00 00 00 08 06 | ÿÿÿÿÿÿÀ.ª ......
0020: 00 01 08 00 06 04 00 01 00 05 3c 03 d9 2f c0 a8 | .........<.Ù/À¨
0030: 00 0b 00 00 00 00 00 00 c0 a8 00 97             | ........À¨.—
```

According to the 802.11b Address Field Logic table, the next address we should encounter will be the source, or transmitting, address in the Address 2 field. The Xircom MAC address identifies the source as SNOOPER. From the sniffer dump in Mini Sniff 11.5 we see that Xircom's OUI is hexadecimal 00 05 3C.

Mini Sniff 11.5

```
      DLC: Basic Service Set ID       = Station Netgea6FD3DA
      DLC: Source Address             = Station Xircom03D92F
      DLC: Destination Address        = BROADCAST FFFFFFFFFFFF, Broadcast
      DLC: Sequence Control           = 0x0CC0
      DLC: ...Sequence Number         = 0x0CC (204)
      DLC: ...Fragment Number         = 0x0    (0)
ADDR  HEX                                       ASCII
0000: 08 01 d5 00 00 09 5b 6f d3 da 00 05 3c 03 d9 2f | ..õ...[oóÚ..<.Ù/
0010: ff ff ff ff ff ff c0 0c aa aa 03 00 00 00 08 06 | ÿÿÿÿÿÿÀ.ª ......
0020: 00 01 08 00 06 04 00 01 00 05 3c 03 d9 2f c0 a8 | .........<.Ù/À¨
0030: 00 0b 00 00 00 00 00 00 c0 a8 00 97             | ........À¨.—
```

This is an ARP request. Therefore, it is a broadcast frame. The destination address is located in the Address 3 field and it should contain all 1's, which it does according to Mini Sniff 11.6.

Mini Sniff 11.6

```
        DLC: Basic Service Set ID            = Station Netgea6FD3DA
        DLC: Source Address                  = Station Xircom03D92F
        DLC: Destination Address             = BROADCAST FFFFFFFFFFFF, Broadcast
        DLC: Sequence Control                = 0x0CC0
        DLC: ...Sequence Number              = 0x0CC (204)
        DLC: ...Fragment Number              = 0x0    (0)
ADDR    HEX                                                  ASCII
0000:   08 01 d5 00 00 09 5b 6f d3 da 00 05 3c 03 d9 2f  |  ..õ...[oóÚ..<.Ù/
0010:   ff ff ff ff ff ff c0 0c aa aa 03 00 00 00 08 06  |  ÿÿÿÿÿÿÀ.ª ª......
0020:   00 01 08 00 06 04 00 01 00 05 3c 03 d9 2f c0 a8  |  ..........<.Ù/À¨
0030:   00 0b 00 00 00 00 00 00 c0 a8 00 97              |  ........À¨.—
```

The AirDrop-P MAC determined that there was no reason to fragment the ARP request frame and that is reflected in the lack of a Fragment Number in the Sequence Control word in Mini Sniff 11.7.

Mini Sniff 11.7

```
        DLC: Sequence Control                = 0x0CC0
        DLC: ...Sequence Number              = 0x0CC (204)
        DLC: ...Fragment Number              = 0x0    (0)
ADDR    HEX                                                  ASCII
0000:   08 01 d5 00 00 09 5b 6f d3 da 00 05 3c 03 d9 2f  |  ..õ...[oóÚ..<.Ù/
0010:   ff ff ff ff ff ff c0 0c aa aa 03 00 00 00 08 06  |  ÿÿÿÿÿÿÀ.ª ª......
0020:   00 01 08 00 06 04 00 01 00 05 3c 03 d9 2f c0 a8  |  ..........<.Ù/À¨
0030:   00 0b 00 00 00 00 00 00 c0 a8 00 97              |  ........À¨.—
```

There's no need to concern yourself with the Sub-Network Attachment Point (SNAP) header information in Mini Sniff 11.8 as it will never change. The Destination Service Access Point (DSAP) and the Source Service Access Point (SSAP) fields will always contain hexadecimal AA. The Control Field that follows will always contain hexadecimal 03, which specifies unnumbered information (UI).

Mini Sniff 11.8

```
LLC:    ----- LLC Header -----
        LLC:
        LLC:  DSAP Address = AA, DSAP IG Bit = 00 (Individual Address)
        LLC:  SSAP Address = AA, SSAP CR Bit = 00 (Command)
        LLC:  Unnumbered frame: UI
ADDR    HEX                                                  ASCII
0000:   08 01 d5 00 00 09 5b 6f d3 da 00 05 3c 03 d9 2f  |  ..õ...[oóÚ..<.Ù/
0010:   ff ff ff ff ff ff c0 0c aa aa 03 00 00 00 08 06  |  ÿÿÿÿÿÿÀ.ª ª......
0020:   00 01 08 00 06 04 00 01 00 05 3c 03 d9 2f c0 a8  |  ..........<.Ù/À¨
0030:   00 0b 00 00 00 00 00 00 c0 a8 00 97              |  ........À¨.—
```

The SNAP header tells a router to look for packet protocol information in the Protocol Identification field, which consists of the five bytes following the UI byte. We issued a PING from SNOOPER. SNOOPER had an empty ARP cache. So, the first thing SNOOPER had to do was use the ARP protocol to issue an ARP request. SNOOPER did just that and we caught every byte of it in Mini Sniff 11.9.

Mini Sniff 11.9

```
SNAP: ----- SNAP Header -----
      SNAP:
      SNAP: Type      = 0806 (ARP)
ADDR  HEX                                                  ASCII
0000: 08 01 d5 00 00 09 5b 6f d3 da 00 05 3c 03 d9 2f | ..õ...[oóÚ..<.Ù/
0010: ff ff ff ff ff ff c0 0c aa aa 03 00 00 00 08 06 | ÿÿÿÿÿÿÀ.ªª.......
0020: 00 01 08 00 06 04 00 01 00 05 3c 03 d9 2f c0 a8 | ..........<.Ù/À¨
0030: 00 0b 00 00 00 00 00 00 c0 a8 00 97             | .......À¨.—
```

In Mini Sniff 11.10, SNOOPER only knows the AirDrop-P's IP address, which I entered in the PING command. SNOOPER, not knowing the hardware (MAC) address of the AirDrop-P, must issue an ARP request. The ARP request is a broadcast message asking for a hardware address from the owner of the IP address in the PING command. The AirDrop-P sees a broadcast hardware address and allows the packet to be retrieved for processing. I've stopped just short of allowing the incoming ARP request to be processed by the Air-Drop-P. Now that you've seen the Netasyst Sniffer view, let's look at the frame from the PIC18LF8621 perspective. While we're at it I'll break down the ARP frame data.

Mini Sniff 11.10

```
ARP: ----- ARP/RARP frame -----
     ARP:
     ARP: Hardware type = 1 (10Mb Ethernet)
     ARP: Protocol type = 0800 (IP)
     ARP: Length of hardware address = 6 bytes
     ARP: Length of protocol address = 4 bytes
     ARP: Opcode 1 (ARP request)
     ARP: Sender's hardware address = 00053C03D92F
     ARP: Sender's protocol address = [192.168.0.11]
     ARP: Target hardware address  = 000000000000
     ARP: Target protocol address  = [192.168.0.151]
ADDR  HEX                                                  ASCII
0000: 08 01 d5 00 00 09 5b 6f d3 da 00 05 3c 03 d9 2f | ..õ...[oóÚ..<.Ù/
0010: ff ff ff ff ff ff c0 0c aa aa 03 00 00 00 08 06 | ÿÿÿÿÿÿÀ.ªª......
0020: 00 01 08 00 06 04 00 01 00 05 3c 03 d9 2f c0 a8 | ..........<.Ù/À¨
0030: 00 0b 00 00 00 00 00 00 c0 a8 00 97             | .......À¨.—
```

What I have just described is the transmission that originated at the SNOOPER laptop and terminated at the AP. The AP then resends the ARP request information to its final destination, which in this case is every station that can hear and interpret the broadcast message. The only differences in the AP's resent message are found in Frame Control Field #2, the duration field, the sequence control field and the way the frame is addressed. When the AP resends the frame, the BSSID in the address fields represents the transmitting address. The physical source and destination stations remain true to form. There's nothing strange going on here. Everything coincides perfectly with the 802.11b Address Field Logic I described earlier. Mini Sniff 11.11 is a look at the differences through the eye of the Netasyst Sniffer.

Mini Sniff 11.11

802.11B ADDRESS FIELD LOGIC						
DEST	ADDR1	ADDR2	ADDR3	ADDR4	To DS	From DS
TO AP	BSSID	SA	DA	NA	1	0
FROM AP	DA	BSSID	SA	NA	0	1
IBSS	DA	SA	BSSID	NA	0	0

```
DEST        = DESTINATION
To DS       = TO DISTRIBUTION SYSTEM BIT
From DS     = FROM DISTRIBUTION SYSTEM BIT
NA          = NOT USED
```

```
DLC: Frame Control Field #2 = 02
DLC:                .... ...0 = Not to Distribution System
DLC:                .... ..1. = From Distribution System
DLC:                .... .0.. = Last fragment
DLC:                .... 0... = Not retry
DLC:                ...0 .... = Active Mode
DLC:                ..0. .... = No more data
DLC:                .0.. .... = Wired Equivalent Privacy is off
DLC:                0... .... = Not ordered
DLC: Duration                       = 149 (in microseconds)
DLC: Destination Address            = BROADCAST FFFFFFFFFFFF, Broadcast
DLC: Basic Service Set ID           = Station Netgea6FD3DA
DLC: Source address                 = Station Xircom03D92F
DLC: Sequence Control               = 0x95F0
DLC: ...Sequence Number             = 0x95F (2399)
DLC: ...Fragment Number             = 0x0   (0)
ADDR  HEX                                           ASCII
0000: 08 02 95 00 ff ff ff ff ff ff 00 09 5b 6f d3 da | .▪•.ÿÿÿÿÿÿ..[oóÚ
0010: 00 05 3c 03 d9 2f f0 95 aa aa 03 00 00 00 08 06 | ..<.Ù/ð•ªª......
0020: 00 01 08 00 06 04 00 01 00 05 3c 03 d9 2f c0 a8 | ..........<.Ù/À¨
0030: 00 0b 00 00 00 00 00 00 c0 a8 00 97             | ........À¨.—
```

Each frame that enters the AirDrop-P data memory area has its header information arranged as shown in the Ethernet Header Layout code snippet in Sub Snippet 11.2. Note that there are no Frame Control or Address fields contained within the header area of the *packet* array.

Sub Snippet 11.2

```
****************************************************************
//**************************************************************
//*     Ethernet Header Layout
//**************************************************************
unsigned int packet[761];
#define enetpacketLen11      0x00
#define enetpacketDest       0x01
#define enetpacketDest01     0x01   //destination mac address
#define enetpacketDest23     0x02
#define enetpacketDest45     0x03
#define enetpacketSrc        0x04
#define enetpacketSrc01      0x04   //source mac address
#define enetpacketSrc23      0x05
#define enetpacketSrc45      0x06
#define enetpacketLen03      0x07
#define enetpacketSnap       0x08
#define enetpacketCntrl      0x09
//                           0x0A
#define enetpacketType       0x0B   //type/length field
#define enetpacketData       0x0C   //IP data area begins here
****************************************************************
```

Let's take a look at how the definitions in the Ethernet Header Layout match up with the 802.11b frame data contained within Screen Capture 11.1.

The six bytes of the destination address begin at slot 0x01 of the *packet* array in Sub Snippet 11.3 and correlate to the highlighted area in Mini Sniff 11.3.

Sub Snippet 11.3

```
****************************************************************
#define enetpacketDest       0x01
#define enetpacketDest01     0x01   //destination mac address
#define enetpacketDest23     0x02
#define enetpacketDest45     0x03
****************************************************************
```

Mini Sniff 11.12

```
      DLC: Basic Service Set ID      = Station Netgea6FD3DA
      DLC: Source Address            = Station Xircom03D92F
      DLC: Destination Address       = BROADCAST FFFFFFFFFFFF, Broadcast
      DLC: Sequence Control          = 0x0CC0
      DLC: ...Sequence Number        = 0x0CC (204)
      DLC: ...Fragment Number        = 0x0   (0)
ADDR  HEX                                            ASCII
0000: 08 01 d5 00 00 09 5b 6f d3 da 00 05 3c 03 d9 2f | ..Õ...[oÓÚ..<.Ù/
0010: ff ff ff ff ff ff c0 0c aa aa 03 00 00 00 08 06 | ÿÿÿÿÿÿÀ.ªª......
0020: 00 01 08 00 06 04 00 01 00 05 3c 03 d9 2f c0 a8 | ..........<.Ù/À¨
0030: 00 0b 00 00 00 00 00 00 c0 a8 00 97             | ........À¨.—
```

The *packet* array partially shown in Sub Snippet 11.4 is laid out in 802.3 format. Recall that we earlier instructed the PRISM chipset to use 802.3 format instead of 802.11 format when handling data. So, the next field in the *packet* array will be the source address field. The source in this instance is the SNOOPER laptop, whose MAC address is highlighted in Mini Sniff 11.13.

Sub Snippet 11.4
```
***************************************************************
#define enetpacketSrc       0x04
#define enetpacketSrc01     0x04   //source mac address
#define enetpacketSrc23     0x05
#define enetpacketSrc45     0x06
***************************************************************
```

Mini Sniff 11.13
```
        DLC: Basic Service Set ID          = Station Netgea6FD3DA
        DLC: Source Address                = Station Xircom03D92F
        DLC: Destination Address           = BROADCAST FFFFFFFFFFFF, Broadcast
        DLC: Sequence Control              = 0x0CC0
        DLC: ...Sequence Number            = 0x0CC (204)
        DLC: ...Fragment Number            = 0x0   (0)
ADDR    HEX                                          ASCII
0000: 08 01 d5 00 00 09 5b 6f d3 da 00 05 3c 03 d9 2f | ..õ...[oÓÚ..<.Ù/
0010: ff ff ff ff ff ff c0 0c aa aa 03 00 00 00 08 06 | ÿÿÿÿÿÿÀ.ªª......
0020: 00 01 08 00 06 04 00 01 00 05 3c 03 d9 2f c0 a8 | .........<.Ù/À¨
0030: 00 0b 00 00 00 00 00 00 c0 a8 00 97             | ........À¨.—
```

We must make room for the SNAP header, control and protocol identifier information in Sub Snippet 11.5 as it is part of the packet that is transmitted. The words associated with Sub Snippet 11.5 are highlighted in the sniffer capture text shown in Mini Sniff 11.14.

Sub Snippet 11.5
```
***************************************************************
#define enetpacketSnap      0x08
#define enetpacketCntrl     0x09
//                          0x0A
#define enetpacketType      0x0B  //type/length field
***************************************************************
```

Mini Sniff 11.14
```
LLC:   ----- LLC Header -----
       LLC:
       LLC:   DSAP Address = AA, DSAP IG Bit = 00 (Individual Address)
       LLC:   SSAP Address = AA, SSAP CR Bit = 00 (Command)
       LLC:   Unnumbered frame: UI
       LLC:
SNAP:  ----- SNAP Header -----
       SNAP:
       SNAP: Type      = 0806 (ARP)
```

```
ADDR   HEX                                                     ASCII
0000:  08 01 d5 00 00 09 5b 6f d3 da 00 05 3c 03 d9 2f  |  ..õ...[oóÚ..<.Ù/
0010:  ff ff ff ff ff ff c0 0c aa aa 03 00 00 00 08 06  |  ÿÿÿÿÿÿÀ.ªª......
0020:  00 01 08 00 06 04 00 01 00 05 3c 03 d9 2f c0 a8  |  ..........<.Ù/À¨
0030:  00 0b 00 00 00 00 00 00 c0 a8 00 97              |  ........À¨.—
```

If you're wondering about the 802.11b packet length words, the packet length word at *packet* array slot [0] is the first word pulled from the 802.11b CompactFlash NIC's receive buffer. The inverted 802.11b packet length word at *packet* array slot [7] is also culled from the 802.11b CompactFlash NIC's receive buffer. To differentiate the inverted 802.11b packet length word, I've referred to it as 802.3 packet length. Both packet length words are stuffed with the correct packet length value buffer slot programmatically before the 802.11b CompactFlash NIC is transmits the 802.11b frame.

That does it for the header/address area of the *packet* array, which I will also refer to as the AirDrop-P packet buffer. So, let's move on and take a look at the data the makes up the body of the ARP request.

As you can see in the highlighted fields of Sub Snippet 11.6, the ARP body resides in the IP data space in the AirDrop-P packet buffer. Ultimately, we want to poke the AirDrop-P's MAC address in the *arp_thaddr* buffer slot, change the type of ARP packet to response and send it along it way back to the station that issued the ARP request.

Sub Snippet 11.6

```
//******************************************************************
//******************************************************************
//*     Ethernet Header Layout
//******************************************************************
unsigned int packet[761];
#define enetpacketLen11      0x00
#define enetpacketDest       0x01
#define enetpacketDest01     0x01   //destination mac address
#define enetpacketDest23     0x02
#define enetpacketDest45     0x03
#define enetpacketSrc        0x04
#define enetpacketSrc01      0x04   //source mac address
#define enetpacketSrc23      0x05
#define enetpacketSrc45      0x06
#define enetpacketLen03      0x07
#define enetpacketSnap       0x08
#define enetpacketCntrl      0x09
//                           0x0A
#define enetpacketType       0x0B   //type/length field
#define enetpacketData       0x0C   //IP data area begins here
//******************************************************************
//*     ARP Layout
//******************************************************************
#define arp_hwtype    0x0C
#define arp_prtype    0x0D
#define arp_hwprlen   0x0E
#define arp_op 0x0F
#define arp_shaddr    0x10   //arp source mac address
//              0x11
```

178

```
//                       0x12
#define arp_sipaddr   0x13    //arp source ip address
//                       0x14
#define arp_thaddr    0x15    //arp target mac address
//                       0x16
//                       0x17
#define arp_tipaddr   0x18    //arp target ip address
//                               0x19
*******************************************************************
```

Sometimes a standard formatted dump isn't the best way to convey the lay of the land as far as microcontroller memory goes. So, I've broken down the ARP request packet that the AirDrop-P received from SNOOPER in Array Table 11.1. This should make it a bit easier to relate the source code layouts to the actual locations of the data.

PACKET ARRAY LAYOUT			
Buffer Address	**Array Slot**	**DATA**	**Description**
0x06FA	[0]	0x0024	802.11b packet data length
0x06FC	[1]	0xFFFF	Destination MAC Address
0x06FE	[2]	0xFFFF	
0x0700	[3]	0xFFFF	
0x0702	[4]	0x0500	Source MAC Address
0x0704	[5]	0x033C	
0x0706	[6]	0x2FD9	
0x0708	[7]	0x2400	802.3 packet data length
0x070A	[8]	0xAAAA	SNAP Header
0x070C	[9]	0x0003	UI/Protocol Identifier
0x070E	[10]	0x0000	Protocol Identifier
0x0710	[11]	0x0608	ARP Protocol
0x0712	[12]	0x0100	Hardware Type
0x0714	[13]	0x0008	Protocol Type
0x0716	[14]	0x0406	Hardware/Protocol Length
0x0718	[15]	0x0100	Opcode
0x071A	[16]	0x0500	Source MAC Address
0x071C	[17]	0x033C	
0x071E	[18]	0x2FD9	
0x0720	[19]	0xA8C0	Source IP Address
0x0722	[20]	0x0B00	
0x0724	[21]	0x0000	Target MAC Address
0x0726	[22]	0x0000	
0x0728	[23]	0x0000	
0x072A	[24]	0xA8C0	Target IP Address
0x072C	[25]	0x9700	

*Array Table 11.1: This is a "vertical" dump of the AirDrop-P's packet buffer array. The Buffer Address is the actual PIC18LF8621 SRAM memory location. A relative reference to the Buffer Address is given as an Array Slot number, which is no more than the position of a word in the **packet** array. The DATA column fields represent the actual contents of the **packet** array location.*

The first ARP body bytes highlighted in Mini Sniff 11.15 begin at offset 0x0712 in the AirDrop-P microcontroller memory, which equates to slot [12] in the *packet* array. The 0x01 value at memory location 0x0718 in the *packet* array denotes the hardware type, which is 10Mb Ethernet. Remember, we're talking 802.3 in this area, not 802.11b.

Mini Sniff 11.15

```
#define arp_hwtype   0x0C

ARP: ----- ARP/RARP frame -----
     ARP:
     ARP: Hardware type = 1 (10Mb Ethernet)
     ARP: Protocol type = 0800 (IP)
     ARP: Length of hardware address = 6 bytes
     ARP: Length of protocol address = 4 bytes
     ARP: Opcode 1 (ARP request)
     ARP: Sender's hardware address = 00053C03D92F
     ARP: Sender's protocol address = [192.168.0.11]
     ARP: Target hardware address  = 000000000000
     ARP: Target protocol address  = [192.168.0.151]
ADDR  HEX                                               ASCII
0000: 08 01 d5 00 00 09 5b 6f d3 da 00 05 3c 03 d9 2f | ..õ...[oóÚ..<.Ù/
0010: ff ff ff ff ff ff c0 0c aa aa 03 00 00 00 08 06 | ÿÿÿÿÿÿÀ.ª.ª......
0020: 00 01 08 00 06 04 00 01 00 05 3c 03 d9 2f c0 a8 | ..........<.Ù/À¨
0030: 00 0b 00 00 00 00 00 00 c0 a8 00 97              | ........À¨.—
```

Since an IP address is being used to locate and contact the AirDrop-P, the next word in Mini Sniff 11.16 identifies the protocol as IP, or Internet Protocol. Also, since the value of the Length/Type field is greater than 0x0600, the field is defined as a Protocol type field.

Mini Sniff 11.16

```
#define arp_prtype   0x0D

ARP: ----- ARP/RARP frame -----
     ARP:
     ARP: Hardware type = 1 (10Mb Ethernet)
     ARP: Protocol type = 0800 (IP)
     ARP: Length of hardware address = 6 bytes
     ARP: Length of protocol address = 4 bytes
     ARP: Opcode 1 (ARP request)
     ARP: Sender's hardware address = 00053C03D92F
     ARP: Sender's protocol address = [192.168.0.11]
     ARP: Target hardware address  = 000000000000
     ARP: Target protocol address  = [192.168.0.151]
ADDR  HEX                                               ASCII
0000: 08 01 d5 00 00 09 5b 6f d3 da 00 05 3c 03 d9 2f | ..õ...[oóÚ..<.Ù/
0010: ff ff ff ff ff ff c0 0c aa aa 03 00 00 00 08 06 | ÿÿÿÿÿÿÀ.ª.ª......
0020: 00 01 08 00 06 04 00 01 00 05 3c 03 d9 2f c0 a8 | ..........<.Ù/À¨
0030: 00 0b 00 00 00 00 00 00 c0 a8 00 97              | ........À¨.—
```

We know that an AirDrop-P MAC, or hardware, address is six bytes in length and an IP address is four bytes in length. This fact is verified by the word at slot [14] of the AirDrop-P packet buffer and in Mini Sniff 11.17. The byte at address 0x0716 represents the length of the hardware address in bytes (6 bytes) and the value of the byte at address 0x0717 is the length of the protocol address, which in the case of IP is four bytes.

Mini Sniff 11.17

```
#define arp_hwprlen  0x0E

ARP: ----- ARP/RARP frame -----
      ARP:
      ARP: Hardware type = 1 (10Mb Ethernet)
      ARP: Protocol type = 0800 (IP)
      ARP: Length of hardware address = 6 bytes
      ARP: Length of protocol address = 4 bytes
      ARP: Opcode 1 (ARP request)
      ARP: Sender's hardware address = 00053C03D92F
      ARP: Sender's protocol address = [192.168.0.11]
      ARP: Target hardware address   = 000000000000
      ARP: Target protocol address   = [192.168.0.151]
ADDR  HEX                                              ASCII
0000: 08 01 d5 00 00 09 5b 6f d3 da 00 05 3c 03 d9 2f | ..õ...[oóÚ..<.Ù/
0010: ff ff ff ff ff ff c0 0c aa aa 03 00 00 00 08 06 | ÿÿÿÿÿÿÀ.ªª......
0020: 00 01 08 00 06 04 00 01 00 05 3c 03 d9 2f c0 a8 | .........<.Ù/À¨
0030: 00 0b 00 00 00 00 00 00 c0 a8 00 97              | ........À¨.—
```

The 0x0001 in slot [15] of the AirDrop-P packet buffer, which is reflected in the highlighted area of Mini Sniff 11.18, tells us the frame is an ARP request. We'll change this value when we reply to signify an ARP reply.

Mini Sniff 11.18

```
#define arp_op 0x0F

ARP: ----- ARP/RARP frame -----
      ARP:
      ARP: Hardware type = 1 (10Mb Ethernet)
      ARP: Protocol type = 0800 (IP)
      ARP: Length of hardware address = 6 bytes
      ARP: Length of protocol address = 4 bytes
      ARP: Opcode 1 (ARP request)
      ARP: Sender's hardware address = 00053C03D92F
      ARP: Sender's protocol address = [192.168.0.11]
      ARP: Target hardware address   = 000000000000
      ARP: Target protocol address   = [192.168.0.151]
ADDR  HEX                                              ASCII
0000: 08 01 d5 00 00 09 5b 6f d3 da 00 05 3c 03 d9 2f | ..õ...[oóÚ..<.Ù/
0010: ff ff ff ff ff ff c0 0c aa aa 03 00 00 00 08 06 | ÿÿÿÿÿÿÀ.ªª......
0020: 00 01 08 00 06 04 00 01 00 05 3c 03 d9 2f c0 a8 | .........<.Ù/À¨
0030: 00 0b 00 00 00 00 00 00 c0 a8 00 97              | ........À¨.—
```

In that this is an ARP request packet, everything is known except the MAC address of the target wireless LAN station, which is our AirDrop-P station. The *arp_thaddr* field is filled with zeros right now. When we craft an ARP reply, we'll stuff the requesting station's MAC address, which is now the Sender's hardware address in Mini Sniff 11.19, into the *arp_thaddr* field. The zeros in the Target hardware address field of Mini Sniff 11.19 say that the remote station doesn't know the AirDrop-P's MAC address at this point. We'll be stuffing the Air-Drop-P's MAC address in the source MAC address field depicted as an array snippet in Sub Snippet 11.7. The source MAC address begins in AirDrop-P packet buffer slot [16].

Sub Snippet 11.7

```
************************************************************
#define arp_shaddr    0x10    //arp source mac address
//                    0x11
//                    0x12
************************************************************
```

Mini Sniff 11.19

```
ARP: ----- ARP/RARP frame -----
      ARP:
      ARP: Hardware type = 1 (10Mb Ethernet)
      ARP: Protocol type = 0800 (IP)
      ARP: Length of hardware address = 6 bytes
      ARP: Length of protocol address = 4 bytes
      ARP: Opcode 1 (ARP request)
      ARP: Sender's hardware address = 00053C03D92F
      ARP: Sender's protocol address = [192.168.0.11]
      ARP: Target hardware address  = 000000000000
      ARP: Target protocol address  = [192.168.0.151]
ADDR  HEX                                              ASCII
0000: 08 01 d5 00 00 09 5b 6f d3 da 00 05 3c 03 d9 2f | ..Õ...[oÓÚ..<.Ù/
0010: ff ff ff ff ff ff c0 0c aa aa 03 00 00 00 08 06 | ÿÿÿÿÿÿÀ.ªª......
0020: 00 01 08 00 06 04 00 01 00 05 3c 03 d9 2f c0 a8 | ..........<.Ù/À¨
0030: 00 0b 00 00 00 00 00 00 c0 a8 00 97             | ........À¨.—
```

AirDrop-P packet buffer slot [19] holds the beginning of the source IP address. At this point in time, the sender is SNOOPER. The sender will become the AirDrop-P module in the ARP reply. Mini Sniff 11.20 backs up the data contained within the AirDrop-P packet buffer that is laid out as described in Sub Snippet 11.8.

Sub Snippet 11.8

```
************************************************************
#define arp_sipaddr 0x13    //arp source ip address
//                  0x14
************************************************************
```

Mini Sniff 11.20

```
ARP: ----- ARP/RARP frame -----
      ARP:
      ARP: Hardware type = 1 (10Mb Ethernet)
```

```
      ARP: Protocol type = 0800 (IP)
      ARP: Length of hardware address = 6 bytes
      ARP: Length of protocol address = 4 bytes
      ARP: Opcode 1 (ARP request)
      ARP: Sender's hardware address = 00053C03D92F
      ARP: Sender's protocol address = [192.168.0.11]
      ARP: Target hardware address   = 000000000000
      ARP: Target protocol address   = [192.168.0.151]
ADDR  HEX                                                      ASCII
0000: 08 01 d5 00 00 09 5b 6f d3 da 00 05 3c 03 d9 2f | ..õ...[oóú..<.Ù/
0010: ff ff ff ff ff ff c0 0c aa aa 03 00 00 00 08 06 | ÿÿÿÿÿÿÀ.ªª......
0020: 00 01 08 00 06 04 00 01 00 05 3c 03 d9 2f c0 a8 | .........<.Ù/À¨
0030: 00 0b 00 00 00 00 00 00 c0 a8 00 97              | ........À¨.—
```

The unknown recipient's MAC address is represented by a train of zeroes beginning at slot [21] of the AirDrop-P packet buffer. Since the AirDrop-P module will become the sender and the requesting station will become the target in the ARP reply, we'll make sure to stuff the highlighted area in Mini Sniff 11.21 with the requesting station's MAC address before sending an ARP reply. Sub Snippet 11.9 shows the relative position of the Target hardware address in the ARP area of the AirDrop-P packet buffer.

Sub Snippet 11.9

```
************************************************************
#define arp_thaddr   0x15    //arp target mac address
//                    0x16
//                    0x17

************************************************************
```

Mini Sniff 11.21

```
ARP: ----- ARP/RARP frame -----
      ARP:
      ARP: Hardware type = 1 (10Mb Ethernet)
      ARP: Protocol type = 0800 (IP)
      ARP: Length of hardware address = 6 bytes
      ARP: Length of protocol address = 4 bytes
      ARP: Opcode 1 (ARP request)
      ARP: Sender's hardware address = 00053C03D92F
      ARP: Sender's protocol address = [192.168.0.11]
      ARP: Target hardware address   = 000000000000
      ARP: Target protocol address   = [192.168.0.151]
ADDR  HEX                                                      ASCII
0000: 08 01 d5 00 00 09 5b 6f d3 da 00 05 3c 03 d9 2f | ..õ...[oóú..<.Ù/
0010: ff ff ff ff ff ff c0 0c aa aa 03 00 00 00 08 06 | ÿÿÿÿÿÿÀ.ªª......
0020: 00 01 08 00 06 04 00 01 00 05 3c 03 d9 2f c0 a8 | .........<.Ù/À¨
0030: 00 0b 00 00 00 00 00 00 c0 a8 00 97              | ........À¨.—
```

AirDrop-P packet buffer slot [24] is the beginning of the end of the ARP body. Right now, this is the AirDrop-P's IP address. When we assemble a reply, this field will be filled with the requesting station's IP address. Sub Snippet 11.10 is the AirDrop 802.11b driver's perspective of the target protocol address and Mini Sniff 11.22 is the Netasyst Sniffer vantage point.

Sub Snippet 11.10

```
************************************************************
#define arp_tipaddr  0x18    //arp target ip address
//                           0x19
************************************************************
```

Mini Sniff 11.22

```
ARP: ----- ARP/RARP frame -----
     ARP:
     ARP: Hardware type = 1 (10Mb Ethernet)
     ARP: Protocol type = 0800 (IP)
     ARP: Length of hardware address = 6 bytes
     ARP: Length of protocol address = 4 bytes
     ARP: Opcode 1 (ARP request)
     ARP: Sender's hardware address = 00053C03D92F
     ARP: Sender's protocol address = [192.168.0.11]
     ARP: Target hardware address  = 000000000000
     ARP: Target protocol address  = [192.168.0.151]
ADDR  HEX                                              ASCII
0000: 08 01 d5 00 00 09 5b 6f d3 da 00 05 3c 03 d9 2f | ..õ...[oóÚ..<.Ù/
0010: ff ff ff ff ff ff c0 0c aa aa 03 00 00 00 08 06 | ÿÿÿÿÿÿÀ.ªª......
0020: 00 01 08 00 06 04 00 01 00 05 3c 03 d9 2f c0 a8 | .........<.Ù/À¨
0030: 00 0b 00 00 00 00 00 00 c0 a8 00 97              | .......À¨.─
```

The BSSID was not by us used here as the 802.11b CompactFlash NIC is handling the ARP frame in 802.3 format. In 802.3 mode, the 802.11b CompactFlash NIC takes care of placing the BSSID in the correct address field of the transmitted frame. Take another look at Array Table 11.1. The first word in Array Table 11.1 tells us that 0x0024 bytes are contained in the data portion of the packet. This is reiterated in 802.3 format in slot [7] of the *packet* array. The data portion of the ARP request begins at slot [8], the SNAP header. Counting from slot [8] to slot [25] gives up 36 (0x24) bytes.

OK...You know how the AirDrop-P packet structure looks and how things fit into it. Let's take an indepth look at the AirDrop 802.11b driver code behind the 802.11b frames and packets. We might as well put together that ARP reply while we're at it.

AirDrop-P Frame Reception

The *get_frame* function pulls a received packet from the 802.11b CompactFlash NIC's receive buffer, interprets the packet header information and calls functions to process the packet information. The entire *get_frame* function source code set and all of the supporting definitions and functions are listed for you in Code Snippet 11.1. Let's take it apart piece by piece to see just how it works.

Code Snippet 11.1

```
******************************
#define rxdatalength_offset 44
//*************************************************************
//*     IP Protocol Types
//*************************************************************
#define PROT_ICMP           0x01
#define PROT_TCP            0x06
#define PROT_UDP            0x11
//*************************************************************
//*     Ethernet Header Layout
//*************************************************************
unsigned int packet[761];
#define enetpacketLen11     0x00
#define enetpacketDest      0x01
#define enetpacketDest01    0x01    //destination mac address
#define enetpacketDest23    0x02
#define enetpacketDest45    0x03
#define enetpacketSrc       0x04
#define enetpacketSrc01     0x04    //source mac address
#define enetpacketSrc23     0x05
#define enetpacketSrc45     0x06
#define enetpacketLen03     0x07
#define enetpacketSnap      0x08
#define enetpacketCntrl     0x09
//                          0x0A
#define enetpacketType      0x0B    //type/length field
#define enetpacketData      0x0C    //IP data area begins here
//*************************************************************
//*     ARP Layout
//*************************************************************
#define arp_hwtype          0x0C
#define arp_prtype          0x0D
#define arp_hwprlen         0x0E
#define arp_op              0x0F
#define arp_shaddr          0x10    //arp source mac address
//                          0x11
//                          0x12
#define arp_sipaddr         0x13    //arp source ip address
//                          0x14
#define arp_thaddr          0x15    //arp target mac address
//                          0x16
//                          0x17
#define arp_tipaddr         0x18    //arp target ip address
//                          0x19
//*************************************************************
//*     IP Header Layout
//*************************************************************
#define ip_vers_len         0x0C    //IP version and header length
//#define ip_tos             0x0F    //IP type of service
#define ip_pktlen           0x0D    //packet length
```

```
#define ip_id              0x0E   //datagram id
#define ip_frag_offset     0x0F   //fragment offset
#define ip_ttlproto        0x10   //time to live
//#define ip_proto                //protocol (ICMP=1, TCP=6, UDP=11)
#define ip_hdr_cksum       0x11   //header checksum
#define ip_srcaddr         0x12   //IP address of source
//                         0X13
#define ip_destaddr        0x14   //IP addess of destination
//                         0X15
#define ip_data            0x16   //IP data area
//*********************************************************
//*    TCP Header Layout
//*********************************************************
#define TCP_srcport        0x16   //TCP source port
#define TCP_destport       0x17   //TCP destination port
#define TCP_seqnum         0x18   //sequence number
//                         0x19
#define TCP_acknum         0x1A   //acknowledgment number
//                         0x1B
#define TCP_hdrflags       0x1C   //4-bit header len and flags
#define TCP_window         0x1D   //window size
#define TCP_cksum          0x1E   //TCP checksum
#define TCP_urgentptr      0x1F   //urgent pointer
#define TCP_data           0x20   //option/data
//*********************************************************
//*    TCP Flags
//*    IN flags represent incoming bits
//*    OUT flags represent outgoing bits
//*********************************************************
#define FIN_IN      (packet[TCP_hdrflags] & 0x0100)
#define SYN_IN      (packet[TCP_hdrflags] & 0x0200)
#define RST_IN      (packet[TCP_hdrflags] & 0x0400)
#define PSH_IN      (packet[TCP_hdrflags] & 0x0800)
#define ACK_IN      (packet[TCP_hdrflags] & 0x1000)
#define URG_IN      (packet[TCP_hdrflags] & 0x2000)
#define FIN_OUT     packet[TCP_hdrflags] |= 0x0100 //0000 0001 0000 0000
#define SYN_OUT     packet[TCP_hdrflags] |= 0x0200 //0000 0010 0000 0000
#define RST_OUT     packet[TCP_hdrflags] |= 0x0400 //0000 0100 0000 0000
#define PSH_OUT     packet[TCP_hdrflags] |= 0x0800 //0000 1000 0000 0000
#define ACK_OUT     packet[TCP_hdrflags] |= 0x1000 //0001 0000 0000 0000
#define URG_OUT     packet[TCP_hdrflags] |= 0x2000 //0010 0000 0000 0000
//*********************************************************
//*    ICMP Header
//*********************************************************
#define ICMP_typecode      ip_data
//#define ICMP_code         ICMP_type+1
#define ICMP_cksum         ICMP_typecode+1
#define ICMP_id            ICMP_cksum+1
#define ICMP_seqnum        ICMP_id+1
#define ICMP_data          ICMP_seqnum+1
```

```
//*****************************************************************
//*     UDP Header
//*****************************************************************
#define UDP_srcport         ip_data
#define UDP_destport        UDP_srcport+1
#define UDP_len             UDP_destport+1
#define UDP_cksum           UDP_len+1
#define UDP_data            UDP_cksum+1


//* COMMANDS
#define   Access_Cmd                0x0021

#define   BAP0Busy_Bit_Mask         0x8000 //busy bit in Offset0 register
#define   BAP0Err_Bit_Mask          0x4000 //Err bit in Offset0 register
#define   BAP1Busy_Bit_Mask         0x8000 //busy bit in Offset1 register
#define   BAP1Err_Bit_Mask          0x4000 //Err bit in Offset1 register
//*****************************************************************
//*     PORT Definitions
//*****************************************************************
#define data_in       PORTD
#define data_out      LATD
#define addr_hi       LATG
#define addr_lo       LATH
//*****************************************************************
//*     PORT TRIS Definitions
//*****************************************************************
#define TO_NICTRISD = 0x00
#define FROM_NIC      TRISD = 0xFF
//*****************************************************************
//*     CF I/O Definitions
//*****************************************************************
#define OE                  0x01
#define IORD                0x02
#define WE                  0x40
#define IOWR                0x80
#define set_OE              PCTL |= OE
#define clr_OE              PCTL &= ~OE
#define set_IORD            PCTL |= IORD
#define clr_IORD            PCTL &= ~IORD
#define set_IOWR            PCTL |= IOWR
#define clr_IOWR            PCTL &= ~IOWR

void wr_cf_data(char bapnum,char* buffer,unsigned int count)
{
 unsigned int addr,i;
 int *byteptr;
 char data_lo;

 if(bapnum)
  addr = Data1_Register;
 else
  addr = Data0_Register;
```

```
byteptr = (int*)buffer;
for(i=0;i<(count&0xFFFE);)
  {
   wr_cf_io16(*byteptr++,addr);
   i+=2;
  }
if(count % 2)
  {
   data_lo = buffer[i++];
   wr_cf_io16(data_lo,addr);
  }
 }

void wr_cf_addr(unsigned int addr)
{
 addr_hi = (make8(addr,1));
 addr_lo = addr & 0x00FF;
}

unsigned int rd_cf_io16(unsigned int addr)
{
 char data_lo,data_hi,i;
 unsigned int data16;

 wr_cf_addr(addr);
 clr_IORD;
 i=3;
 while(--i);
 data_lo=data_in;
 set_IORD;
 NOP();

 wr_cf_addr(addr+1);
 clr_IORD;
 i=3;
 while(--i);
 data_hi=data_in;
 set_IORD;
 NOP();
 data16 = make16(data_hi,data_lo);
 return(data16);
}

void wr_cf_io16(unsigned int data16,unsigned int addr)
{
 char i;

 wr_cf_addr(addr);
 data_out = LOW_BYTE(data16);
 TO_NIC;
 clr_IOWR;
 i=1;
```

```
  while(--i);
  set_IOWR;
  NOP();
  data_out = 0xFF;

  wr_cf_addr(addr+1);
  data_out = HIGH_BYTE(data16);
  clr_IOWR;
  i=1;
  while(--i);
  set_IOWR;
  NOP();
  data_out = 0xFF;
  FROM_NIC;
}
//***********************************************************
//* get a free bap
//***********************************************************
char get_free_bap(void)
{
  unsigned int temp;
  char rc;

  rc = 2;
  temp = rd_cf_io16(Offset1_Register);
  if(!(temp & BAP1Busy_Bit_Mask))
   rc = 1;
  temp = rd_cf_io16(Offset0_Register);
  if(!(temp & BAP0Busy_Bit_Mask))
   rc = 0;
  return(rc);
}
****************************

//***********************************************************
//*    Get A Frame From the CF Receive Buffer
//*    This routine removes an Ethernet frame from the receive buffer.
//***********************************************************
void get_frame()
{
  unsigned int temp,i;
  char rc,bapnum;

  rc = 0;
  RxFID = rd_cf_io16(RxFID_Register);     // Get the FID of the allocated buf-
fer
  bapnum = get_free_bap();
  switch(bapnum)
   {
    case 0:
     wr_cf_io16(RxFID,Select0_Register);
     wr_cf_io16(rxdatalength_offset,Offset0_Register);
```

```
    do{
        temp = rd_cf_io16(Offset0_Register);
        }while(temp & BAP0Busy_Bit_Mask);
    packet[enetpacketLen11] = rd_cf_io16(Data0_Register);    //get data
length
    if(packet[enetpacketLen11] > 0x05DC)
     packet[enetpacketLen11] = 0x05DC;
    for(i=0;i<3;++i)
     packet[enetpacketDest+i] = rd_cf_io16(Data0_Register);
    for(i=0;i<3;++i)
     packet[enetpacketSrc+i] = rd_cf_io16(Data0_Register);
    for(i=0;i<packet[enetpacketLen11];++i)
     packet[enetpacketLen03+i] = rd_cf_io16(Data0_Register);
    // Acknowledge that we now own the FID
    wr_cf_io16(EvStat_Rx_Bit_Mask,EvAck_Register );
    break;

  case 1:
   wr_cf_io16(RxFID,Select1_Register);
   wr_cf_io16(rxdatalength_offset,Offset1_Register);
   do{
       temp = rd_cf_io16(Offset1_Register);
       }while(temp & BAP0Busy_Bit_Mask);
    packet[enetpacketLen11] = rd_cf_io16(Data1_Register);    //get data
length
    if(packet[enetpacketLen11] > 0x05DC)
     packet[enetpacketLen11] = 0x05DC;
    for(i=0;i<3;++i)
     packet[enetpacketDest+i] = rd_cf_io16(Data1_Register);
    for(i=0;i<3;++i)
     packet[enetpacketSrc+i] = rd_cf_io16(Data1_Register);
    for(i=0;i<packet[enetpacketLen11];++i)   //removed +1
     packet[enetpacketLen03+i] = rd_cf_io16(Data1_Register);
    // Acknowledge that we now own the FID
    wr_cf_io16(EvStat_Rx_Bit_Mask,EvAck_Register );
    break;
  default:
    rc = 1;
    break;
 }
 if(!rc)
 {
  //process an ARP packet
  if(packet[enetpacketType] == 0x0608)
  {
   if(packet[arp_hwtype] == 0x0100 &&
    packet[arp_prtype] == 0x0008 &&
   packet[arp_hwprlen] == 0x0406 &&
   packet[arp_op] == 0x0100 &&
   ipaddri[0] == packet[arp_tipaddr] &&
   ipaddri[1] == packet[arp_tipaddr+1])
   arp();
```

```
    }

    //process an IP packet
    else if(packet[enetpacketType] == 0x0008
     && packet[ip_destaddr] == ipaddri[0]
     && packet[ip_destaddr+1] == ipaddri[1])
    {
       if(HIGH_BYTE(packet[ip_ttlproto]) == PROT_ICMP)
        icmp();
       else if(HIGH_BYTE(packet[ip_ttlproto]) == PROT_UDP)
        udp();
       else if(HIGH_BYTE(packet[ip_ttlproto]) == PROT_TCP)
        tcp();
    }
  }
  else
   do_event_housekeeping();
}
```
★★★★★★★★★★★★★★★★★★★★★★★★★★★★★

Code Snippet 11.1: I didn't include the function calls to ARP, ICMP, TCP, UDP and do_event_ housekeeping because we'll examine them all in detail as we move forward. The breakpoint for the ARP request segement we just talked about was placed at the shaded area of this code snippet.

We're going to continue working with the original ARP request frame we've been working with all along in this chapter. After initializing the return code value in Sub Snippet 11.11, a read is performed to obtain the FID of the receive buffer that is holding the incoming ARP request frame. We'll need a BAP to transfer the data out of the 802.11b CompactFlash NIC receive buffer and into the PIC18LF8621's AirDrop-P packet buffer. Our code executed and selected BAP0. So, the *bapnum* variable value becomes 0x00. As you can see in Code Snippet 11.1, just in case BAP1 was selected instead of BAP0, the BAP0 and BAP1 code segments are identical with the exception of the BAP number and associated registers that are used.

To begin the read of the 802.11b CompactFlash NIC's receive buffer we must place our newly obtained RxFID value in the Select0_Register. You'll have to trust me on this one as there are no graphics to support my next statement. As stated in the definition statement for the *rxdatalength_offset* in Code Snippet 11.1, there are 44 bytes from beginning of the AirDrop 802.11b receive frame structure to the 802.11 length field, which is slot [0] of the AirDrop-P packet buffer. The mystery byte area preceding slot [0] of the AirDrop-P packet buffer is used internally by the PRISM chipset and consists of addressing, length and control data. You don't see any scoop on the mystery space because the PRISM chipset is in control of the memory area and you will never have to deal with it as long as you use the PRISM chipset in 802.3 format. All of the work we'll ever do on the mystery memory area has already been done in the *create_comframestructure* function. As it turns out, the 802.3 packet data length field is really not the odd field in the bunch. The 802.11b packet data length field at slot [0] is the odd one as it is formatted as big endian and has nothing at all to do with the the transmission or reception of a frame. The slot [0] value is a product of the 802.11b CompactFlash NIC firmware and is not transmitted or received. It's there for us to use as a value. Isn't that nice? Recall that little endian format puts the least significant byte of a word at the lower address in memory and the most significant byte of that word at the next higher memory address. Big endian is the opposite format of little endian. The AirDrop-P's 802.11b CompactFlash NIC wants to see everything we give to it in little endian format. The 802.11b CompactFlash NIC also gives us data in little endian format. The proof of this lies in the order that the *rd_cf_io16* and *wr_cf_io16* I/O routines read and write the 802.11b CompactFlash NIC's memory areas. The big endian value at slot [0] is simply the 802.11b CompactFlash NIC's version of the packet data length value. OK...Let's get back to the main thread of our conversation, retrieving data from the 802.11b CompactFlash NIC receive buffer. To access the 802.11 length field the offset value to the location of the 802.3 packet data length field must be loaded into the offset register associated with BAP0, the Offset0_Register. When the Offset0_Register is loaded, a BAP0 read process is kicked off. As soon as the BAP0 busy bit clears, we read in the 802.3 length word into the appropriate array slot (slot [7] or *packet[enetpacketLen11]*) of our AirDrop-P packet buffer. The AirDrop-P packet buffer is only setup to do 1500-byte packets and if the 802.11 packet data length is greater than 1500, we simply force the 802.11 packet data length to be exactly 1500 bytes, which dumps the extra bytes as they are never read from the 802.11b CompactFlash NIC's receive buffer.

The BAP process automatically increments the memory pointer and the next words we read are the destination address words, which are stored in array slots [1]–[3] of the Air-Drop-P packet buffer. The destination words are followed by the source address words that get stored in array slots [4]-[6] of the AirDrop-P packet buffer. After reading and storing the remaining packet data words in the 802.11b CompactFlash NIC receive buffer identified by our RxFID, we tell the PRISM chipset that we have retrieved all of the contents of the Rx-FID we are interested in by acknowledging the receive bit in the Event Acknowledgment Register. Every header, length and data word in Array Table 11.1 is now contained with the PIC18LF8621's *packet* array (the AirDrop-P packet buffer). Once we acknowledge ownership of the RxFID, we can't go back to get anything else regarding our newly acquired RxFID from the 802.11b CompactFlash NIC receive buffer.

Sub Snippet 11.11
****
```
rc = 0;
RxFID = rd_cf_io16(RxFID_Register);   // Get the FID of the allocated buffer
 bapnum = get_free_bap();
 switch(bapnum)
   {
   case 0:
    wr_cf_io16(RxFID,Select0_Register);
    wr_cf_io16(rxdatalength_offset,Offset0_Register);
    do{
         temp = rd_cf_io16(Offset0_Register);
         }while(temp & BAP0Busy_Bit_Mask);
   //READ 802.11B PACKET DATA LENGTH
    packet[enetpacketLen11] = rd_cf_io16(Data0_Register);
```

```
//CHECK FOR PACKET DATA LENGTH > 1500 BYTES

 If(packet[enetpacketLen11] > 0x05DC)
  packet[enetpacketLen11] = 0x05DC;
//GET DESTINATION MAC ADDRESS
  for(i=0;i<3;++i)
   packet[enetpacketDest+i] = rd_cf_io16(Data0_Register);
//GET SOURCE MAC ADDRESS
  for(i=0;i<3;++i)
   packet[enetpacketSrc+i] = rd_cf_io16(Data0_Register);
//GET PACKET DATA
  for(i=0;i<packet[enetpacketLen11];++i)
   packet[enetpacketLen03+i] = rd_cf_io16(Data0_Register);
// ACKNOWLEDGE THAT WE NOW OWN THE FID
  wr_cf_io16(EvStat_Rx_Bit_Mask,EvAck_Register );
  break;
**************************************************************
```

Now that we have all of the packet information, what do we do with it? Well, first we must determine what type of packet we have in the AirDrop-P packet buffer. Assuming we had no troubles and the return code is still 0x00, the first order of business is to parse the protocol type word contained within array slot [11] of the AirDrop-P packet buffer. We happen to know ahead of time that we are dealing with an ARP request frame. It doesn't matter about our early warning as the 0x0806 (little endian 0x0608) in slot [11] represents ARP. If you follow down the list in Array Table 11.1 and compare the values to the *if* statements under the *//process an ARP packet* comment, you'll see that everything matches. The results of the *if* exchanges lead to a call to the *arp* function. Don't worry, we'll get to the rest of the IP protocols in Sub Snippet 11.12 later.

Sub Snippet 11.12
```
**************************************************************

//************************************************************
//*    IP ADDRESS DEFINITION
//*    This is the Ethernet Module IP address.
//*    You may change this to any valid address.
//************************************************************
char ipaddrc[4] = { 192,168,0,151};

***************************
if(!rc)
 {
  //process an ARP packet
  if(packet[enetpacketType] == 0x0608)
  {
   if(packet[arp_hwtype] == 0x0100 &&
    packet[arp_prtype] == 0x0008 &&
   packet[arp_hwprlen] == 0x0406 &&
   packet[arp_op] == 0x0100 &&
   ipaddri[0] == packet[arp_tipaddr] &&
```

```
    ipaddri[1] == packet[arp_tipaddr+1])
    arp();
  }

  //process an IP packet
  else if(packet[enetpacketType] == 0x0008
   && packet[ip_destaddr] == ipaddri[0]
   && packet[ip_destaddr+1] == ipaddri[1])
  {
    if(HIGH_BYTE(packet[ip_ttlproto]) == PROT_ICMP)
     icmp();
    else if(HIGH_BYTE(packet[ip_ttlproto]) == PROT_UDP)
     udp();
    else if(HIGH_BYTE(packet[ip_ttlproto]) == PROT_TCP)
     tcp();
  }
 }
 else
  do_event_housekeeping();
}
**************************************************************
```

All of the source code contained within Code Snippet 11.2 collectively builds an ARP reply. Most of it you're already familiar with. So, let's walk through the process of building and transmitting an ARP reply using the AirDrop-P. By the way, this will be first shot of RF fired directly by the AirDrop 802.11b driver.

Code Snippet 11.2

```
******************************
#define rxdatalength_offset 44
//**********************************************************
//*    Ethernet Header Layout
//**********************************************************
unsigned int packet[761];
#define enetpacketLen11      0x00
#define enetpacketDest       0x01
#define enetpacketDest01     0x01  //destination mac address
#define enetpacketDest23     0x02
#define enetpacketDest45     0x03
#define enetpacketSrc        0x04
#define enetpacketSrc01      0x04  //source mac address
#define enetpacketSrc23      0x05
#define enetpacketSrc45      0x06
#define enetpacketLen03      0x07
#define enetpacketSnap       0x08
#define enetpacketCntrl      0x09
//                           0x0A
#define enetpacketType       0x0B  //type/length field
#define enetpacketData       0x0C  //IP data area begins here
//**********************************************************
//*    ARP Layout
```

```
//**************************************************************
#define arp_hwtype          0x0C
#define arp_prtype          0x0D
#define arp_hwprlen         0x0E
#define arp_op              0x0F
#define arp_shaddr          0x10    //arp source mac address
//                          0x11
//                          0x12
#define arp_sipaddr         0x13    //arp source ip address
//                          0x14
#define arp_thaddr          0x15    //arp target mac address
//                          0x16
//                          0x17
#define arp_tipaddr         0x18    //arp target ip address
//                          0x19

#define   BAP0Busy_Bit_Mask         0x8000 //busy bit in Offset0 register
#define   BAP0Err_Bit_Mask          0x4000 //Err bit in Offset0 register
#define   BAP1Busy_Bit_Mask         0x8000 //busy bit in Offset1 register
#define   BAP1Err_Bit_Mask          0x4000 //Err bit in Offset1 register


//**************************************************************
//*    PORT Definitions
//**************************************************************
#define data_in       PORTD
#define data_out      LATD
#define addr_hi       LATG
#define addr_lo       LATH
//**************************************************************
//*    PORT TRIS Definitions
//**************************************************************
#define TO_NICTRISD = 0x00
#define FROM_NIC      TRISD = 0xFF
//**************************************************************
//*    CF I/O Definitions
//**************************************************************
#define OE                0x01
#define IORD              0x02
#define WE                0x40
#define IOWR              0x80
#define set_OE            PCTL |= OE
#define clr_OE            PCTL &= ~OE
#define set_IORD          PCTL |= IORD
#define clr_IORD          PCTL &= ~IORD
#define set_IOWR          PCTL |= IOWR
#define clr_IOWR          PCTL &= ~IOWR

void wr_cf_data(char bapnum,char* buffer,unsigned int count)
{
 unsigned int addr,i;
 int *byteptr;
 char data_lo;
```

```
 if(bapnum)
   addr = Data1_Register;
 else
   addr = Data0_Register;

 byteptr = (int*)buffer;
 for(i=0;i<(count&0xFFFE);)
   {
     wr_cf_io16(*byteptr++,addr);
     i+=2;
   }
 if(count % 2)
   {
     data_lo = buffer[i++];
     wr_cf_io16(data_lo,addr);
   }
 }

void wr_cf_addr(unsigned int addr)
{
 addr_hi = (make8(addr,1));
 addr_lo = addr & 0x00FF;
}

unsigned int rd_cf_io16(unsigned int addr)
{
 char data_lo,data_hi,i;
 unsigned int data16;

 wr_cf_addr(addr);
 clr_IORD;
 i=3;
 while(--i);
 data_lo=data_in;
 set_IORD;
 NOP();

 wr_cf_addr(addr+1);
 clr_IORD;
 i=3;
 while(--i);
 data_hi=data_in;
 set_IORD;
 NOP();
 data16 = make16(data_hi,data_lo);
 return(data16);
}

void wr_cf_io16(unsigned int data16,unsigned int addr)
{
 char i;
```

```
wr_cf_addr(addr);
data_out = LOW_BYTE(data16);
TO_NIC;
clr_IOWR;
i=1;
while(--i);
set_IOWR;
NOP();
data_out = 0xFF;

wr_cf_addr(addr+1);
data_out = HIGH_BYTE(data16);
clr_IOWR;
i=1;
while(--i);
set_IOWR;
NOP();
data_out = 0xFF;
FROM_NIC;
}

char  send_command(unsigned int cmd, unsigned int parm0)
{
char cmd_code,rc;
unsigned int cmd_status,evstat_data;

 do{
      cmd_data = rd_cf_io16(Command_Register);
    }while(cmd_data &  CmdBusy_Bit_Mask);

 wr_cf_io16(parm0, Param0_Register);
 wr_cf_io16(cmd, Command_Register);

 do{
      cmd_data = rd_cf_io16(Command_Register);
    }while(cmd_data &  CmdBusy_Bit_Mask);

 do{
      evstat_data = rd_cf_io16(EvStat_Register);
    }while(!(evstat_data & EvStat_Cmd_Bit_Mask));

 cmd_status = rd_cf_io16( Status_Register);
 rc = cmd_status &  Status_Result_Mask;
 wr_cf_io16( EvStat_Cmd_Bit_Mask, EvAck_Register);

 switch(rc)
  {
   case 0x00:
    rc = 0;          //command completed OK
   break;
   case 0x01:
    rc = 1;          //Card failure
```

```
   break;
 case 0x05:
   rc = 2;           //no buffer space
   break;
 case 0x7F:
   rc = 3;           //command error
   break;
 }
 return(rc);
}


//***********************************************************
//* get a free bap
//***********************************************************
char get_free_bap(void)
{
 unsigned int temp;
 char rc;

 rc = 2;
 temp = rd_cf_io16(Offset1_Register);
 if(!(temp & BAP1Busy_Bit_Mask))
  rc = 1;
 temp = rd_cf_io16(Offset0_Register);
 if(!(temp & BAP0Busy_Bit_Mask))
  rc = 0;
 return(rc);
}


//***********************************************************
//* bap_write
//***********************************************************
char bap_write(unsigned int fidrid,unsigned int offset,char* buffer, unsigned
int count)
{
 unsigned int temp;
 char rc,bapnum;

 bapnum = get_free_bap();
 rc = 0;
 switch(bapnum)
 {
  case 0:
   wr_cf_io16(fidrid,Select0_Register);
   wr_cf_io16(offset,Offset0_Register);
   do{
        temp = rd_cf_io16(Offset0_Register);
      }while(temp & BAP0Busy_Bit_Mask);
   temp = rd_cf_io16(Offset0_Register);
   if(temp & BAP0Err_Bit_Mask)
    rc=1;
   else
```

```
   wr_cf_data(BAP0,buffer,count);
  break;

  case 1:
   wr_cf_io16(fidrid,Select1_Register);
   wr_cf_io16(offset,Offset1_Register);
   do{
        temp = rd_cf_io16(Offset1_Register);
      }while(temp & BAP1Busy_Bit_Mask);
   temp = rd_cf_io16(Offset1_Register);
   if(temp & BAP1Err_Bit_Mask)
    rc=1;
   else
    wr_cf_data(BAP1,buffer,count);
 break;
}
 return(rc);
}

unsigned int get_free_TxFID(void)
{
 unsigned int TxFID;

 TxFID = 0x0000;
 if(free_TxFIDs & 0x01)
 {
  bitclr(free_TxFIDs,0);
  TxFID = TxFID_buffers[0];
 }
 else if(free_TxFIDs & 0x02)
 {
  bitclr(free_TxFIDs,1);
  TxFID = TxFID_buffers[1];
 }
 else if(free_TxFIDs & 0x04)
 {
  bitclr(free_TxFIDs,2);
  TxFID = TxFID_buffers[2];
 }
 return(TxFID);
 }
**************************

//***********************************************************
//*    Perform ARP Response
//*    This routine supplies a requesting computer with the
//*    Ethernet modules's MAC (hardware) address.
//***********************************************************
char arp(void)
{
 unsigned int TxFID,i;
 char rc;
```

```
rc = 0;
TxFID = get_free_TxFID();
if(TxFID == 0)
 rc = 1;
else
{
 for(i=0;i<2;++i)
 {
  packet[arp_tipaddr+i] = packet[arp_sipaddr+i];
  packet[arp_sipaddr+i] = ipaddri[i];
 }
 for(i=0;i<3;++i)
 {
   packet[enetpacketDest+i] = packet[enetpacketSrc+i];
   packet[arp_thaddr+i] = packet[enetpacketSrc+i];
   packet[enetpacketSrc+i] = macaddri[i];
   packet[arp_shaddr+i] = macaddri[i];
 }
 packet[arp_op] = 0x0200; //arp reply
 rc = bap_write(TxFID,rxdatalength_offset,(char*)packet,
 packet[enetpacketLen11]+16);
 rc=send_command(TransmitReclaim_Cmd,TxFID);

 do_event_housekeeping();
 }
 return(rc);
}
```

****

Code Snippet 11.1: Replying to an ARP request requires us to fetch and ready one of those preallocated transmit FIDs we created earlier. After we've stuffed the TxFID and swapped some source and destination addresses, the frame is ready to be transmitted.

In the early going, we preallocated three TxFIDs that were intended to be pooled and used for frame transmissions. Since we are crafting an ARP reply, we'll need to pull one of those TxFIDs out of the pool and prepare it for transmission. Each of the three TxFIDs is associated with a *free_TxFID* flag bit. The first TxFID we encounter with its *free_TxFID* flag bit set is ours to use. To claim the TxFID for use, we clear the TxFID's associated *free_TxFID* flag bit. If all of the TxFIDs are busy, something is very wrong as this should never happen. In fact, the PRISM chipset memory allocation engine is so fast that we'll never use more than one of the three preallocated TxFIDs stored in the TxFID buffer area described by the code in Sub Snippet 11.13.

Sub Snippet 11.13
****

```
//********************************************************
//*     Macros
//********************************************************
```

```
#define bitset(var, bitno) ((var) |= 1 << (bitno))
#define bitclr(var, bitno) ((var) &= ~(1 << (bitno)))

unsigned int get_free_TxFID(void)
{
 unsigned int TxFID;

 TxFID = 0x0000;
 if(free_TxFIDs & 0x01)
 {
  bitclr(free_TxFIDs,0);
  TxFID = TxFID_buffers[0];
 }
 else if(free_TxFIDs & 0x02)
 {
  bitclr(free_TxFIDs,1);
  TxFID = TxFID_buffers[1];
 }
 else if(free_TxFIDs & 0x04)
 {
  bitclr(free_TxFIDs,2);
  TxFID = TxFID_buffers[2];
 }
 return(TxFID);
 }

****************************
 rc = 0;
 TxFID = get_free_TxFID();
 if(TxFID == 0)
  rc = 1;
*************************************************************
```

If everything goes as planned, and we are able to grab a free TxFID from the TxFID pool, we can start reshuffling the ARP request frame into an ARP reply frame. Since we're simply turning the ARP request packet around, we must address the ARP reply packet we're building to the station that issued the ARP request. That's easily done by simply loading the ARP reply destination IP address with the ARP request source IP address and loading the ARP reply source IP address with the AirDrop-P IP address as shown in Sub Snippet 11.14.

Sub Snippet 11.14

```
*************************************************************
else
 {
  for(i=0;i<2;++i)
  {
   packet[arp_tipaddr+i] = packet[arp_sipaddr+i];
   packet[arp_sipaddr+i] = ipaddri[i];
  }
*************************************************************
```

To complete the ARP reply return address, we must also swap the ARP request source MAC address information with the ARP request destination MAC address. Then there's the matter of filling in those zeroes with the ARP requesting station's MAC address, now that the station that issued the ARP request is now the target station. The station that issued the ARP request will get the unknown MAC address (our AirDrop-P 802.11b CompactFlash NIC MAC address in this case) from the ARP reply's source MAC address field, which is populated by the code in Sub Snippet 11.15.

Sub Snippet 11.15

```
*************************************************************
  for(i=0;i<3;++i)
  {
    packet[enetpacketDest+i] = packet[enetpacketSrc+i];
    packet[arp_thaddr+i] = packet[enetpacketSrc+i];
    packet[enetpacketSrc+i] = macaddri[i];
    packet[arp_shaddr+i] = macaddri[i];
  }
*************************************************************
```

All of the destination MAC and IP addresses have been altered to point back at the station that requested the ARP. We've also made sure that the receiving station knows where the ARP reply came from. Now all that's left to do to complete the ARP reply packet is to identify our newly reshuffled and address-swapped packet as an ARP reply in Sub Snippet 11.16.

Sub Snippet 11.16

```
*************************************************************
  packet[arp_op] = 0x0200; //arp reply
*************************************************************
```

Let's check our work against what we see in Figure 11.2. Since we didn't add any data or delete any data, the packet data length words should still be as they were in the ARP request packet. We're replying to the station that issued the ARP request. So, the new destination MAC address is now that of the station that issued the ARP request. The source MAC address array slots are filled with the AirDrop-P's 802.11b NIC's MAC address. No changes were made to the SNAP header and protocol identification fields and they remain the same as they were in the ARP request packet. There were also no changes made to the hardware type, protocol type and hardware/protocol length AirDrop-P packet buffer slots. We did change the opcode *packet* array slot to reflect the fact that the new ARP packet is an ARP response packet. To satisfy the remote station's ARP request, in the body of the ARP reply we filled in the source MAC address fields with our AirDrop-P NIC's MAC address and included the AirDrop-P's IP address in the ARP reply's source IP address array slots. The target hardware is now the remote station that issued the ARP request and we filled both the target MAC address and target IP address fields with the requesting station's MAC and IP addresses. This ARP reply packet is ready for transmission.

PACKET ARRAY LAYOUT			
Buffer Address	**Array Slot**	**DATA**	**Description**
0x06FA	[0]	0x0024	802.11b packet data length
0x06FC	[1]	0x0500	Destination MAC Address
0x06FE	[2]	0x033C	
0x0700	[3]	0x2FD9	
0x0702	[4]	0xE000	Source MAC Address
0x0704	[5]	0xBC98	
0x0706	[6]	0x6759	
0x0708	[7]	0x2400	802.3 packet data length
0x070A	[8]	0xAAAA	SNAP Header
0x070C	[9]	0x0003	UI/Protocol Identifier
0x070E	[10]	0x0000	Protocol Identifier
0x0710	[11]	0x0608	ARP Protocol
0x0712	[12]	0x0100	Hardware Type
0x0714	[13]	0x0008	Protocol Type
0x0716	[14]	0x0406	Hardware/Protocol Length
0x0718	[15]	0x0200	Opcode
0x071A	[16]	0xE000	Source MAC Address
0x071C	[17]	0xBC98	
0x071E	[18]	0x6759	
0x0720	[19]	0xA8C0	Source IP Address
0x0722	[20]	0x9700	
0x0724	[21]	0x0500	Target MAC Address
0x0726	[22]	0x033C	
0x0728	[23]	0x2FD9	
0x072A	[24]	0xA8C0	Target IP Address
0x072C	[25]	0x0B00	

Figure 11.2: This is what the AirDrop-P packet buffer looks like after all of the swapping and reshuffling of source and destination IP and MAC addresses. Note that we haven't changed the length of the ARP reply packet in the process.

Byte by byte we fill the voids in the TxFID that is standing by to be transmitted. The sixteen bytes added just before transmission time to the 802.11b packet data length array slot (*packet[enetpacketLen11]+16*) represent the 802.11b packet data length (2 bytes), the destination MAC address (6 bytes), the source MAC address (6 bytes) and the 802.3 packet data length (2 bytes). Those collective 16 bytes need to be written to the waiting TxFID by the BAP but are not included in the 802.11b or 802.3 packet data length counts. Once all of the bytes are loaded into the waiting TxFID, a transmit command is issued to the 802.11b CompactFlash NIC. The *TransmitReclaim_Cmd* in Sub Snippet 11.17 is special in that it not only performs the transmission, it also automatically reclaims a free TxFID for replacement into our TxFID pool.

Sub Snippet 11.17

```
**********************************************************
  rc = bap_write(TxFID,rxdatalength_offset,(char*)packet,
  packet[enetpacketLen11]+16);
  rc=send_command(TransmitReclaim_Cmd,TxFID);
**********************************************************
```

As you would imagine, the PRISM chipset is pretty busy doing all sorts of things. Some of these "things" generate events that set an associated bit in the Event Status Register. The *do_event_housekeeping* function in Sub Snippet 11.18 is responsible for going about and cleaning up events that occur during the time the 802.11b CompactFlash NIC is operating. The event we're looking for right now occurs when the TxFID we just transmitted is reclaimed. Since the reclamation of the TxFID is technically no more than a buffer allocation, The *do_event_housekeeping* function is looking for the buffer allocation bit to be set.

Sub Snippet 11.18

```
**********************************************************

//**********************************************************
//* Perform Housekeeping on Events
//**********************************************************
void do_event_housekeeping(void)
{
        unsigned int evstat_data,FID;

        evstat_data = rd_cf_io16(EvStat_Register);

        if(evstat_data & EvStat_Tx_Bit_Mask)
        {
         //printf("\r\nTransmit Event Occurred");
         wr_cf_io16(EvStat_Tx_Bit_Mask,EvAck_Register);
        }

        if(evstat_data & EvStat_TxExc_Bit_Mask)
        {
         //printf("\r\nTransmit Event Exception Occurred");
         wr_cf_io16(EvStat_TxExc_Bit_Mask,EvAck_Register);
        }

        if(evstat_data & EvStat_Alloc_Bit_Mask)
        {
         //printf("\r\nBuffer Allocation Event Occurred");
         buffer_alloc_event_handler();
         wr_cf_io16(EvStat_Tx_Bit_Mask,EvAck_Register);
        }

        if(evstat_data & EvStat_Cmd_Bit_Mask)
        {
         //printf("\r\nCommand Completed Event Occurred");
         wr_cf_io16(EvStat_Cmd_Bit_Mask,EvAck_Register);
        }
```

```
        if(evstat_data & EvStat_Info_Bit_Mask)
        {
         //printf("\r\nInformation Frame Received Event Occurred");
         FID = rd_cf_io16(InfoFID_Register);
         //printf("\r\nInformation FID: %04X", FID);
         wr_cf_io16(EvStat_Info_Bit_Mask,EvAck_Register);
        }

        if(evstat_data & EvStat_InfDrop_Bit_Mask)
        {
         //printf("\r\nUnsolicited Information Frame Dropped Event Occurred");
         wr_cf_io16(EvStat_InfDrop_Bit_Mask,EvAck_Register);
        }

        if(evstat_data & EvStat_WTErr_Bit_Mask)
        {
         //printf("\r\nWait Error Event Occurred");
         wr_cf_io16(EvStat_WTErr_Bit_Mask,EvAck_Register);
        }
}
****************************
  do_event_housekeeping();
   }
   return(rc);
   }
*************************************************************
```

If all goes well, the buffer allocation bit will set and the *do_event_housekeeping* function will call the *buffer_alloc_event_handler* in Sub Snippet 11.19. As I mentioned earlier, the reclamation of the TxFID is really an allocation process that is kicked off automatically by a bit set in the transmit command. The FID reclamation process is so efficient that in almost every case the TxFID that was just used is reclaimed. Once the TxFID is reallocated, the FID is ready for us to retrieve from the AllocFID Register. The newly acquired FID, which is most likely the same FID we just used, is reentered into the TxFID pool by simply setting its *free_TxFIDs* flag bit. Just in some weird case the reallocated TxFID is different, we poke it into the position of the previously used TxFID and arm it by setting its associated *free_Tx-FIDs* flag bit.

Sub Snippet 11.19

```
*************************************************************
void buffer_alloc_event_handler(void)
{
        unsigned int FID;
        char i,j;

        FID = rd_cf_io16(AllocFID_Register);
        j = 1;
        for(i=0;i<3;++i)
        {
```

```
          if(TxFID_buffers[i] == FID)
           bitset(free_TxFIDs, i);
           j = 0;
           //printf("\r\nReclaimed Buffer %04X", FID);
            }
       if(j)
        {
         if(!(free_TxFIDs & 0x01))
         {
            TxFID_buffers[0] = FID;
            bitset(free_TxFIDs, 0);
         }
         else if(!(free_TxFIDs & 0x02))
         {
            TxFID_buffers[1] = FID;
            bitset(free_TxFIDs, 1);
         }
         else if(!(free_TxFIDs & 0x04))
         {
            TxFID_buffers[2] = FID;
            bitset(free_TxFIDs, 2);
         }
        }
       }
  }
```
****

This book exists. So, everything transmitted just fine. Screen Capture 11.2 is a graphical rendition of what SNOOPER saw as an ARP reply from the AirDrop-P. I've also pulled the text from Screen Capture 11.2 for your reading pleasure.

There's absolutely nothing that I can tell you about the ARP reply in Sniffer Text 11.2 that you don't already know. Note that the AirDrop-P is transmitting the ARP reply at full throttle (11 Mbps).

Sniffer Text 11.2

```
DLC: ----- DLC Header -----
     DLC:
     DLC: Frame 10 arrived at  15:05:01.3627; frame size is 60 (003C hex)
bytes.
     DLC: Signal level                = 100%
     DLC: Channel                     = 1
     DLC: Data rate                   = 22 (11.0 Megabits per second)
     DLC:
     DLC: Frame Control Field #1 = 08
     DLC:                 .... ..00 = 0x0 Protocol Version
     DLC:                 .... 10.. = 0x2 Data Frame
     DLC:                 0000 .... = 0x0 Data (Subtype)
     DLC: Frame Control Field #2 = 02
     DLC:                 .... ...0 = Not to Distribution System
     DLC:                 .... ..1. = From Distribution System
     DLC:                 .... .0.. = Last fragment
```

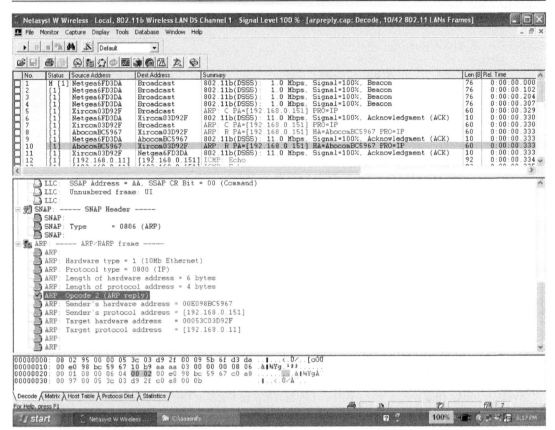

Screen Capture 11.2: Here's a Netasyst Sniffer view of the ARP reply that was assembled and transmitted by the AirDrop-P. Once you turn on the radio, 802.11b is just another way of flinging Internet protocols around.

```
      DLC:                   .... 0... = Not retry
      DLC:                   ...0 .... = Active Mode
      DLC:                   ..0. .... = No more data
      DLC:                   .0.. .... = Wired Equivalent Privacy is off
      DLC:                   0... .... = Not ordered
      DLC: Duration                    = 149 (in microseconds)
      DLC: Destination Address         = Station Xircom03D92F
      DLC: Basic Service Set ID        = Station Netgea6FD3DA
      DLC: Source address              = Station AbocomBC5967
      DLC: Sequence Control            = 0xB910
      DLC: ...Sequence Number          = 0xB91 (2961)
      DLC: ...Fragment Number          = 0x0   (0)
      DLC:
LLC:  ----- LLC Header -----
      LLC:
      LLC:  DSAP Address = AA, DSAP IG Bit = 00 (Individual Address)
      LLC:  SSAP Address = AA, SSAP CR Bit = 00 (Command)
```

```
        LLC:  Unnumbered frame: UI
        LLC:
SNAP: ----- SNAP Header -----
        SNAP:
        SNAP: Type       = 0806 (ARP)
        SNAP:
ARP: ----- ARP/RARP frame -----
        ARP:
        ARP: Hardware type = 1 (10Mb Ethernet)
        ARP: Protocol type = 0800 (IP)
        ARP: Length of hardware address = 6 bytes
        ARP: Length of protocol address = 4 bytes
        ARP: Opcode 2 (ARP reply)
        ARP: Sender's hardware address = 00E098BC5967
        ARP: Sender's protocol address = [192.168.0.151]
        ARP: Target hardware address  = 00053C03D92F
        ARP: Target protocol address  = [192.168.0.11]
        ARP:
        ARP:
ADDR  HEX                                                ASCII
0000: 08 02 95 00 00 05 3c 03 d9 2f 00 09 5b 6f d3 da | ..•...<.Ù/..[oóÚ
0010: 00 e0 98 bc 59 67 10 b9 aa aa 03 00 00 00 08 06 | .à~¼Yg.¹ªª......
0020: 00 01 08 00 06 04 00 02 00 e0 98 bc 59 67 c0 a8 | .......ã.à~¼YgÀ¨
0030: 00 97 00 05 3c 03 d9 2f c0 a8 00 0b              | .—..<.Ù/À¨..
```

****

Well, it looks like we've gone full circle and we're back at the never-ending loop in Sub Snippet 11.20 that's looking for a frame to process and another adventure. Think about this. Although we looked at assembling an ARP reply, you now know how to build an ARP request packet as well.

Sub Snippet 11.20
****
```
    while(1){
       do{
            evstat_data = rd_cf_io16(EvStat_Register);
          }while(!(evstat_data & EvStat_Rx_Bit_Mask));
       get_frame();
    }
```
****

SNOOPER didn't issue an ARP request for nothing. When a station requests another stations MAC address, you can be sure that data is about to be transferred. So, let's see if we can sniff around and find out what was behind SNOOPER's ARP request that was aimed at our AirDrop-P.

PINGING the AirDrop

It didn't take long to figure out what the SNOOPER laptop was up to. The Netasyst Sniffer's Summary window in Screen Capture 12.1 shows that the next operation initiated by SNOOPER is and ICMP Echo, better known to all of us as a PING.

ICMP messages serve many purposes. For instance, we're using ICMP to echo a packet from one host to another and back. An ICMP message may also be sent when a datagram cannot reach its destination. ICMP messages are primarily used to provide feedback about problems that exist with datagrams in the communication environment. The only thing an ICMP message can't do is tell on itself when it's bad.

A "PING" is really an application that issues an ICMP echo request packet. Basically a ping sends some data to a remote host and expects the remote host to echo it back. The ICMP header and its data are encapsulated within the IP data area.

Everything contained within Sniffer Text 12.1 is familiar until we get past the SNAP header area. This ICMP Echo message originates at SNOOPER (Xircom03D92F) and is looking to be delivered via the AP (Netgea6FD3DA) to the AirDrop-P module (AbocomBC5967). An 0x0800 is in the SNAP header protocol identification field indicating that this packet is an IP packet. The IP address information confirms the MAC address information. SNOOPER (192.168.0.11) wants to talk to the AirDrop-P module (192.168.0.151). The IP packet payload is an ICMP Echo package.

Sniffer Text 12.1

```
DLC: ----- DLC Header -----
     DLC:
     DLC: Frame 35 arrived at  14:02:23.6343; frame size is 92 (005C hex)
bytes.
     DLC: Signal level            = 100%
     DLC: Channel                 = 1
     DLC: Data rate               = 22 (11.0 Megabits per second)
     DLC:
     DLC: Frame Control Field #1 = 08
     DLC:                  .... ..00 = 0x0 Protocol Version
     DLC:                  .... 10.. = 0x2 Data Frame
     DLC:                  0000 .... = 0x0 Data (Subtype)
     DLC: Frame Control Field #2 = 01
     DLC:                  .... ...1 = To Distribution System
     DLC:                  .... ..0. = Not from Distribution System
```

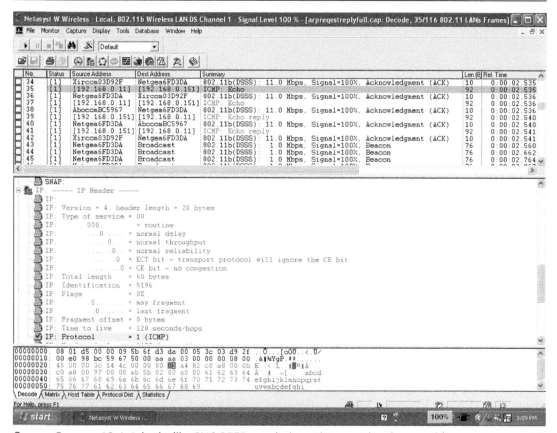

Screen Capture 12.1: It looks like SNOOPER needed an AirDrop-P hardware address to get on with the PING operation. Although you see a pair of ICMP Echo request and a pair of ICMP Echo reply frames, only a single ICMP Echo request/reply exchange occurred between the physical source and destination stations. The first ICMP Echo request message is sent to the AP (Frame 35) and the second ICMP Echo request message is the AP redirecting the ICMP Echo request message to the AirDrop-P module. The same process holds for the ICMP Echo reply transmitted by the AirDrop-P module in Frame 39 and echoed by the AP in Frame 41.

```
            DLC:                   .... .0.. = Last fragment
            DLC:                   .... 0... = Not retry
            DLC:                   ...0 .... = Active Mode
            DLC:                   ..0. .... = No more data
            DLC:                   .0.. .... = Wired Equivalent Privacy is off
            DLC:                   0... .... = Not ordered
            DLC: Duration                    = 213 (in microseconds)
            DLC: Basic Service Set ID        = Station Netgea6FD3DA
            DLC: Source Address              = Station Xircom03D92F
            DLC: Destination Address         = Station AbocomBC5967
            DLC: Sequence Control            = 0x0050
            DLC: ...Sequence Number          = 0x005 (5)
            DLC: ...Fragment Number          = 0x0   (0)
            DLC:
LLC:        ----- LLC Header -----
            LLC:
            LLC:  DSAP Address = AA, DSAP IG Bit = 00 (Individual Address)
            LLC:  SSAP Address = AA, SSAP CR Bit = 00 (Command)
            LLC:  Unnumbered frame: UI
            LLC:
SNAP:       ----- SNAP Header -----
            SNAP:
            SNAP: Type     = 0800 (IP)
            SNAP:
IP: ----- IP Header -----
            IP:
            IP: Version = 4, header length = 20 bytes
            IP: Type of service = 00
            IP:       000. ....   = routine
            IP:       ...0 ....   = normal delay
            IP:       .... 0...   = normal throughput
            IP:       .... .0..   = normal reliability
            IP:       .... ..0. = ECT bit - transport protocol will ignore the CE bit
            IP:       .... ...0 = CE bit - no congestion
            IP: Total length  = 60 bytes
            IP: Identification = 5196
            IP: Flags         = 0X
            IP:       .0.. .... = may fragment
            IP:       ..0. .... = last fragment
            IP: Fragment offset = 0 bytes
            IP: Time to live  = 128 seconds/hops
            IP: Protocol      = 1 (ICMP)
            IP: Header checksum = A482 (correct)
            IP: Source address    = [192.168.0.11]
            IP: Destination address = [192.168.0.151]
            IP: No options
            IP:
ICMP:       ----- ICMP header -----
            ICMP:
            ICMP: Type = 8 (Echo)
```

```
         ICMP: Code = 0
         ICMP: Checksum = AB5B (correct)
         ICMP: Identifier = 512
         ICMP: Sequence number = 40960
         ICMP: [32 bytes of data]
         ICMP:
         ICMP: [Normal end of "ICMP header".]
         ICMP:
ADDR  HEX                                                ASCII
0000: 08 01 d5 00 00 09 5b 6f d3 da 00 05 3c 03 d9 2f | ..õ...[oóÚ..<.Ù/
0010: 00 e0 98 bc 59 67 50 00 aa aa 03 00 00 00 08 00 | .à˜¼YgP.ªª......
0020: 45 00 00 3c 14 4c 00 00 80 01 a4 82 c0 a8 00 0b | E..<.L..€.¤‚À¨..
0030: c0 a8 00 97 08 00 ab 5b 02 00 a0 00 61 62 63 64 | À¨.—..«[.. .abcd
0040: 65 66 67 68 69 6a 6b 6c 6d 6e 6f 70 71 72 73 74 | efghijklmnopqrst
0050: 75 76 77 61 62 63 64 65 66 67 68 69             | uvwabcdefghi
```

Sniffer Text 12.1: The appearance of an IP packet in this sniffer capture text is new to us. However, the way we will manipulate the IP packet is old hat. We may be transmitting the packet via RF, but we handle the packet just as if it arrived on a wire.

The 802.11b CompactFlash NIC will receive the ICMP Echo frame and process it as described in the code segments you see in Code Snippet 12.1. However, instead of taking the ARP path, the code will break right and dive down the IP packet path.

Code Snippet 12.1
****

```
#define rxdatalength_offset 44
//***********************************************************
//*     IP Protocol Types
//***********************************************************
#define PROT_ICMP          0x01
#define PROT_TCP           0x06
#define PROT_UDP           0x11
//***********************************************************
//*     Ethernet Header Layout
//***********************************************************
unsigned int packet[761];
#define enetpacketLen11     0x00
#define enetpacketDest      0x01
#define enetpacketDest01    0x01   //destination mac address
#define enetpacketDest23    0x02
#define enetpacketDest45    0x03
#define enetpacketSrc       0x04
#define enetpacketSrc01     0x04   //source mac address
#define enetpacketSrc23     0x05
#define enetpacketSrc45     0x06
#define enetpacketLen03     0x07
#define enetpacketSnap      0x08
#define enetpacketCntrl     0x09
//                          0x0A
#define enetpacketType      0x0B   //type/length field
```

```
#define enetpacketData        0x0C  //IP data area begins here
//***********************************************************
//*    ARP Layout
//***********************************************************
#define arp_hwtype            0x0C
#define arp_prtype            0x0D
#define arp_hwprlen           0x0E
#define arp_op                0x0F
#define arp_shaddr            0x10    //arp source mac address
//                            0x11
//                            0x12
#define arp_sipaddr           0x13    //arp source ip address
//                            0x14
#define arp_thaddr            0x15    //arp target mac address
//                            0x16
//                            0x17
#define arp_tipaddr           0x18    //arp target ip address
//                            0x19
//***********************************************************
//*    IP Header Layout
//***********************************************************
#define ip_vers_len           0x0C    //IP version and header length
//#define ip_tos                      //IP type of service
#define ip_pktlen             0x0D    //packet length
#define ip_id                 0x0E    //datagram id
#define ip_frag_offset        0x0F    //fragment offset
#define ip_ttlproto           0x10    //time to live
//#define ip_proto                    //protocol (ICMP=1, TCP=6, UDP=11)
#define ip_hdr_cksum          0x11    //header checksum
#define ip_srcaddr            0x12    //IP address of source
//                            0X13
#define ip_destaddr           0x14    //IP addess of destination
//                            0X15
#define ip_data               0x16    //IP data area
//***********************************************************
//*    TCP Header Layout
//***********************************************************
#define TCP_srcport           0x16    //TCP source port
#define TCP_destport          0x17    //TCP destination port
#define TCP_seqnum            0x18    //sequence number
//                            0x19
#define TCP_acknum            0x1A    //acknowledgment number
//                            0x1B
#define TCP_hdrflags          0x1C    //4-bit header len and flags
#define TCP_window            0x1D    //window size

#define TCP_cksum             0x1E    //TCP checksum
#define TCP_urgentptr         0x1F    //urgent pointer
#define TCP_data              0x20    //option/data
//***********************************************************
//*    TCP Flags
//*    IN flags represent incoming bits
```

```
//*    OUT flags represent outgoing bits
//***********************************************************
#define FIN_IN (packet[TCP_hdrflags] & 0x0100)
#define SYN_IN (packet[TCP_hdrflags] & 0x0200)
#define RST_IN (packet[TCP_hdrflags] & 0x0400)
#define PSH_IN              (packet[TCP_hdrflags] & 0x0800)
#define ACK_IN              (packet[TCP_hdrflags] & 0x1000)
#define URG_IN              (packet[TCP_hdrflags] & 0x2000)
#define FIN_OUT     packet[TCP_hdrflags] |= 0x0100 //0000 0001 0000 0000
#define SYN_OUT     packet[TCP_hdrflags] |= 0x0200 //0000 0010 0000 0000
#define RST_OUT     packet[TCP_hdrflags] |= 0x0400 //0000 0100 0000 0000
#define PSH_OUT     packet[TCP_hdrflags] |= 0x0800 //0000 1000 0000 0000
#define ACK_OUT     packet[TCP_hdrflags] |= 0x1000 //0001 0000 0000 0000
#define URG_OUT     packet[TCP_hdrflags] |= 0x2000 //0010 0000 0000 0000
//***********************************************************
//*    ICMP Header
//***********************************************************
#define ICMP_typecode     ip_data
//#define ICMP_code        ICMP_type+1
#define ICMP_cksum         ICMP_typecode+1
#define ICMP_id            ICMP_cksum+1
#define ICMP_seqnum  ICMP_id+1
#define ICMP_data          ICMP_seqnum+1
//***********************************************************
//*    UDP Header
//***********************************************************
#define UDP_srcport        ip_data
#define UDP_destport UDP_srcport+1
#define UDP_len            UDP_destport+1
#define UDP_cksum          UDP_len+1
#define UDP_data           UDP_cksum+1

//* COMMANDS
#define  Access_Cmd              0x0021

#define  BAP0Busy_Bit_Mask       0x8000 //busy bit in Offset0 register
#define  BAP0Err_Bit_Mask            0x4000 //Err bit in Offset0 register
#define  BAP1Busy_Bit_Mask       0x8000 //busy bit in Offset1 register
#define  BAP1Err_Bit_Mask            0x4000 //Err bit in Offset1 register


//***********************************************************
//*    PORT Definitions
//***********************************************************
#define data_in       PORTD
#define data_out      LATD
#define addr_hi       LATG
#define addr_lo       LATH
//***********************************************************
//*    PORT TRIS Definitions
//***********************************************************
#define TO_NICTRISD = 0x00
#define FROM_NIC     TRISD = 0xFF
```

```
//*************************************************************
//*    CF I/O Definitions
//*************************************************************
#define OE                   0x01
#define IORD                 0x02
#define WE                   0x40
#define IOWR                 0x80
#define set_OE               PCTL |= OE
#define clr_OE               PCTL &= ~OE
#define set_IORD             PCTL |= IORD
#define clr_IORD             PCTL &= ~IORD
#define set_IOWR             PCTL |= IOWR
#define clr_IOWR             PCTL &= ~IOWR

void wr_cf_data(char bapnum,char* buffer,unsigned int count)
{
 unsigned int addr,i;
 int *byteptr;
 char data_lo;

 if(bapnum)
  addr = Data1_Register;
 else
  addr = Data0_Register;

 byteptr = (int*)buffer;
 for(i=0;i<(count&0xFFFE);)
  {
   wr_cf_io16(*byteptr++,addr);
   i+=2;
  }
 if(count % 2)
  {
   data_lo = buffer[i++];
   wr_cf_io16(data_lo,addr);
  }
 }

void wr_cf_addr(unsigned int addr)
{
 addr_hi = (make8(addr,1));
 addr_lo = addr & 0x00FF;
}

unsigned int rd_cf_io16(unsigned int addr)
{
 char data_lo,data_hi,i;
 unsigned int data16;

 wr_cf_addr(addr);
 clr_IORD;
 i=3;
```

```
 while(--i);
 data_lo=data_in;
 set_IORD;
 NOP();

 wr_cf_addr(addr+1);
 clr_IORD;
 i=3;
 while(--i);
 data_hi=data_in;
 set_IORD;
 NOP();
 data16 = make16(data_hi,data_lo);
 return(data16);
}

void wr_cf_io16(unsigned int data16,unsigned int addr)
{
 char i;

 wr_cf_addr(addr);
 data_out = LOW_BYTE(data16);
 TO_NIC;
 clr_IOWR;
 i=1;
 while(--i);
 set_IOWR;
 NOP();
 data_out = 0xFF;

 wr_cf_addr(addr+1);
 data_out = HIGH_BYTE(data16);
 clr_IOWR;
 i=1;
 while(--i);
 set_IOWR;
 NOP();
 data_out = 0xFF;
 FROM_NIC;
}

//*********************************************************
//* get a free bap
//*********************************************************
char get_free_bap(void)
{
 unsigned int temp;
 char rc;

 rc = 2;
 temp = rd_cf_io16(Offset1_Register);
 if(!(temp & BAP1Busy_Bit_Mask))
```

```
   rc = 1;
  temp = rd_cf_io16(Offset0_Register);
  if(!(temp & BAP0Busy_Bit_Mask))
   rc = 0;
  return(rc);
}
****************************

//*************************************************************
//*    Get A Frame From the CF Receive Buffer
//*    This routine removes an Ethernet frame from the receive buffer.
//*************************************************************
void get_frame()
{
 unsigned int temp,i;
 char rc,bapnum;

 rc = 0;
 RxFID = rd_cf_io16(RxFID_Register);      // Get the FID of the allocated buf-
fer
 bapnum = get_free_bap();
 switch(bapnum)
   {
    case 0:
     wr_cf_io16(RxFID,Select0_Register);
     wr_cf_io16(rxdatalength_offset,Offset0_Register);
     do{
          temp = rd_cf_io16(Offset0_Register);
         }while(temp & BAP0Busy_Bit_Mask);
     packet[enetpacketLen11] = rd_cf_io16(Data0_Register);    //get data
length
     if(packet[enetpacketLen11] > 0x05DC)
      packet[enetpacketLen11] = 0x05DC;
     for(i=0;i<3;++i)
      packet[enetpacketDest+i] = rd_cf_io16(Data0_Register);
     for(i=0;i<3;++i)
      packet[enetpacketSrc+i] = rd_cf_io16(Data0_Register);
     for(i=0;i<packet[enetpacketLen11];++i)
      packet[enetpacketLen03+i] = rd_cf_io16(Data0_Register);
     // Acknowledge that we now own the FID
     wr_cf_io16(EvStat_Rx_Bit_Mask,EvAck_Register );
     break;

    case 1:
     wr_cf_io16(RxFID,Select1_Register);
     wr_cf_io16(rxdatalength_offset,Offset1_Register);
     do{
          temp = rd_cf_io16(Offset1_Register);
         }while(temp & BAP0Busy_Bit_Mask);
     packet[enetpacketLen11] = rd_cf_io16(Data1_Register);    //get data
length
     if(packet[enetpacketLen11] > 0x05DC)
```

```
     packet[enetpacketLen11] = 0x05DC;
    for(i=0;i<3;++i)
     packet[enetpacketDest+i] = rd_cf_io16(Data1_Register);
    for(i=0;i<3;++i)
     packet[enetpacketSrc+i] = rd_cf_io16(Data1_Register);
    for(i=0;i<packet[enetpacketLen11];++i)   //removed +1
     packet[enetpacketLen03+i] = rd_cf_io16(Data1_Register);
    // Acknowledge that we now own the FID
    wr_cf_io16(EvStat_Rx_Bit_Mask,EvAck_Register );
    break;
  default:
    rc = 1;
    break;
  }
 if(!rc)
 {
 //process an ARP packet
 if(packet[enetpacketType] == 0x0608)
 {
  if(packet[arp_hwtype] == 0x0100 &&
   packet[arp_prtype] == 0x0008 &&
  packet[arp_hwprlen] == 0x0406 &&
  packet[arp_op] == 0x0100 &&
  ipaddri[0] == packet[arp_tipaddr] &&
  ipaddri[1] == packet[arp_tipaddr+1])
  arp();
 }

 //process an IP packet
 else if(packet[enetpacketType] == 0x0008
  && packet[ip_destaddr] == ipaddri[0]
  && packet[ip_destaddr+1] == ipaddri[1])
 {
   if(HIGH_BYTE(packet[ip_ttlproto]) == PROT_ICMP)
    icmp();
   else if(HIGH_BYTE(packet[ip_ttlproto]) == PROT_UDP)
    udp();
   else if(HIGH_BYTE(packet[ip_ttlproto]) == PROT_TCP)
    tcp();
   }
 }
 else
  do_event_housekeeping();
}
```
****

Code Snippet 12.1: We walked through the details of obtaining the contents of the 802.11b CompactFlash NIC receive buffer in the previous chapter. The thread in this code snippet will lead us through the AirDrop 802.11b driver's IP handler and into the AirDrop 802.11b driver's ICMP handler, which will respond to the PING from SNOOPER.

Array Table 12.1 is a representation of the ICMP Echo frame that the *get_frame* function retrieved from the 802.11b CompactFlash NIC receive buffer. For the AirDrop-P module, the idea is to swap stuff around, recalculate the checksums and bounce the frame back to the sender.

PACKET BUFFER LAYOUT			
Buffer Address	**Array Slot**	**DATA**	**Description**
0x06FA	[0]	0x0044	802.11b packet data length
0x06FC	[1]	0xE000	Destination MAC Address
0x06FE	[2]	0xBC98	
0x0700	[3]	0x6759	
0x0702	[4]	0x0500	Source MAC Address
0x0704	[5]	0x033C	
0x0706	[6]	0x2FD9	
0x0708	[7]	0x4400	802.3 packet data length
0x070A	[8]	0xAAAA	SNAP Header
0x070C	[9]	0x0003	UI/Protocol Identifier
0x070E	[10]	0x0000	Protocol Identifier
0x0710	[11]	0x0008	IP Protocol
0x0712	[12]	0x0045	IP version/header length/tos
0x0714	[13]	0x3C00	IP packet length
0x0716	[14]	0x4C14	IP datagram id
0x0718	[15]	0x0000	fragment offset
0x071A	[16]	0x0180	time to live/protocol
0x071C	[17]	0x82A4	IP header checksum
0x071E	[18]	0xA8C0	Source IP Address
0x0720	[19]	0x0B00	
0x0722	[20]	0xA8C0	Destination IP Address
0x0724	[21]	0x9700	
0x0726	[22]	0x0008	type (08=Echo)/code
0x0728	[23]	0x5BAB	ICMP checksum
0x072A	[24]	0x0002	identifier
0x072C	[25]	0x00A0	sequence number
0x072E	[26]	0x6261	data
0x0730	[27]	0x6463	
0x0732	[28]	0x6665	
0x0734	[29]	0x6867	
0x0736	[30]	0x6A69	
0x0738	[31]	0x6C6B	
0x073A	[32]	0x6E6D	
0x073C	[33]	0x706F	
0x073E	[34]	0x7271	
0x0740	[35]	0x7473	

PACKET BUFFER LAYOUT			
Buffer Address	**Array Slot**	**DATA**	**Description**
0x0742	[36]	0x7675	
0x0744	[37]	0x6177	
0x0746	[38]	0x6362	
0x0748	[39]	0x6564	
0x074A	[40]	0x6766	
0x074C	[41]	0x6968	

Array Table 12.1: Let's count the packet data bytes in this ICMP Echo request frame. Let's see, beginning at array slot [8] and counting the bytes through slot [41] gives us 34 words, which converted to bytes is 68 (0x44) bytes. Remember to send this frame just as it is, we must add 16 bytes to the transmission total to include the destination MAC address, the source MAC address, the 802.11b packet data length word and the 802.3 packet data length word. See, I told you this 802.11b stuff is easy.

The AirDrop 802.11b driver ICMP Echo handler code shown in Code Snippet 12.2 is pretty simple in itself. Despite its apparent simplicity, as you can see, the ICMP Echo handler depends heavily on other AirDrop 802.11b driver utilities.

Code Snippet 12.2
```
****************************

//***********************************************************
//*    Macros
//***********************************************************
#define bitset(var, bitno) ((var) |= 1 << (bitno))
#define bitclr(var, bitno) ((var) &= ~(1 << (bitno)))
#define make8(var,offset)   ((unsigned int)var >> (offset * 8)) & 0x00FF
#define make16(varhigh,varlow)    (((unsigned int)varhigh & 0xFF)* 0x100) +
((unsigned int)varlow & 0x00FF)


//***********************************************************
//*    IP Header Layout
//***********************************************************
#define ip_vers_len      0x0C   //IP version and header length
//#define ip_tos          0x0F   //IP type of service
#define ip_pktlen        0x0D   //packet length
#define ip_id            0x0E   //datagram id
#define ip_frag_offset   0x0F   //fragment offset
#define ip_ttlproto      0x10   //time to live
//#define ip_proto                //protocol (ICMP=1, TCP=6, UDP=11)
#define ip_hdr_cksum     0x11   //header checksum
#define ip_srcaddr       0x12   //IP address of source
//                       0X13
#define ip_destaddr      0x14   //IP addess of destination
//                       0X15
#define ip_data          0x16   //IP data area
```

```
//*************************************************************
//*    ICMP Header
//*************************************************************
#define ICMP_typecode      ip_data
//#define ICMP_code        ICMP_type+1
#define ICMP_cksum         ICMP_typecode+1
#define ICMP_id            ICMP_cksum+1
#define ICMP_seqnum        ICMP_id+1
#define ICMP_data          ICMP_seqnum+1


//*************************************************************
//*    SETIPADDRS
//*    This function builds the IP header.
//*************************************************************
void setipaddrs(void)
{
   //move IP source address to destination address
   packet[ip_destaddr]=packet[ip_srcaddr];
   packet[ip_destaddr+1]=packet[ip_srcaddr+1];
   //make ethernet module IP address source address
   packet[ip_srcaddr]=ipaddri[0];
   packet[ip_srcaddr+1]=ipaddri[1];
   //move hardware source address to destinatin address
   packet[enetpacketDest01]=packet[enetpacketSrc01];
   packet[enetpacketDest23]=packet[enetpacketSrc23];
   packet[enetpacketDest45]=packet[enetpacketSrc45];
   //make ethernet module mac address the source address
   packet[enetpacketSrc01]=macaddri[0];
   packet[enetpacketSrc23]=macaddri[1];
   packet[enetpacketSrc45]=macaddri[2];

   //calculate the IP header checksum
   packet[ip_hdr_cksum]=0x00;

   hdr_chksum =0;
   hdrlen = (LOW_BYTE(packet[ip_vers_len]) & 0x0F) *4;
   addr = &packet[ip_vers_len];
   cksum();
   packet[ip_hdr_cksum]= ~(hdr_chksum + ((hdr_chksum & 0xFFFF0000) >> 16));
 }

//*************************************************************
//*    CHECKSUM CALCULATION ROUTINE
//*************************************************************
void cksum()
{
        //hdr_chksum = 0;
      while(hdrlen > 1)
      {
          hdr_chksum = hdr_chksum + *addr++;
        hdrlen -=2;
```

```
                }
        if(hdrlen > 0)
            {
            hdr_chksum = hdr_chksum + (*addr & 0x00FF);
        --hdrlen;
            }
}


//************************************************************
//*    Echo Packet Function
//*    This routine does not modify the incoming packet size and
//*    thus echoes the original packet structure.
//************************************************************
char echo_packet()
{
        char rc;
        unsigned int TxFID;

        TxFID = get_free_TxFID();
        if(TxFID == 0)
         rc = 1;
        else
         {
     rc = bap_write(TxFID,rxdatalength_offset,(char*)packet,packet[enetpacke
tLen11]+16);
           if(rc=send_command(TransmitReclaim_Cmd,TxFID))
                printf("\r\nTransmit failed");

     do_event_housekeeping();
         }
         return(rc);
}

unsigned int swapbytes(unsigned int val)
{
        char temphi,templo;

        temphi = make8(val,1);
        templo = make8(val,0);
        return(make16(templo,temphi));
}
****************************

//************************************************************
//*    Perform ICMP Function
//*    This routine responds to a ping.
//************************************************************
    unsigned int iphdrlen,ippktlen;
    char tempcH,tempcL;

void icmp()
{
```

```
       char rc;

       //set echo reply
       packet[ICMP_typecode]=0x00;

       //clear the ICMP checksum
       packet[ICMP_cksum]=0x00;

       //setup the IP header
       setipaddrs();

       //calculate the ICMP checksum
       hdr_chksum =0;
       ippktlen=swapbytes(packet[ip_pktlen]);
       iphdrlen =(LOW_BYTE(packet[ip_vers_len]) & 0x0F) *4;

       hdrlen = ippktlen - iphdrlen;
       addr = &packet[ICMP_typecode];
       cksum();
       packet[ICMP_cksum]= ~(hdr_chksum + ((hdr_chksum & 0xFFFF0000) >> 16));
       //send the ICMP packet along on its way
       rc = echo_packet();
}
```
****

Code Snippet 12.2: The ICMP Echo handler in the AirDrop 802.11b driver only responds to PINGS. The jist of the ICMP code is similar to the AirDrop 802.11b driver ARP response code. However, because IP is involved, the ICMP code must recalculate checksums in addition to reshuffling MAC and IP addresses.

Even though we will flop around here and there to supporting AirDrop 802.11b driver functions, the ICMP Echo handler code is rather easy to follow. The first tasks performed by the ICMP Echo handler are to label the echo reply packet as just that, an echo reply. Checksums are very important in the function and the AirDrop-P packet buffer *ICMP_cksum* slot is cleared in Sub Snippet 12.1 to make sure we get an accurate checksum calculation.

Sub Snippet 12.1
****
```
   //set echo reply
     packet[ICMP_typecode]=0x00;

   //clear the ICMP checksum
     packet[ICMP_cksum]=0x00;
```
****

The *setipaddrs* function in Sub Snippet 12.2 points the ICMP Echo request packet back at the station that issued the ICMP Echo request. This is much like what we had to do to redirect the ARP request packet. The *ipaddri* array values are hard-coded in the AirDrop-P packet buffer. We culled the macaddri array values from the 802.11b CompactFlash NIC in the early stages of the AirDrop-P CompactFlash card initialization.

Sub Snippet 12.2

```
***********************************************************

//***********************************************************
//*    Ethernet Header Layout
//***********************************************************
unsigned int packet[761];
#define enetpacketLen11      0x00
#define enetpacketDest       0x01
#define enetpacketDest01     0x01   //destination mac address
#define enetpacketDest23     0x02
#define enetpacketDest45     0x03
#define enetpacketSrc        0x04
#define enetpacketSrc01      0x04   //source mac address
#define enetpacketSrc23      0x05
#define enetpacketSrc45      0x06
#define enetpacketLen03      0x07
#define enetpacketSnap       0x08
#define enetpacketCntrl      0x09
//                           0x0A
#define enetpacketType       0x0B   //type/length field
#define enetpacketData       0x0C   //IP data area begins here


//***********************************************************
//*    IP Header Layout
//***********************************************************
#define ip_vers_len          0x0C   //IP version and header length
//#define ip_tos             0x0F   //IP type of service
#define ip_pktlen            0x0D   //packet length
#define ip_id                0x0E   //datagram id
#define ip_frag_offset       0x0F   //fragment offset
#define ip_ttlproto          0x10   //time to live
//#define ip_proto                  //protocol (ICMP=1, TCP=6, UDP=11)
#define ip_hdr_cksum         0x11   //header checksum
#define ip_srcaddr           0x12   //IP address of source
//                           0X13
#define ip_destaddr          0x14   //IP addess of destination
//                           0X15
#define ip_data              0x16   //IP data area

void setipaddrs(void)
{
   //move IP source address to IP destination address
   packet[ip_destaddr]=packet[ip_srcaddr];
   packet[ip_destaddr+1]=packet[ip_srcaddr+1];
   //make AirDrop module IP address source address
   packet[ip_srcaddr]=ipaddri[0];
   packet[ip_srcaddr+1]=ipaddri[1];
   //move hardware source address to hardware destination address
   packet[enetpacketDest01]=packet[enetpacketSrc01];
   packet[enetpacketDest23]=packet[enetpacketSrc23];
```

```
packet[enetpacketDest45]=packet[enetpacketSrc45];
//make AirDrop module mac address the source address
packet[enetpacketSrc01]=macaddri[0];
packet[enetpacketSrc23]=macaddri[1];
packet[enetpacketSrc45]=macaddri[2];

*****************************

//setup the IP header
setipaddrs();
*************************************************************
```

Everybody makes a big deal out of computing checksums. There's absolutely nothing to checksum calculations if you have a good definition of what must be done to arrive at the correct checksum value. With that, the ICMP checksum that is ultimately calculated in Sub Snippet 12.3 is defined in this way:

The ICMP checksum is the 16-bit one's complement of the one's complement sum of the ICMP message starting with the ICMP Type. If the total length is odd, the received data is padded with one octet of zeros for computing the checksum.

There's nothing in the ICMP header to tell us how long the ICMP header is. So, we calculate the ICMP message length by simply subtracting the IP header length from the total length of the frame. Using the low byte of AirDrop-P packet buffer slot [12] (0x45), the expression *(LOW_BYTE(packet[ip_vers_len]) & 0x0F)* evaluates to 0x05. The IP header length is computed by multiplying *(LOW_BYTE(packet[ip_vers_len]) & 0x0F)* by 4, which results in an IP header length of 20 bytes. AirDrop-P packet buffer slot [13] of the AirDrop-P packet buffer contains the length of the IP datagram, which is (0x003C) 60 bytes. So, 60 bytes of IP datagram – 20 bytes of IP header leaves 40 bytes of ICMP message. Let's check our work. Beginning at AirDrop-P packet buffer slot [22] and counting ICMP header bytes through AirDrop-P packet buffer slot [25] results in 8 bytes of ICMP header. ICMP data begins at AirDrop-P packet buffer slot [26] and continues through AirDrop-P packet buffer slot [41] for 32 bytes. So, 32 bytes of ICMP data plus 8 bytes of ICMP header is equal to 40 bytes of ICMP message. The variable *hdrlen* holds the length of the ICMP message and passes this value to the *cksum* function. After pointing to the ICMP type slot in the AirDrop-P packet buffer (slot [22]), the intermediate ICMP checksum is calculated with a call to the *cksum* function.

Sub Snippet 12.3

```
*************************************************************
//calculate the ICMP checksum
hdr_chksum =0;
ippktlen=swapbytes(packet[ip_pktlen]);
iphdrlen =(LOW_BYTE(packet[ip_vers_len]) & 0x0F) *4;

hdrlen = ippktlen - iphdrlen;
addr = &packet[ICMP_typecode];
cksum();
packet[ICMP_cksum]= ~(hdr_chksum + ((hdr_chksum & 0xFFFF0000) >> 16));
*************************************************************
```

Maybe everyone just gets caught up with the check part and forget that a checksum is really just what it says it is, a sum. The *hdr_chksum* value computed by the AirDrop 802.11b driver *cksum* function in Sub Snippet 112.4 is no more than a 32-bit sum of a specified range of bytes.

Sub Snippet 12.4
****

```
//**********************************************************
//*    CHECKSUM CALCULATION ROUTINE
//**********************************************************
void cksum()
{
        //hdr_chksum = 0;
      while(hdrlen > 1)
      {
            hdr_chksum = hdr_chksum + *addr++;
         hdrlen -=2;
         }
      if(hdrlen > 0)
         {
         hdr_chksum = hdr_chksum + (*addr & 0x00FF);
        --hdrlen;
         }
}
```
****

The final ICMP checksum computed in Sub Snippet 12.5 is formulated by simply adding all of the bits that overflowed out of the least significant 16 bits of the *cksum* function's calculated 32-bit *hdr_chksum* value back into the least significant 16-bits of the calculated *hdr_chksum* value and then logically complementing the modified-by-overflow-bit-addition *hdr_chksum* value. Since the ICMP_cksum AirDrop-P packet buffer slot is only 16-bits wide, only the lower 16-bits of the *hdr_chksum* value is realized.

Sub Snippet 12.5
****

```
    packet[ICMP_cksum]= ~(hdr_chksum + ((hdr_chksum & 0xFFFF0000) >> 16));
```
****

We're ready to transmit our ICMP Echo reply, but first let's check our work against the AirDrop-P packet buffer contents in Array Table 12.2. In an ARP-like fashion, we first turn around the ICMP Echo reply destination and source MAC addresses. Nothing in the IP header area changes until we get to the source and destination IP address AirDrop-P packet buffer slots. We turn the IP addresses around to point back at the station that issued the ICMP Echo request and identify the AirDrop-P module as the ICMP Echo reply sender. The only real data value change is made in AirDrop-P packet buffer slot [22]. The ICMP Echo request identifier (0x08) is changed to an ICMP Echo reply identifier (0x00). Making the ICMP Echo identifier change requires us to recalculate the ICMP checksum value. The new ICMP checksum value is stored in AirDrop-P packet buffer slot [23]. The rest of the ICMP message remains the same as it appeared in the original ICMP Echo request.

PACKET BUFFER LAYOUT			
Buffer Address	**Array Slot**	**DATA**	**Description**
0x06FA	[0]	0x0044	802.11b packet data length
0x06FC	[1]	0x0500	Destination MAC Address
0x06FE	[2]	0x033C	
0x0700	[3]	0x2FD9	
0x0702	[4]	0xE000	Source MAC Address
0x0704	[5]	0xBC98	
0x0706	[6]	0x6759	
0x0708	[7]	0x4400	802.3 packet data length
0x070A	[8]	0xAAAA	SNAP Header
0x070C	[9]	0x0003	UI/Protocol Identifier
0x070E	[10]	0x0000	Protocol Identifier
0x0710	[11]	0x0008	IP Protocol
0x0712	[12]	0x0045	IP version/header length/tos
0x0714	[13]	0x3C00	IP packet length
0x0716	[14]	0x4C14	IP datagram id
0x0718	[15]	0x0000	fragment offset
0x071A	[16]	0x0180	time to live/protocol
0x071C	[17]	0x82A4	IP header checksum
0x071E	[18]	0xA8C0	Source IP Address
0x0720	[19]	0x9700	
0x0722	[20]	0xA8C0	Destination IP Address
0x0724	[21]	0x0B00	
0x0726	[22]	0x0000	type (00=Echo reply)/code
0x0728	[23]	0x5BB3	ICMP checksum
0x072A	[24]	0x0002	identifier
0x072C	[25]	0x00A0	sequence number
0x072E	[26]	0x6261	data
0x0730	[27]	0x6463	
0x0732	[28]	0x6665	
0x0734	[29]	0x6867	
0x0736	[30]	0x6A69	
0x0738	[31]	0x6C6B	
0x073A	[32]	0x6E6D	
0x073C	[33]	0x706F	
0x073E	[34]	0x7271	
0x0740	[35]	0x7473	
0x0742	[36]	0x7675	
0x0744	[37]	0x6177	
0x0746	[38]	0x6362	

PACKET BUFFER LAYOUT			
Buffer Address	**Array Slot**	**DATA**	**Description**
0x0748	[39]	0x6564	
0x074A	[40]	0x6766	
0x074C	[41]	0x6968	

Array Table 12.2: There was no need to recalculate the IP header checksum as we didn't change anything in the IP header area. In fact, the only real change we made was to the type field in AirDrop-P packet buffer slot [22], which forced a recalculation of the ICMP checksum value.

Everything looks good. Now we can pass the new frame containing our ICMP Echo reply to the 802.11b CompactFlash NIC for transmission. We do that by calling the *echo_packet* function.

There's nothing new in Sub Snippet 12.6. We're in transmit mode and have a new frame to send. So, we must seek out a free TxFID from our pool of preallocated TxFIDs. A free BAP was found and selected at the beginning of the *get_frame* function. Using the services of the free BAP, which is now our BAP, we write the contents of Array Table 12.2 to our waiting TxFID. A transmit/reclaim command is issued and the TxFID leaves our air space and now belongs to the AirDrop-P 802.11b CompactFlash NIC's MAC engine. A TxFID is reclaimed and reintroduced into our TxFID pool by calling the *do_event_housekeeping* function.

Sub Snippet 12.6
```
//*************************************************************

//*************************************************************
//*    Echo Packet Function
//*    This routine does not modify the incoming packet size and
//*    thus echoes the original packet structure.
//*************************************************************
char echo_packet()
{
        char rc;
        unsigned int TxFID;

        TxFID = get_free_TxFID();
        if(TxFID == 0)
         rc = 1;
        else
         {
     rc = bap_write(TxFID,rxdatalength_offset,(char*)packet,packet[enetpacke
tLen11]+16);
          if(rc=send_command(TransmitReclaim_Cmd,TxFID))
               printf("\r\nTransmit failed");

       do_event_housekeeping();
         }
        return(rc);
}
```

```
****************************

//send the ICMP packet along on its way
   rc = echo_packet();

**************************************************************
```

The results of our work on the ICMP Echo reply can be seen in Screen Capture 12.2 and the related sniffer text shown in Sniffer Text 12.2.

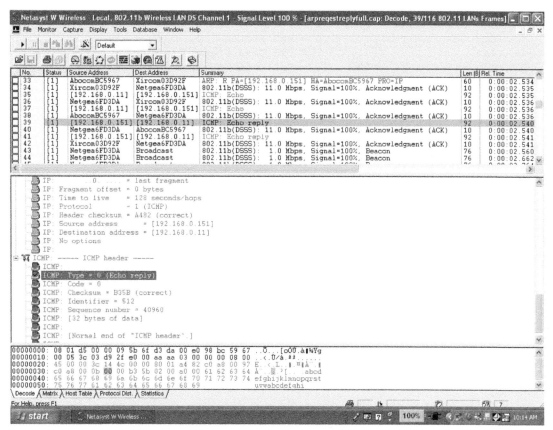

Screen Capture 12.2: Now that you know how to respond to a PING and how to respond to an ARP request, hopefully you can visualize how you could easily turn the ARP and ICMP Echo code around and issue an ARP request followed by PING request just as the SNOOPER laptop did.

The AP's primary job is to deliver 802.11b frames to their destinations. Once the AP receives the ICMP Echo reply frame contents shown in Sniffer Text 12.2, it alters the addressing scheme in the DLC (Data Link Control) Header area to direct the incoming frame to its final destination, the SNOOPER laptop. The original payload must not be altered by the AP. So, nothing is changed from the LLC (Logical Link Control) Header to the end of the ICMP Echo reply frame.

Sniffer Text 12.2

```
DLC: ----- DLC Header -----
      DLC:
      DLC: Frame 39 arrived at  14:02:23.6390; frame size is 92 (005C hex)
bytes.
      DLC: Signal level              = 100%
      DLC: Channel                   = 1
      DLC: Data rate                 = 22 (11.0 Megabits per second)
      DLC:
      DLC: Frame Control Field #1 = 08
      DLC:                  .... ..00 = 0x0 Protocol Version
      DLC:                  .... 10.. = 0x2 Data Frame
      DLC:                  0000 .... = 0x0 Data (Subtype)
      DLC: Frame Control Field #2 = 01
      DLC:                  .... ...1 = To Distribution System
      DLC:                  .... ..0. = Not from Distribution System
      DLC:                  .... .0.. = Last fragment
      DLC:                  .... 0... = Not retry
      DLC:                  ...0 .... = Active Mode
      DLC:                  ..0. .... = No more data
      DLC:                  .0.. .... = Wired Equivalent Privacy is off
      DLC:                  0... .... = Not ordered
      DLC: Duration                  = 213 (in microseconds)
      DLC: Basic Service Set ID      = Station Netgea6FD3DA
      DLC: Source Address            = Station AbocomBC5967
      DLC: Destination Address       = Station Xircom03D92F
      DLC: Sequence Control          = 0x00E0
      DLC: ...Sequence Number        = 0x00E (14)
      DLC: ...Fragment Number        = 0x0   (0)
      DLC:
LLC:  ----- LLC Header -----
      LLC:
      LLC:  DSAP Address = AA, DSAP IG Bit = 00 (Individual Address)
      LLC:  SSAP Address = AA, SSAP CR Bit = 00 (Command)
      LLC:  Unnumbered frame: UI
      LLC:
SNAP: ----- SNAP Header -----
      SNAP:
      SNAP: Type      = 0800 (IP)
      SNAP:
IP: ----- IP Header -----
      IP:
      IP: Version = 4, header length = 20 bytes
      IP: Type of service = 00
      IP:      000. ....   = routine
      IP:      ...0 ....   = normal delay
      IP:      .... 0...   = normal throughput
      IP:      .... .0..   = normal reliability
      IP:      .... ..0.   = ECT bit - transport protocol will ignore the CE bit
      IP:      .... ...0   = CE bit - no congestion
```

```
        IP: Total length     = 60 bytes
        IP: Identification   = 5196
        IP: Flags            = 0X
        IP:         .0.. .... = may fragment
        IP:         ..0. .... = last fragment
        IP: Fragment offset  = 0 bytes
        IP: Time to live     = 128 seconds/hops
        IP: Protocol         = 1 (ICMP)
        IP: Header checksum  = A482 (correct)
        IP: Source address      = [192.168.0.151]
        IP: Destination address = [192.168.0.11]
        IP: No options
        IP:
ICMP: ----- ICMP header -----
        ICMP:
        ICMP: Type = 0 (Echo reply)
        ICMP: Code = 0
        ICMP: Checksum = B35B (correct)
        ICMP: Identifier = 512
        ICMP: Sequence number = 40960
        ICMP: [32 bytes of data]
        ICMP:
        ICMP: [Normal end of "ICMP header".]
        ICMP:
ADDR  HEX                                                  ASCII
0000: 08 01 d5 00 00 09 5b 6f d3 da 00 e0 98 bc 59 67 | ..õ...[oóÚ.à˜¼Yg
0010: 00 05 3c 03 d9 2f e0 00 aa aa 03 00 00 00 08 00 | ..<.Ù/à.ªª......
0020: 45 00 00 3c 14 4c 00 00 80 01 a4 82 c0 a8 00 97 | E..<.L..€.¤‚À¨.—
0030: c0 a8 00 0b 00 00 b3 5b 02 00 a0 00 61 62 63 64 | À¨....³[.. .abcd
0040: 65 66 67 68 69 6a 6b 6c 6d 6e 6f 70 71 72 73 74 | efghijklmnopqrst
0050: 75 76 77 61 62 63 64 65 66 67 68 69             | uvwabcdefghi
```

Sniffer Text 12.2: Even though the processes are very similar, putting together an ICMP Echo reply is just a bit more work than assembling an ARP reply.

Examining the IP Header

I introduced IP in this chapter with its inclusion in the ICMP Echo packet. A good understanding of what's inside the IP header will make things a bit easier when we get to the subject of TCP. So, let's stop here and take a very close look at the IP header.

Figure 12.1 is a graphical depiction of the contents of AN IP header. The top row of Figure 12.1 represents the bit positions within each 32-bit field of the IP header. As you might imagine, we have to accommodate the IP header structure in the AirDrop 802.11b driver. Code Snippet 12.3 shows how the AirDrop 802.11b driver defines the fields of an IP header. Now that you can see the IP header for what it is, a group of 32-bit fields with subfields, let's go about defining an IP packet and all of the fields that make it up.

IP HEADER LAYOUT						
BIT	0-3	4-7	8-15	16-18	19-23	24-31
	Version/Header Length	IP Header Length	Type of Service	Total Length		
	Identification			Flags	Fragment Offset	
	Time To Live		Protocol	Header Checksum		
	Source IP Address					
	Destination IP Address					
	Options					Padding
	Data					

Figure 12.1: The IP header isn't hard to deal with if you break it down into its component parts.

Code Snippet 12.3
****

```
//***********************************************************
//*     IP Header Layout
//***********************************************************
#define ip_vers_len        0x0C   //IP version and header length
//#define ip_tos                   //IP type of service
#define ip_pktlen          0x0D   //packet length
#define ip_id              0x0E   //datagram id
#define ip_frag_offset     0x0F   //fragment offset
#define ip_ttlproto        0x10   //time to live
//#define ip_proto                 //protocol (ICMP=1, TCP=6, UDP=11)
#define ip_hdr_cksum 0x11          //header checksum
#define ip_srcaddr         0x12   //IP address of source
//                         0X13
#define ip_destaddr        0x14   //IP addess of destination
//                         0X15
#define ip_data            0x16   //IP data area
```

****
Code Snippet 12.3: The definitions in this snippet match up one-to-one and in perfect order with the fields of a textbook IP header.

IP packets are called datagrams. The definition of a datagram implies that a datagram is an independent entity that can carry a message on a network but cannot guarantee the safe arrival of that message. For instance, UDP and IP don't have any internal mechanisms that automatically check packet delivery status.

The first four bits of the IP header in Mini Sniff 12.1 contain the datagram's IP version, which is currently 4 (0x4X). The next four bits of the IP header (0xX5)represent the length of the IP header. The IP header length is calculated by multiplying the IP header length by 4. The IP header length value is actually the number of 32-bit words in the IP header. The Options/Padding and Data fields are not part of the IP header.

Mini Sniff 12.1

```
IP: ----- IP Header -----
      IP:
      IP: Version = 4, header length = 20 bytes
      IP: Type of service = 00
      IP:       000. ....   = routine
      IP:       ...0 ....   = normal delay
      IP:       .... 0...   = normal throughput
      IP:       .... .0..   = normal reliability
      IP:       .... ..0.   = ECT bit - transport protocol will ignore the CE
bit
      IP:       .... ...0   = CE bit - no congestion
      IP: Total length    = 60 bytes
      IP: Identification  = 5196
      IP: Flags           = 0X
      IP:       .0.. ....   = may fragment
      IP:       ..0. ....   = last fragment
      IP: Fragment offset = 0 bytes
      IP: Time to live    = 128 seconds/hops
      IP: Protocol        = 1 (ICMP)
      IP: Header checksum = A482 (correct)
      IP: Source address      = [192.168.0.11]
      IP: Destination address = [192.168.0.151]
      IP: No options
ADDR  HEX                                                ASCII
0000: 08 01 d5 00 00 09 5b 6f d3 da 00 05 3c 03 d9 2f | ..Õ...[oÓÚ..<.Ù/
0010: 00 e0 98 bc 59 67 50 00 aa aa 03 00 00 00 08 00 | .à˜¼YgP.ªª......
0020: 45 00 00 3c 14 4c 00 00 80 01 a4 82 c0 a8 00 0b | E..<.L..€.¤,À¨..
0030: c0 a8 00 97 08 00 ab 5b 02 00 a0 00 61 62 63 64 | À¨.—..«[.. .abcd
0040: 65 66 67 68 69 6a 6b 6c 6d 6e 6f 70 71 72 73 74 | efghijklmnopqrst
0050: 75 76 77 61 62 63 64 65 66 67 68 69             | uvwabcdefghi
```

The Type of Service field in Mini Sniff 12.2 is one of those fields that makes owning a wireless network sniffer worth every penny you pay for it. Instead of having to rumble through pages and pages of IP documentation to find out about the bits within the Type of Service field, it only takes a quick glance at the Netasyst Sniffer capture to get the total breakdown of the Type of Service field. The AirDrop 802.11b driver doesn't care about any of the bits in the Type of Service field. That doesn't mean you can twiddle them if you want. As you can see from the Netasyst Sniffer capture bit descriptions, the Type of Service field specifies how upper-layer protocols want the datagram to be handled. The Type of Service field value is 0x00, which says that there aren't any special handling instructions for our datagram.

Mini Sniff 12.2

```
IP: ----- IP Header -----
      IP:
      IP: Version = 4, header length = 20 bytes
      IP: Type of service = 00
      IP:       000. ....   = routine
      IP:       ...0 ....   = normal delay
```

```
IP:              .... 0... = normal throughput
IP:              .... .0.. = normal reliability
IP:              .... ..0. = ECT bit - transport protocol will ignore the CE bit
IP:              .... ...0 = CE bit - no congestion
IP: Total length      = 60 bytes
IP: Identification    = 5196
IP: Flags             = 0X
IP:          .0.. .... = may fragment
IP:          ..0. .... = last fragment
IP: Fragment offset = 0 bytes
IP: Time to live      = 128 seconds/hops
IP: Protocol          = 1 (ICMP)
IP: Header checksum = A482 (correct)
IP: Source address       = [192.168.0.11]
IP: Destination address = [192.168.0.151]
IP: No options
```

```
ADDR   HEX                                                    ASCII
0000:  08 01 d5 00 00 09 5b 6f d3 da 00 05 3c 03 d9 2f  | ..õ...[oóÚ..<.Ù/
0010:  00 e0 98 bc 59 67 50 00 aa aa 03 00 00 00 08 00  | .à~¼YgP.ªª......
0020:  45 00 00 3c 14 4c 00 00 80 01 a4 82 c0 a8 00 0b  | E..<.L..€.¤,À"..
0030:  c0 a8 00 97 08 00 ab 5b 02 00 a0 00 61 62 63 64  | À".—..«[.. .abcd
0040:  65 66 67 68 69 6a 6b 6c 6d 6e 6f 70 71 72 73 74  | efghijklmnopqrst
0050:  75 76 77 61 62 63 64 65 66 67 68 69              | uvwabcdefghi
```

As you saw earlier in the ICMP driver, the Total Length field of the IP header, which is highlighted in Mini Sniff 12.3, is a field that is used by the AirDrop 802.11b driver. This field's value represents the length of the entire IP packet including the data and the IP header.

Mini Sniff 12.3

```
IP: ----- IP Header -----
     IP:
     IP: Version = 4, header length = 20 bytes
     IP: Type of service = 00
     IP:          000. .... = routine
     IP:          ...0 .... = normal delay
     IP:          .... 0... = normal throughput
     IP:          .... .0.. = normal reliability
     IP:          .... ..0. = ECT bit - transport protocol will ignore the CE bit
     IP:          .... ...0 = CE bit - no congestion
     IP: Total length      = 60 bytes
     IP: Identification    = 5196
     IP: Flags             = 0X
     IP:          .0.. .... = may fragment
     IP:          ..0. .... = last fragment
     IP: Fragment offset = 0 bytes
     IP: Time to live      = 128 seconds/hops
     IP: Protocol          = 1 (ICMP)
     IP: Header checksum = A482 (correct)
     IP: Source address       = [192.168.0.11]
     IP: Destination address = [192.168.0.151]
     IP: No options
```

```
ADDR  HEX                                                     ASCII
0000: 08 01 d5 00 00 09 5b 6f d3 da 00 05 3c 03 d9 2f  | ..Õ...[oÓÚ..<.Ù/
0010: 00 e0 98 bc 59 67 50 00 aa aa 03 00 00 00 08 00  | .à˜¼YgP.ªª......
0020: 45 00 00 3c 14 4c 00 00 80 01 a4 82 c0 a8 00 0b  | E..<.L..€.¤,À¨..
0030: c0 a8 00 97 08 00 ab 5b 02 00 a0 00 61 62 63 64  | À¨.—..«[.. .abcd
0040: 65 66 67 68 69 6a 6b 6c 6d 6e 6f 70 71 72 73 74  | efghijklmnopqrst
0050: 75 76 77 61 62 63 64 65 66 67 68 69              | uvwabcdefghi
```

The Identification field is a number that represents the current datagram. It's used to reorder fragments. The Identification number, highlighted in Mini Sniff 12.4, will increment by one for each ICMP Echo request issued by the SNOOPER laptop.

Mini Sniff 12.4

```
IP: ----- IP Header -----
     IP:
     IP: Version = 4, header length = 20 bytes
     IP: Type of service = 00
     IP:      000. ....  = routine
     IP:      ...0 ....  = normal delay
     IP:      .... 0...  = normal throughput
     IP:      .... .0..  = normal reliability
     IP:      .... ..0.  = ECT bit - transport protocol will ignore the CE
bit
     IP:      .... ...0  = CE bit - no congestion
     IP: Total length     = 60 bytes
     IP: Identification   = 5196
     IP: Flags            = 0X
     IP:      .0.. ....   = may fragment
     IP:      ..0. ....   = last fragment
     IP: Fragment offset  = 0 bytes
     IP: Time to live     = 128 seconds/hops
     IP: Protocol         = 1 (ICMP)
     IP: Header checksum  = A482 (correct)
     IP: Source address       = [192.168.0.11]
     IP: Destination address  = [192.168.0.151]
     IP: No options
ADDR  HEX                                                     ASCII
0000: 08 01 d5 00 00 09 5b 6f d3 da 00 05 3c 03 d9 2f  | ..Õ...[oÓÚ..<.Ù/
0010: 00 e0 98 bc 59 67 50 00 aa aa 03 00 00 00 08 00  | .à˜¼YgP.ªª......
0020: 45 00 00 3c 14 4c 00 00 80 01 a4 82 c0 a8 00 0b  | E..<.L..€.¤,À¨..
0030: c0 a8 00 97 08 00 ab 5b 02 00 a0 00 61 62 63 64  | À¨.—..«[.. .abcd
0040: 65 66 67 68 69 6a 6b 6c 6d 6e 6f 70 71 72 73 74  | efghijklmnopqrst
0050: 75 76 77 61 62 63 64 65 66 67 68 69              | uvwabcdefghi
```

The Flags field is used to determine if the packet can be fragmented. The low order bit controls fragmentation while the middle bit signals if this is the last fragment in a series of fragmented packets. The high order bit of the Flags field is not used. According to our Mini Sniff 12.5 capture, fragmentation of this packet is not allowed.

Mini Sniff 12.5

```
IP: ----- IP Header -----
     IP:
     IP: Version = 4, header length = 20 bytes
     IP: Type of service = 00
     IP:        000. ....  = routine
     IP:        ...0 ....  = normal delay
     IP:        .... 0...  = normal throughput
     IP:        .... .0..  = normal reliability
     IP:        .... ..0.  = ECT bit - transport protocol will ignore the CE bit
     IP:        .... ...0  = CE bit - no congestion
     IP: Total length    = 60 bytes
     IP: Identification  = 5196
     IP: Flags           = 0X
     IP:        .0.. .... = may fragment
     IP:        ..0. .... = last fragment
     IP: Fragment offset = 0 bytes
     IP: Time to live    = 128 seconds/hops
     IP: Protocol        = 1 (ICMP)
     IP: Header checksum = A482 (correct)
     IP: Source address      = [192.168.0.11]
     IP: Destination address = [192.168.0.151]
     IP: No options
```

```
ADDR  HEX                                                      ASCII
0000: 08 01 d5 00 00 09 5b 6f d3 da 00 05 3c 03 d9 2f  | ..õ...[oóÚ..<.Ù/
0010: 00 e0 98 bc 59 67 50 00 aa aa 03 00 00 00 08 00  | .à˜¼YgP.ªª......
0020: 45 00 00 3c 14 4c 00 00 80 01 a4 82 c0 a8 00 0b  | E..<.L..€.¤,À¨..
0030: c0 a8 00 97 08 00 ab 5b 02 00 a0 00 61 62 63 64  | À¨.—..«[.. .abcd
0040: 65 66 67 68 69 6a 6b 6c 6d 6e 6f 70 71 72 73 74  | efghijklmnopqrst
0050: 75 76 77 61 62 63 64 65 66 67 68 69              | uvwabcdefghi
```

Even though the process of sending datagrams is reliable to a great extent, there still looms the possibility of having a "bad" datagram bouncing around uncontrolled on a network. The Time to Live value in Mini Sniff 12.6 is really a counter that gradually decrements until its value reaches zero. When Time to Live equals zero, the life of the packet ends and it is discarded.

Mini Sniff 12.6

```
IP: ----- IP Header -----
     IP:
     IP: Version = 4, header length = 20 bytes
     IP: Type of service = 00
     IP:        000. ....  = routine
     IP:        ...0 ....  = normal delay
     IP:        .... 0...  = normal throughput
     IP:        .... .0..  = normal reliability
     IP:        .... ..0.  = ECT bit - transport protocol will ignore the CE
bit
     IP:        .... ...0  = CE bit - no congestion
     IP: Total length    = 60 bytes
```

```
IP: Identification      = 5196
IP: Flags               = 0X
IP:         .0.. ....   = may fragment
IP:         ..0. ....   = last fragment
IP: Fragment offset     = 0 bytes
IP: Time to live        = 128 seconds/hops
IP: Protocol            = 1 (ICMP)
IP: Header checksum      = A482 (correct)
IP: Source address      = [192.168.0.11]
IP: Destination address = [192.168.0.151]
IP: No options
ADDR  HEX                                               ASCII
0000: 08 01 d5 00 00 09 5b 6f d3 da 00 05 3c 03 d9 2f | ..õ...[oÓÚ..<.Ù/
0010: 00 e0 98 bc 59 67 50 00 aa aa 03 00 00 00 08 00 | .à~¼YgP.ªª......
0020: 45 00 00 3c 14 4c 00 00 80 01 a4 82 c0 a8 00 0b | E..<.L..€.¤,À"..
0030: c0 a8 00 97 08 00 ab 5b 02 00 a0 00 61 62 63 64 | À".−..«[.. .abcd
0040: 65 66 67 68 69 6a 6b 6c 6d 6e 6f 70 71 72 73 74 | efghijklmnopqrst
0050: 75 76 77 61 62 63 64 65 66 67 68 69             | uvwabcdefghi
```

The Protocol field that is singled out in Mini Sniff 12.7 is a very important field to the AirDrop 802.11b driver. The value in this field represents an upper layer protocol that will be the recipient of the IP packet after it is processed. Look back at Code Snippet 12.1 and you'll see that the AirDrop 802.11b driver will use this field to distinguish between and service the ICMP, TCP and UDP protocols. A value of 0x01 in this field indicates that the payload of this packet contains an ICMP message.

Mini Sniff 12.7

```
IP: ----- IP Header -----
    IP:
    IP: Version = 4, header length = 20 bytes
    IP: Type of service = 00
    IP:         000. ....   = routine
    IP:         ...0 ....   = normal delay
    IP:         .... 0...   = normal throughput
    IP:         .... .0..   = normal reliability
    IP:         .... ..0.   = ECT bit - transport protocol will ignore the CE bit
    IP:         .... ...0   = CE bit - no congestion
    IP: Total length     = 60 bytes
    IP: Identification   = 5196
    IP: Flags            = 0X
    IP:         .0.. ....   = may fragment
    IP:         ..0. ....   = last fragment
    IP: Fragment offset  = 0 bytes
    IP: Time to live     = 128 seconds/hops
    IP: Protocol         = 1 (ICMP)
    IP: Header checksum  = A482 (correct)
    IP: Source address      = [192.168.0.11]
    IP: Destination address = [192.168.0.151]
    IP: No options
```

```
ADDR  HEX                                                 ASCII
0000: 08 01 d5 00 00 09 5b 6f d3 da 00 05 3c 03 d9 2f | ..õ...[oóÚ..<.Ù/
0010: 00 e0 98 bc 59 67 50 00 aa aa 03 00 00 00 08 00 | .à~¼YgP.ª ª......
0020: 45 00 00 3c 14 4c 00 00 80 01 a4 82 c0 a8 00 0b | E..<.L..€.¤,À¨..
0030: c0 a8 00 97 08 00 ab 5b 02 00 a0 00 61 62 63 64 | À¨.—..«[.. .abcd
0040: 65 66 67 68 69 6a 6b 6c 6d 6e 6f 70 71 72 73 74 | efghijklmnopqrst
0050: 75 76 77 61 62 63 64 65 66 67 68 69             | uvwabcdefghi
```

The IP header checksum is very important as it is a way of checking the integrity of the IP header. The IP header checksum you see in Mini Sniff 12.8 is defined as follows:

The IP checksum is defined as the 16-bit one's complement of the one's complement sum of all 16-bit words in the header.

Although the IP header information remained the same in the ICMP Echo reply packet, the *setippaddrs* function that was called from the *icmp* function actually recalculated the IP header checksum anyway.

Mini Sniff 12.8

```
IP: ----- IP Header -----
     IP:
     IP: Version = 4, header length = 20 bytes
     IP: Type of service = 00
     IP:      000. ....  = routine
     IP:      ...0 ....  = normal delay
     IP:      .... 0...  = normal throughput
     IP:      .... .0..  = normal reliability
     IP:      .... ..0.  = ECT bit - transport protocol will ignore the CE bit
     IP:      .... ...0  = CE bit - no congestion
     IP: Total length    = 60 bytes
     IP: Identification   = 5196
     IP: Flags           = 0X
     IP:      .0.. ....  = may fragment
     IP:      ..0. ....  = last fragment
     IP: Fragment offset = 0 bytes
     IP: Time to live    = 128 seconds/hops
     IP: Protocol        = 1 (ICMP)
     IP: Header checksum = A482 (correct)
     IP: Source address      = [192.168.0.11]
     IP: Destination address = [192.168.0.151]
     IP: No options
```

```
ADDR  HEX                                                 ASCII
0000: 08 01 d5 00 00 09 5b 6f d3 da 00 05 3c 03 d9 2f | ..õ...[oóÚ..<.Ù/
0010: 00 e0 98 bc 59 67 50 00 aa aa 03 00 00 00 08 00 | .à~¼YgP.ª ª......
0020: 45 00 00 3c 14 4c 00 00 80 01 a4 82 c0 a8 00 0b | E..<.L..€.¤,À¨..
0030: c0 a8 00 97 08 00 ab 5b 02 00 a0 00 61 62 63 64 | À¨.—..«[.. .abcd
0040: 65 66 67 68 69 6a 6b 6c 6d 6e 6f 70 71 72 73 74 | efghijklmnopqrst
0050: 75 76 77 61 62 63 64 65 66 67 68 69             | uvwabcdefghi
```

Calculating the IP header checksum with the code in Sub Snippet 12.7 is very similar to putting the math to the ICMP checksum. About the only difference between the ICMP checksum and IP header checksum procedures is where the sum is accumulated. First, the IP header checksum AirDrop-P packet buffer slot is cleared to zero. Then, the IP header length is calculated using the IP header's Version/Header Length field. Finally, just like the ICMP checksum calculation, all of the IP header words are accumulated by the *cksum* function. Any bits that overflowed into the upper 16-bits of the 32-bit *hdr_chksum* variable are added to the lower 16-bits of the 32-bit *hdr_chksum* variable and the 32-bit *hdr_chksum* variable is then one's complemented. The checksum result is a 16-bit word deposited in the *ip_hdr_cksum* AirDrop-P packet buffer slot. For an ICMP operation the *ip_hdr_cksum* AirDrop-P packet buffer slot is slot [17].

Sub Snippet 12.7
```
//calculate the IP header checksum
packet[ip_hdr_cksum]=0x00;

hdr_chksum =0;
hdrlen = (LOW_BYTE(packet[ip_vers_len]) & 0x0F) *4;
addr = &packet[ip_vers_len];
cksum();
packet[ip_hdr_cksum]= ~(hdr_chksum + ((hdr_chksum & 0xFFFF0000) >> 16));
```

The source and destination addresses you see in Mini Sniff 12.9 and Mini Sniff 12.10 respectively are the protocol addresses of the players in the ICMP echo communications session. Note that the IP header is the only place you will see an IP address in an IP datagram. No options have been specified for our IP datagram. Therefore the Option field does not exist and the destination IP address field gives way directly to the IP data area.

Mini Sniff 12.9
```
IP: ----- IP Header -----
    IP:
    IP: Version = 4, header length = 20 bytes
    IP: Type of service = 00
    IP:         000. ....    = routine
    IP:         ...0 ....  = normal delay
    IP:         .... 0...  = normal throughput
    IP:         .... .0..  = normal reliability
    IP:         .... ..0.  = ECT bit - transport protocol will ignore the CE bit
    IP:         .... ...0  = CE bit - no congestion
    IP: Total length     = 60 bytes
    IP: Identification   = 5196
    IP: Flags            = 0X
    IP:         .0.. ....  = may fragment
    IP:         ..0. ....  = last fragment
    IP: Fragment offset  = 0 bytes
    IP: Time to live     = 128 seconds/hops
```

```
      IP: Protocol            = 1 (ICMP)
      IP: Header checksum = A482 (correct)
      IP: Source address       = [192.168.0.11]
      IP: Destination address = [192.168.0.151]
      IP: No options
ADDR  HEX                                                   ASCII
0000: 08 01 d5 00 00 09 5b 6f d3 da 00 05 3c 03 d9 2f | ..õ...[oóÚ..<.Ù/
0010: 00 e0 98 bc 59 67 50 00 aa aa 03 00 00 00 08 00 | .à~¼YgP.ª ª......
0020: 45 00 00 3c 14 4c 00 00 80 01 a4 82 c0 a8 00 0b | E..<.L..€.¤,À¨..
0030: c0 a8 00 97 08 00 ab 5b 02 00 a0 00 61 62 63 64 | À¨.—..«[.. .abcd
0040: 65 66 67 68 69 6a 6b 6c 6d 6e 6f 70 71 72 73 74 | efghijklmnopqrst
0050: 75 76 77 61 62 63 64 65 66 67 68 69             | uvwabcdefghi
```

Mini Sniff 12.10

```
IP: ----- IP Header -----
      IP:
      IP: Version = 4, header length = 20 bytes
      IP: Type of service = 00
      IP:        000. .... = routine
      IP:        ...0 .... = normal delay
      IP:        .... 0... = normal throughput
      IP:        .... .0.. = normal reliability
      IP:        .... ..0. = ECT bit - transport protocol will ignore the CE bit
      IP:        .... ...0 = CE bit - no congestion
      IP: Total length    = 60 bytes
      IP: Identification  = 5196
      IP: Flags           = 0X
      IP:       .0.. .... = may fragment
      IP:       ..0. .... = last fragment
      IP: Fragment offset = 0 bytes
      IP: Time to live    = 128 seconds/hops
      IP: Protocol        = 1 (ICMP)
      IP: Header checksum = A482 (correct)
      IP: Source address      = [192.168.0.11]
      IP: Destination address = [192.168.0.151]
      IP: No options
ADDR  HEX                                                   ASCII
0000: 08 01 d5 00 00 09 5b 6f d3 da 00 05 3c 03 d9 2f | ..õ...[oóÚ..<.Ù/
0010: 00 e0 98 bc 59 67 50 00 aa aa 03 00 00 00 08 00 | .à~¼YgP.ª ª......
0020: 45 00 00 3c 14 4c 00 00 80 01 a4 82 c0 a8 00 0b | E..<.L..€.¤,À¨..
0030: c0 a8 00 97 08 00 ab 5b 02 00 a0 00 61 62 63 64 | À¨.—..«[.. .abcd
0040: 65 66 67 68 69 6a 6b 6c 6d 6e 6f 70 71 72 73 74 | efghijklmnopqrst
0050: 75 76 77 61 62 63 64 65 66 67 68 69             | uvwabcdefghi
```

The IP header data area, shown from the Netasyst Sniffer point of view in Mini Sniff 12.11, holds the entire ICMP message including the ICMP header. This process of embedding the ICMP message inside the IP header data area inside the 802.11b frame is called encapsulation.

Mini Sniff 12.11

```
ICMP: ----- ICMP header -----
      ICMP:
      ICMP: Type = 8 (Echo)
      ICMP: Code = 0
      ICMP: Checksum = AB5B (correct)
      ICMP: Identifier = 512
      ICMP: Sequence number = 40960
      ICMP: [32 bytes of data]
      ICMP:
      ICMP: [Normal end of "ICMP header".]
      ICMP:
ADDR  HEX                                                  ASCII
0000: 08 01 d5 00 00 09 5b 6f d3 da 00 05 3c 03 d9 2f  |  ..õ...[oÓÚ..<.Ù/
0010: 00 e0 98 bc 59 67 50 00 aa aa 03 00 00 00 08 00  |  .à˜¼YgP.ªª......
0020: 45 00 00 3c 14 4c 00 00 80 01 a4 82 c0 a8 00 0b  |  E..<.L..€.¤‚À¨..
0030: c0 a8 00 97 08 00 ab 5b 02 00 a0 00 61 62 63 64  |  À¨.—..«[...abcd
0040: 65 66 67 68 69 6a 6b 6c 6d 6e 6f 70 71 72 73 74  |  efghijklmnopqrst
0050: 75 76 77 61 62 63 64 65 66 67 68 69              |  uvwabcdefghi
```

You're getting better and better at flying the AirDrop. You fired your first "shot" by responding to an ARP request. Now, you're flying wing with other stations on the wireless LAN using IP, ICMP and the PING application. You've also gained some valuable checksum experience, which you will be able to apply again later. In the next chapter, you'll be flying a cargo mission as we apply UDP to the AirDrop.

Now for something entirely different…Do you know the nickname for the studio musicians that helped create the Motown Sound?

Flying Cargo with UDP and the AirDrop

UDP is the closest thing to standard serial communication techniques you'll get in an internet protocol. Like RS-232, UDP can be considered as a "I don't care" protocol. Each protocol will send and receive data but neither of them care if it really gets where it's going or not. UDP and RS-232 leave the data integrity tasks up to the application. You can send just about anything with UDP at any time you wish with very little effort. A very useful file transfer protocol, TFTP (Trivial File Transfer Protocol) is built on top of UDP. Of all of the Internet protocols in my opinion UDP is the easiest to understand and implement. Just like all of the other internet protocols, UDP datagrams are handled the same whether they emanate from an RF source like the AirDrop-P 802.11b CompactFlash NIC or are transported over ether in a wire. In this chapter we'll put UDP thorough its paces via our AirDrop-P.

Running a UDP Application on the AirDrop-P

UDP is short for User Datagram Protocol. UDP is also unofficially known in some circles as the Unreliable Delivery Protocol. Like IP, UDP has absolutely no means of insuring that a data packet will arrive in one piece or even arrive at all. However, you'll find that it is reliable enough for most tasks it's used for.

As I mentioned earlier, UDP is a very simple and easy to understand protocol. UDP modifies an application-generated message by tagging on a checksum, a source port number and a destination port number of its own before passing the UDP segment to IP for encapsulation. IP does its best to deliver the UDP segment since there is nothing within IP or the UDP segment it is carrying to guarantee that the UDP segment will arrive intact.

Logically, a UDP host transmits a UDP datagram through a source port to a UDP recipient's destination port. The destination port number and destination IP address are used to route the UDP segment to the correct application once the segment arrives at its destination. By using port numbers, various applications can be using the services of UDP simultaneously. This is called multiplexing. The combination of the IP address and the port number is called a socket.

Like RS-232, a UDP transmission can occur at any time without the need to establish a communications session with the remote host. Since there is no handshaking or predetermined contact between UDP hosts, UDP is defined as a connectionless protocol. Again, this is similar to the way RS-232 stations operate.

Despite the shortcomings that UDP appears to emanate, UDP does have advantages over its cousin TCP. For instance, UDP does not have to establish a formal connection and as a result is a faster and more efficient way to send a message. As you will see later, TCP uses a three-way handshake to establish a communications session before transmitting any data and TCP does an awful lot of housekeeping compared to none for UDP.

UDP is able to send messages as fast as the microcontroller and application it is involved with can run. The only thing that slows UDP down are the limitations of the hardware it is running on and the bandwidth of the LAN it is riding on. Unlike UDP, TCP has built in rev limiters that throttle the data rate to relieve congestion on the LAN segment. UDP segments with any kind of problems are simply discarded by the devices on the LAN segment.

The layout of the UDP header in Code Snippet 13.1 shows us that the UDP segment rides in the IP data area. Thus the UDP datagram is encapsulated within the IP datagram. The UDP source and destination ports are 16-bits in length, which allows port numbers to range from 0 to 65535. The UDP header is only 8 bytes long compared to the 20 bytes that make up the IP header. The UDP datagram length or size is simply the total number of bytes in the UDP header plus the number of bytes in the UDP payload (data area).

Code Snippet 13.1
****

```
//************************************************************
//*    IP Header Layout
//************************************************************
#define ip_vers_len         0x0C   //IP version and header length
//#define ip_tos                   //IP type of service
#define ip_pktlen           0x0D   //packet length
#define ip_id               0x0E   //datagram id
#define ip_frag_offset      0x0F   //fragment offset
#define ip_ttlproto         0x10   //time to live
//#define ip_proto                 //protocol (ICMP=1, TCP=6, UDP=11)
#define ip_hdr_cksum        0x11   //header checksum
#define ip_srcaddr          0x12   //IP address of source
//                          0X13
#define ip_destaddr         0x14   //IP addess of destination
//                          0X15
#define ip_data             0x16   //IP data area


//************************************************************
//*    UDP Header Layout
//************************************************************
#define UDP_srcport         ip_data
#define UDP_destport        UDP_srcport+1
#define UDP_len             UDP_destport+1
#define UDP_cksum           UDP_len+1
#define UDP_data            UDP_cksum+1
*****************************
```

Code Snippet 13.1: Once you understand what encapsulation is and how it works, deciphering protocols like UDP is as easy as peeling off layers of an onion.

UDP checksumming is optional. However, the AirDrop 802.11b driver performs a checksum on every UDP datagram. The UDP checksum is put there for use by the application as UDP itself doesn't care about it at all.

UDP and TCP have their unique advantages (and disadvantages) depending on what your application needs to accomplish. Let's move on and use the EDTP Internet Test Panel and the AirDrop-P to explore some of the things that UDP does well.

The EDTP Internet Test Panel and the Code Behind It

When it comes to writing something quick and dirty to run on a personal computer, I prefer using Microsoft's Visual Basic 6.0. VB6 is one of the quickest and easiest ways to write simple applications like the Internet Test Panel. I realize that there's a big commotion being tossed over the new .NET stuff, but VB6 is still good enough for writing small applications like the Internet Test Panel as it includes network applications modules for UDP and TCP/IP. The inclusion of the networking modules makes VB6 a perfect UDP datagram generator for machines that run Windows.

The Internet Test Panel shown in Screen Capture 13.1 is a Visual Basic 6.0 application that uses UDP socket programming to send a UDP datagram to a well-know port or a socket of your choice. The Internet Test Panel also includes programming to send a UDP message to an LCD-equipped UDP host.

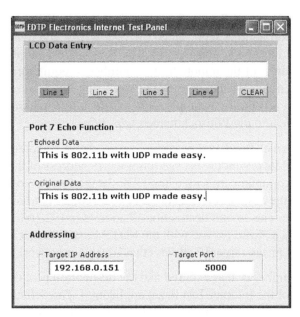

Screen Capture 13.1: The internet protocols don't really care about what medium they ride on. So, you can use this little utility program to test the operation of the AirDrop-P or any of the wired EDTP Ethernet devices in conjunction with any other Windows-based station on the LAN that can run the Internet Test Panel application.

The VB6 code behind the Internet Test Panel is easy to grasp. Each window in the Internet Test Panel is associated with a name and a variable that represents the text within the window. For example, the LCD Data Entry window in Screen Capture 13.1 is named *txtlcdin* and the text within *txtlcdin* is recognized by the program as *txtlcdin.Text*.

The *udp_PC* mnemonic in Sub Snippet 13.1 represents the Visual Basic Winsock module that provides the UDP protocol and UDP socket services. In the Internet Test Panel application our Winsock module, *udp_PC*, is bound to local port 5002. The binding precludes any other application or Winsock module from using port 5002.

Sub Snippet 13.1
****

```
Private Sub Form_Load()
On Error Resume Next
txtip.Text = "192.168.1.150"
txtport.Text = "5000"
With udp_PC
.RemoteHost = txtip.Text
.RemotePort = Val(txtport.Text)
.Bind 5002
End With
frm_B.Show
End Sub
```
****

Prefixing the *RemoteHost* and *RemotePort* variables with the Winsock designator associates the IP address and port address represented by *RemoteHost* and *RemotePort* with the Winsock in the prefix and its bound port number. The *RemoteHost* variable is actually a string that represents the IP address of the remote host. A number is required for the *RemotePort* variable and the *Val* function is used to convert the text string into a numeric value that the *RemotePort* variable will accept. For example, *udp_PC.RemoteHost* associates the IP address represented by *RemoteHost* to the Winsock named *udp_PC*.

The Visual Basic Internet Test Panel program is event driven. Text entered in the Internet Test Panel's Target IP Address window is used to set the IP address of the remote host the Internet Test Panel will communicate with using UDP datagrams. As you can see in Screen Capture 13.1, I have set the destination IP address to 192.168.0.151 to match the IP address of the AirDrop-P module. The default destination IP address is 192.168.1.150 as the Internet Test Panel was originally realized for use with the EDTP Easy Ethernet series of wired Ethernet devices. You can key in any target IP address that suits your network.

A destination port number is also required and the destination port value is entered using the Target Port window of the Internet Test Panel application. The default Internet Test Panel destination port number is 5000. The Internet Test Panel's source port number is fixed at 5002. The AirDrop 802.11b driver is coded to respond to ports 5000 and 7 with an IP address of 192.168.0.151.

Simply typing the desired numbers into the appropriate Internet Test Panel windows will change the default values for the Target IP Address and the Target Port. The Visual Basic functions txtip_Change and txtport_Change in Sub Snippet 13.2 will sense the changes in the text within their respective text boxes (txtip and txtport) and load the Internet Test Panel application's *RemoteHost* variable with the new remote IP address and the *RemotePort* variable with the new destination port address.

Sub Snippet 13.2
****
```
Private Sub txtip_Change()
On Error Resume Next
udp_PC.RemoteHost = txtip.Text
End Sub

Private Sub txtport_Change()
On Error Resume Next
udp_PC.RemotePort = Val(txtport.Text)
End Sub
```
****

A well-known port is one that has a standard function associated with it. On today's Internet, Port 7 is the well-known echo port. If a UDP message has a destination port of 7, the data transmitted to the remote host should be echoed from the remote host back to the sender.

The txsend.Text Visual Basic variable listed in Sub Snippet 13.3 is actually text data entered in the Original Data window (txtsend) of the Internet Test Panel. The entering of text into the Original Data window triggers the txtsend_Change function, which sends the contents of the Original Data window to port 7 of the of the remote host with the IP address that is listed in the Target IP Address window of the Internet Test Panel.

Sub Snippet 13.3
****
```
Private Sub txtsend_Change()
On Error Resume Next
udp_PC.RemoteHost = txtip.Text
udp_PC.RemotePort = 7
udp_PC.SendData txtsend.Text
End Sub
```
****

When using the echo function of the Internet Test Panel, the UDP message size increases for every byte entered in the Original Data window. For instance, if the letter 'T' is entered it is immediately sent and echoed. Entering the letter "h" would send "Th", entering "i" would send "Thi" and so forth. If all works as planned, the remote host echoes whatever is in the Original Data window back to the Internet Test Panel application.

The incoming bytes of echoed data trigger another event that calls the udp_PC_Data-Arrival function, which is shown in Sub Snippet 13.4. The echoed data (*txtreceive.Text*) is retrieved using the Visual Basic GetData method and displayed in the Echoed Data window (txtreceive). After executing *udp_PC.GetData strData*, the Visual Basic program knows to route the incoming echoed data from the AirDrop-P to the port address and IP address tied to Winsock *udp_PC*.

Sub Snippet 13.4

```
***********************************************************
Private Sub udp_PC_DataArrival(ByVal bytesTotal As Long)
On Error Resume Next
udp_PC.RemoteHost = txtip.Text
Dim strData As String
udp_PC.GetData strData
txtreceive.Text = strData
End Sub
***********************************************************
```

The LCD Data Entry window (txtlcdin) has a dual-purpose role. Primarily, the data entered into the LCD Data Entry window (*txtlcdin.Text*) is designed to be interpreted by a remote host that is driving a standard 4-line LCD module. The workings of the Visual Basic event mechanism for the LCD Data Entry window are no different than when sending data from the Original Data window. Only the socket addressing is changed. The data entered into the LCD Data Entry window is aimed at port 5000 by default and the data doesn't accumulate like it does in the Port 7 Echo Function windows. The data shown in the LCD Data Entry window will seem to accumulate visually but only one character is sent per event and that's always the last character in the window. The LCD Data Entry window's addressing is controlled by the values of the Target IP Address and Target Port windows.

The AirDrop-P hardware has all of the PIC18LF8621 I/O extended out to header areas to allow you to wire in an LCD of your choice. Also, there is no LCD application code to directly support the VB6 code in Sub Snippet 13.5 in the AirDrop 802.11b driver. A working example that applies a 4 x 16 LCD module is included on the CDROM that is included with this book.

Sub Snippet 13.5

```
***********************************************************
Private Sub txtlcdin_Change()
On Error Resume Next
udp_PC.RemoteHost = txtip.Text
udp_PC.RemotePort = Val(txtport.Text)
udp_PC.SendData Right(txtlcdin.Text, 1)
***********************************************************
```

Just because the AirDrop-P module doesn't have an on-board LCD module doesn't mean we can't use the Internet Test Panel's GUI pushbuttons. The pushbuttons under the LCD Data Entry window that are normally used to switch from line to line on a multi-lined LCD module can be used to control selected I/O pins on the AirDrop-P module. The Internet Test Panel buttons Line 1, Line 2, Line 3, Line 4 and CLEAR send raw hex values of 0x0100, 0x0200, 0x0300, 0x0400 and 0x0000 respectively. The AirDrop 802.11b driver is coded to only pick up the high byte of each word sent by the Internet Test Panel button code in Sub Snippet 13.6.

Sub Snippet 13.6
```
***********************************************************
Private Sub btnclear_Click()
On Error Resume Next
udp_PC.RemotePort = txtport.Text
udp_PC.SendData &H0
txtlcdin = ""
End Sub

Private Sub btnline1_Click()
On Error Resume Next
udp_PC.RemotePort = txtport.Text
udp_PC.SendData &H1
txtlcdin = ""
End Sub

Private Sub btnline2_Click()
On Error Resume Next
udp_PC.RemotePort = txtport.Text
udp_PC.SendData &H2
txtlcdin = ""
End Sub

Private Sub btnline3_Click()
On Error Resume Next
udp_PC.RemotePort = txtport.Text
udp_PC.SendData &H3
txtlcdin = ""
End Sub

Private Sub btnline4_Click()
On Error Resume Next
udp_PC.RemotePort = txtport.Text
udp_PC.SendData &H4
txtlcdin = ""
End Sub
***********************************************************
```

The EDTP Internet Test Panel VB6 source code package is included on the CDROM that accompanies this book. Providing the VB6 source code for the Internet Test Panel allows you to tailor the code to your needs.

Exercising the AirDrop-P with the EDTP Internet Test Panel

The best way to show you how UDP works with the AirDrop-P is to initiate a UDP transmission from SNOOPER's EDTP Internet Test Panel and follow the UDP datagram into and out of the AirDrop-P's PIC18LF8621 and 802.11b CompactFlash NIC. Let's begin with my keying a "T" into the Internet Test Panel. The "T" is processed by the VB6 *txtsend_Change* subroutine in Sub Snippet 13.7, which sends the lone character along its way wrapped inside a UDP datagram, which is encapsulated in an IP datagram.

Sub Snippet 13.7
```
*************************************************************
Private Sub txtsend_Change()
On Error Resume Next
udp_PC.RemoteHost = txtip.Text
udp_PC.RemotePort = 7
udp_PC.SendData txtsend.Text
End Sub
*************************************************************
```

The DLC Header area of Sniffer Text 13.1 should pose no suprises. Note that Sniffer Text 13.1 is composed of the AP's resend frame destined for the AirDrop-P module. Only the addressing in the DLC Header area changes when the AP turns the incoming frame around.

Sniffer Text 13.1
```
DLC: ----- DLC Header -----
      DLC:
      DLC: Frame 77 arrived at 13:51:28.9729; frame size is 61 (003D hex) bytes.
      DLC: Signal level                  = 100%
      DLC: Channel                       = 1
      DLC: Data rate                     = 22 (11.0 Megabits per second)
      DLC:
      DLC: Frame Control Field #1 = 08
      DLC:                   .... ..00 = 0x0 Protocol Version
      DLC:                   .... 10.. = 0x2 Data Frame
      DLC:                   0000 .... = 0x0 Data (Subtype)
      DLC: Frame Control Field #2 = 02
      DLC:                   .... ...0 = Not to Distribution System
      DLC:                   .... ..1. = From Distribution System
      DLC:                   .... .0.. = Last fragment
      DLC:                   .... 0... = Not retry
      DLC:                   ...0 .... = Active Mode
      DLC:                   ..0. .... = No more data
      DLC:                   .0.. .... = Wired Equivalent Privacy is off
      DLC:                   0... .... = Not ordered
      DLC: Duration                      = 149 (in microseconds)
      DLC: Destination Address           = Station AbocomBC5967
      DLC: Basic Service Set ID          = Station Netgea6FD3DA
      DLC: Source address                = Station Xircom03D92F
      DLC: Sequence Control              = 0x7D90
      DLC: ...Sequence Number            = 0x7D9 (2009)
      DLC: ...Fragment Number            = 0x0   (0)
      DLC:
LLC: ----- LLC Header -----
      LLC:
      LLC: DSAP Address = AA, DSAP IG Bit = 00 (Individual Address)
      LLC: SSAP Address = AA, SSAP CR Bit = 00 (Command)
      LLC: Unnumbered frame: UI
      LLC:
```

```
SNAP: ----- SNAP Header -----
      SNAP:
      SNAP: Type        = 0800 (IP)
      SNAP:
IP: ----- IP Header -----
      IP:
      IP: Version = 4, header length = 20 bytes
      IP: Type of service = 00
      IP:       000. ....  = routine
      IP:       ...0 ....  = normal delay
      IP:       .... 0...  = normal throughput
      IP:       .... .0..  = normal reliability
      IP:       .... ..0.  = ECT bit - transport protocol will ignore the CE bit
      IP:       .... ...0  = CE bit - no congestion
      IP: Total length    = 29 bytes
      IP: Identification  = 6638
      IP: Flags           = 0X
      IP:       .0.. ....  = may fragment
      IP:       ..0. ....  = last fragment
      IP: Fragment offset = 0 bytes
      IP: Time to live    = 128 seconds/hops
      IP: Protocol        = 17 (UDP)
      IP: Header checksum = 9EEF (correct)
      IP: Source address      = [192.168.0.11]
      IP: Destination address = [192.168.0.151]
      IP: No options
      IP:
UDP: ----- UDP Header -----
      UDP:
      UDP: Source port      =  5002
      UDP: Destination port =     7 (Echo)
      UDP: Length           =    9
      UDP: Checksum         = 1658 (correct)
      UDP: [1 byte(s) of data]
      UDP:
ADDR  HEX                                               ASCII
0000: 08 02 95 00 00 e0 98 bc 59 67 00 09 5b 6f d3 da | ..•..à~¼Yg..[oÓÚ
0010: 00 05 3c 03 d9 2f 90 7d aa aa 03 00 00 00 08 00 | ..<.Ù/□}ªª......
0020: 45 00 00 1d 19 ee 00 00 80 11 9e ef c0 a8 00 0b | E....î..€.žïÀ¨..
0030: c0 a8 00 97 13 8a 00 07 00 09 16 58 54           | À¨.−.Š.....XT
```

Sniffer Text 13.1: This rendition of sniffer text shows the UDP packet leaving the AP and heading towards the AirDrop-P module. It's the AP's job to deliver the mail and the original data payload headers and all must remain intact.

The AirDrop 802.11b driver's *get_frame* function picks up on the incoming UDP frame and uses the code package shown in Code Snippet 13.2 to determine that the packet has a UDP payload.

Code Snippet 13.2

```
******************************
#define rxdatalength_offset 44
//***********************************************************
//*    IP Protocol Types
//***********************************************************
#define PROT_ICMP           0x01
#define PROT_TCP            0x06
#define PROT_UDP            0x11
//***********************************************************
//*    Ethernet Header Layout
//***********************************************************
unsigned int packet[761];
#define enetpacketLen11     0x00
#define enetpacketDest      0x01
#define enetpacketDest01    0x01  //destination mac address
#define enetpacketDest23    0x02
#define enetpacketDest45    0x03
#define enetpacketSrc       0x04
#define enetpacketSrc01     0x04  //source mac address
#define enetpacketSrc23     0x05
#define enetpacketSrc45     0x06
#define enetpacketLen03     0x07
#define enetpacketSnap      0x08
#define enetpacketCntrl     0x09
//                          0x0A
#define enetpacketType      0x0B  //type/length field
#define enetpacketData      0x0C  //IP data area begins here
//***********************************************************
//*    IP Header Layout
//***********************************************************
#define ip_vers_len         0x0C  //IP version and header length
//#define ip_tos                  //IP type of service
#define ip_pktlen           0x0D  //packet length
#define ip_id               0x0E  //datagram id
#define ip_frag_offset      0x0F  //fragment offset
#define ip_ttlproto         0x10  //time to live
//#define ip_proto                //protocol (ICMP=1, TCP=6, UDP=11)
#define ip_hdr_cksum        0x11  //header checksum
#define ip_srcaddr          0x12  //IP address of source
//                          0X13
#define ip_destaddr         0x14  //IP addess of destination
//                          0X15
#define ip_data             0x16  //IP data area
//***********************************************************
//*    UDP Header
//***********************************************************
#define UDP_srcport         ip_data
#define UDP_destport        UDP_srcport+1
#define UDP_len             UDP_destport+1
#define UDP_cksum           UDP_len+1
#define UDP_data            UDP_cksum+1
```

```
//* COMMANDS
#define   Access_Cmd                  0x0021

#define   BAP0Busy_Bit_Mask           0x8000 //busy bit in Offset0 register
#define   BAP0Err_Bit_Mask            0x4000 //Err bit in Offset0 register
#define   BAP1Busy_Bit_Mask           0x8000 //busy bit in Offset1 register
#define   BAP1Err_Bit_Mask            0x4000 //Err bit in Offset1 register

//*************************************************************
//*    PORT Definitions
//*************************************************************
#define data_in      PORTD
#define data_out     LATD
#define addr_hi      LATG
#define addr_lo      LATH
//*************************************************************
//*    PORT TRIS Definitions
//*************************************************************
#define TO_NICTRISD = 0x00
#define FROM_NIC     TRISD = 0xFF
//*************************************************************
//*    CF I/O Definitions
//*************************************************************
#define OE              0x01
#define IORD            0x02
#define WE              0x40
#define IOWR            0x80
#define set_OE          PCTL |= OE
#define clr_OE          PCTL &= ~OE
#define set_IORD        PCTL |= IORD
#define clr_IORD        PCTL &= ~IORD
#define set_IOWR        PCTL |= IOWR
#define clr_IOWR        PCTL &= ~IOWR

void wr_cf_data(char bapnum,char* buffer,unsigned int count)
{
 unsigned int addr,i;
 int *byteptr;
 char data_lo;

 if(bapnum)
  addr = Data1_Register;
 else
  addr = Data0_Register;

 byteptr = (int*)buffer;
 for(i=0;i<(count&0xFFFE);)
 {
   wr_cf_io16(*byteptr++,addr);
   i+=2;
 }
```

```
 if(count % 2)
  {
   data_lo = buffer[i++];
   wr_cf_io16(data_lo,addr);
  }
 }

void wr_cf_addr(unsigned int addr)
{
 addr_hi = (make8(addr,1));
 addr_lo = addr & 0x00FF;
}

unsigned int rd_cf_io16(unsigned int addr)
{
 char data_lo,data_hi,i;
 unsigned int data16;

 wr_cf_addr(addr);
 clr_IORD;
 i=3;
 while(--i);
 data_lo=data_in;
 set_IORD;
 NOP();

 wr_cf_addr(addr+1);
 clr_IORD;
 i=3;
 while(--i);
 data_hi=data_in;
 set_IORD;
 NOP();
 data16 = make16(data_hi,data_lo);
 return(data16);
}

void wr_cf_io16(unsigned int data16,unsigned int addr)
{
 char i;

 wr_cf_addr(addr);
 data_out = LOW_BYTE(data16);
 TO_NIC;
 clr_IOWR;
 i=1;
 while(--i);
 set_IOWR;
 NOP();
 data_out = 0xFF;

 wr_cf_addr(addr+1);
```

```
 data_out = HIGH_BYTE(data16);
 clr_IOWR;
 i=1;
 while(--i);
 set_IOWR;
 NOP();
 data_out = 0xFF;
 FROM_NIC;
}


//***********************************************************
//* get a free bap
//***********************************************************
char get_free_bap(void)
{
 unsigned int temp;
 char rc;

 rc = 2;
 temp = rd_cf_io16(Offset1_Register);
 if(!(temp & BAP1Busy_Bit_Mask))
  rc = 1;
 temp = rd_cf_io16(Offset0_Register);
 if(!(temp & BAP0Busy_Bit_Mask))
  rc = 0;
 return(rc);
}
***************************


//***********************************************************
//*    Get A Frame From the CF Receive Buffer
//*    This routine removes an Ethernet frame from the receive buffer.
//***********************************************************
void get_frame()
{
 unsigned int temp,i;
 char rc,bapnum;

 rc = 0;
 RxFID = rd_cf_io16(RxFID_Register);     // Get the FID of the allocated buf-
fer
 bapnum = get_free_bap();
 switch(bapnum)
  {
   case 0:
    wr_cf_io16(RxFID,Select0_Register);
    wr_cf_io16(rxdatalength_offset,Offset0_Register);
    do{
         temp = rd_cf_io16(Offset0_Register);
       }while(temp & BAP0Busy_Bit_Mask);
    packet[enetpacketLen11] = rd_cf_io16(Data0_Register);//get data length
    if(packet[enetpacketLen11] > 0x05DC)
```

```
      packet[enetpacketLen11] = 0x05DC;
     for(i=0;i<3;++i)
      packet[enetpacketDest+i] = rd_cf_io16(Data0_Register);
     for(i=0;i<3;++i)
      packet[enetpacketSrc+i] = rd_cf_io16(Data0_Register);
     for(i=0;i<packet[enetpacketLen11];++i)
      packet[enetpacketLen03+i] = rd_cf_io16(Data0_Register);
     // Acknowledge that we now own the FID
     wr_cf_io16(EvStat_Rx_Bit_Mask,EvAck_Register );
     break;

   case 1:
    wr_cf_io16(RxFID,Select1_Register);
    wr_cf_io16(rxdatalength_offset,Offset1_Register);
    do{
         temp = rd_cf_io16(Offset1_Register);
        }while(temp & BAP0Busy_Bit_Mask);
    packet[enetpacketLen11] = rd_cf_io16(Data1_Register);//get data length
    if(packet[enetpacketLen11] > 0x05DC)
     packet[enetpacketLen11] = 0x05DC;
    for(i=0;i<3;++i)
     packet[enetpacketDest+i] = rd_cf_io16(Data1_Register);
    for(i=0;i<3;++i)
     packet[enetpacketSrc+i] = rd_cf_io16(Data1_Register);
    for(i=0;i<packet[enetpacketLen11];++i)   //removed +1
     packet[enetpacketLen03+i] = rd_cf_io16(Data1_Register);
    // Acknowledge that we now own the FID
    wr_cf_io16(EvStat_Rx_Bit_Mask,EvAck_Register );
     break;
  default:
    rc = 1;
    break;
 }
if(!rc)
{
 //process an ARP packet
 if(packet[enetpacketType] == 0x0608)
 {
  if(packet[arp_hwtype] == 0x0100 &&
   packet[arp_prtype] == 0x0008 &&
  packet[arp_hwprlen] == 0x0406 &&
  packet[arp_op] == 0x0100 &&
  ipaddri[0] == packet[arp_tipaddr] &&
  ipaddri[1] == packet[arp_tipaddr+1])
  arp();
 }

 //process an IP packet
 else if(packet[enetpacketType] == 0x0008
  && packet[ip_destaddr] == ipaddri[0]
  && packet[ip_destaddr+1] == ipaddri[1])
 {
```

```
        if(HIGH_BYTE(packet[ip_ttlproto]) == PROT_ICMP)
          icmp();
        else if(HIGH_BYTE(packet[ip_ttlproto]) == PROT_UDP)
          udp();
        else if(HIGH_BYTE(packet[ip_ttlproto]) == PROT_TCP)
          tcp();
     }
  }
  else
    do_event_housekeeping();
}
```
****

Code Snippet 13.2: We're walking the internet protocol chain and you're getting smarter with every step. Three down (ARP, ICMP, UDP) and one to go.

The first layer of the UDP "onion" is seen in Mini Sniff 13.1 contains the SNAP Header, which holds the packet type, IP. There should be nothing new to you here. The AirDrop 802.11b driver *get_frame* function parses the packet type and the destination IP address. If the destination IP address matches up to the AirDrop-P module's assigned IP address (which it does here), the *get_frame* function moves on to parse for the packet payload protocol type.

Mini Sniff 13.1

```
LLC:   ----- LLC Header -----
       LLC:
       LLC:  DSAP Address = AA, DSAP IG Bit = 00 (Individual Address)
       LLC:  SSAP Address = AA, SSAP CR Bit = 00 (Command)
       LLC:  Unnumbered frame: UI
       LLC:
SNAP:  ----- SNAP Header -----
       SNAP:
       SNAP: Type      = 0800 (IP)
       SNAP:
ADDR   HEX                                               ASCII
0000:  08 02 95 00 00 e0 98 bc 59 67 00 09 5b 6f d3 da | ..•...à~¼Yg..[oÓÚ
0010:  00 05 3c 03 d9 2f 90 7d aa aa 03 00 00 00 08 00 | ..<.Ù/ }ªª....░
0020:  45 00 00 1d 19 ee 00 00 80 11 9e ef c0 a8 00 0b | E....î..□.žïÀ¨..
0030:  c0 a8 00 97 13 8a 00 07 00 09 16 58 54          | À¨.−.Š.....XT
```

Looking at the IP Header area in Mini Sniff 13.2 we see that the incoming IP packet is normal in that its header length is 20 bytes in length. Also, we can see that the packet has no abnormal requirements. A total IP header length of 29 bytes tells us that 9 bytes of the IP header contain the packet payload.

Mini Sniff 13.2

```
IP: ----- IP Header -----
    IP:
    IP: Version = 4, header length = 20 bytes
    IP: Type of service = 00
    IP:       000. ....  = routine
```

```
IP:         ...0 .... = normal delay
IP:         .... 0... = normal throughput
IP:         .... .0.. = normal reliability
IP:         .... ..0. = ECT bit - transport protocol will ignore the CE bit
IP:         .... ...0 = CE bit - no congestion
IP: Total length      = 29 bytes
IP: Identification    = 6638
IP: Flags             = 0X
IP:         .0.. .... = may fragment
IP:         ..0. .... = last fragment
IP: Fragment offset = 0 bytes
IP: Time to live      = 128 seconds/hops
IP: Protocol          = 17 (UDP)
IP: Header checksum = 9EEF (correct)
IP: Source address       = [192.168.0.11]
IP: Destination address = [192.168.0.151]
IP: No options
ADDR  HEX                                                   ASCII
0000: 08 02 95 00 00 e0 98 bc 59 67 00 09 5b 6f d3 da  | ..•..à~¼Yg..[oÓÚ
0010: 00 05 3c 03 d9 2f 90 7d aa aa 03 00 00 00 08 00  | ..<.Ù/□}ªª......
0020: 45 00 00 1d 19 ee 00 00 80 11 9e ef c0 a8 00 0b  | E...î..€.žïÀ¨..
0030: c0 a8 00 97 13 8a 00 07 00 09 16 58 54           | À¨.−.Š.....XT
```

The four IP header fields beyond the Total length field (Identification, Flags, Fragment offset and Time to live) are of no concern to us at the moment. The AirDrop 802.11b driver keys on the Protocol field highlighted in Mini Sniff 13.3 and drives the AirDrop 802.11b driver thread to the *udp* function.

Mini Sniff 13.3

```
IP: ----- IP Header -----
    IP:
    IP: Version = 4, header length = 20 bytes
    IP: Type of service = 00
    IP:         000. .... = routine
    IP:         ...0 .... = normal delay
    IP:         .... 0... = normal throughput
    IP:         .... .0.. = normal reliability
    IP:         .... ..0. = ECT bit - transport protocol will ignore the CE bit
    IP:         .... ...0 = CE bit - no congestion
    IP: Total length      = 29 bytes
    IP: Identification    = 6638
    IP: Flags             = 0X
    IP:         .0.. .... = may fragment
    IP:         ..0. .... = last fragment
    IP: Fragment offset = 0 bytes
    IP: Time to live      = 128 seconds/hops
    IP: Protocol          = 17 (UDP)
    IP: Header checksum = 9EEF (correct)
    IP: Source address       = [192.168.0.11]
    IP: Destination address = [192.168.0.151]
```

```
      IP: No options
      IP:
ADDR  HEX                                                      ASCII
0000: 08 02 95 00 00 e0 98 bc 59 67 00 09 5b 6f d3 da | ..•..à~¼Yg..[oÓÚ
0010: 00 05 3c 03 d9 2f 90 7d aa aa 03 00 00 00 08 00 | ..<.Ù/□}ªª......
0020: 45 00 00 1d 19 ee 00 00 80 11 9e ef c0 a8 00 0b | E....î..€.žïÀ"..
0030: c0 a8 00 97 13 8a 00 07 00 09 16 58 54          | À".−.Š.....XT
```

The IP destination address brought to your attention in Mini Sniff 13.4 was checked by the AirDrop 802.11b driver up front to get us to this point. And, since there are no options in the IP header, the next field is the IP data area.

Mini Sniff 13.4

```
IP: ----- IP Header -----
      IP:
      IP: Version = 4, header length = 20 bytes
      IP: Type of service = 00
      IP:       000. ....  = routine
      IP:       ...0 ....  = normal delay
      IP:       .... 0...  = normal throughput
      IP:       .... .0..  = normal reliability
      IP:       .... ..0.  = ECT bit - transport protocol will ignore the CE bit
      IP:       .... ...0  = CE bit - no congestion
      IP: Total length   = 29 bytes
      IP: Identification = 6638
      IP: Flags          = 0X
      IP:       .0.. ....  = may fragment
      IP:       ..0. ....  = last fragment
      IP: Fragment offset = 0 bytes
      IP: Time to live   = 128 seconds/hops
      IP: Protocol       = 17 (UDP)
      IP: Header checksum = 9EEF (correct)
      IP: Source address      = [192.168.0.11]
      IP: Destination address = [192.168.0.151]
      IP: No options
      IP:
ADDR  HEX                                                      ASCII
0000: 08 02 95 00 00 e0 98 bc 59 67 00 09 5b 6f d3 da | ..•..à~¼Yg..[oÓÚ
0010: 00 05 3c 03 d9 2f 90 7d aa aa 03 00 00 00 08 00 | ..<.Ù/□}ªª......
0020: 45 00 00 1d 19 ee 00 00 80 11 9e ef c0 a8 00 0b | E....î.€.žïÀ"..
0030: c0 a8 00 97 13 8a 00 07 00 09 16 58 54          | À".−.Š.....XT
```

At this point, the AirDrop 802.11b driver *udp* function is called to process the UDP payload. Everything needed or called by the *udp* function is shown in Code Snippet 13.3.

Code Snippet 13.3

```
****************************

//************************************************************
//*    SETIPADDRS
//*    This function builds the IP header.
```

259

```
//****************************************************************
void setipaddrs(void)
{
    //move IP source address to destination address
    packet[ip_destaddr]=packet[ip_srcaddr];
    packet[ip_destaddr+1]=packet[ip_srcaddr+1];
    //make ethernet module IP address source address
    packet[ip_srcaddr]=ipaddri[0];
    packet[ip_srcaddr+1]=ipaddri[1];
    //move hardware source address to destinatin address
    packet[enetpacketDest01]=packet[enetpacketSrc01];
    packet[enetpacketDest23]=packet[enetpacketSrc23];
    packet[enetpacketDest45]=packet[enetpacketSrc45];
    //make ethernet module mac address the source address
    packet[enetpacketSrc01]=macaddri[0];
    packet[enetpacketSrc23]=macaddri[1];
    packet[enetpacketSrc45]=macaddri[2];

    //calculate the IP header checksum
    packet[ip_hdr_cksum]=0x00;

    hdr_chksum =0;
    hdrlen = (LOW_BYTE(packet[ip_vers_len]) & 0x0F) *4;
    addr = &packet[ip_vers_len];
    cksum();
    packet[ip_hdr_cksum]= ~(hdr_chksum + ((hdr_chksum & 0xFFFF0000) >> 16));
 }
//****************************************************************
//*    CHECKSUM CALCULATION ROUTINE
//****************************************************************
void cksum()
{
        //hdr_chksum = 0;
      while(hdrlen > 1)
      {
            hdr_chksum = hdr_chksum + *addr++;
        hdrlen -=2;
        }
      if(hdrlen > 0)
        {
        hdr_chksum = hdr_chksum + (*addr & 0x00FF);
      --hdrlen;
        }
}

unsigned int swapbytes(unsigned int val)
{
      char temphi,templo;

      temphi = make8(val,1);
      templo = make8(val,0);
```

```
        return(make16(templo,temphi));
}


//*************************************************************
//*    Echo Packet Function
//*    This routine does not modify the incoming packet size and
//*    thus echoes the original packet structure.
//*************************************************************
char echo_packet()
{
        char rc;
        unsigned int TxFID;

        TxFID = get_free_TxFID();
        if(TxFID == 0)
         rc = 1;
        else
         {
      rc = bap_write(TxFID,rxdatalength_offset,(char*)packet,packet[enetpacke
tLen11]+16);
          if(rc=send_command(TransmitReclaim_Cmd,TxFID))
                printf("\r\nTransmit failed");

        do_event_housekeeping();
         }
         return(rc);
}


****************************


//*************************************************************
//*    UDP Function
//*    This function uses a Visual Basic UDP program to echo the
//*    data back to the VB program and set or reset bit on PORT A
//*    under control of the VB program LCD pushbuttons.
//*************************************************************
void udp()
{
    unsigned int temp;
    char rc;

    //port 7 is the well-known echo port
    if(packet[UDP_destport] == 0x0700)
    {
        //build the IP header
        setipaddrs();

        //swap the UDP source and destination ports
        temp = packet[UDP_srcport];
        packet[UDP_srcport] = packet[UDP_destport];
        packet[UDP_destport] = temp;
```

```
    //calculate the UDP checksum
    packet[UDP_cksum] = 0x00;

    hdr_chksum =0;
    hdrlen = 0x08;
    addr = &packet[ip_srcaddr];
    cksum();
    hdr_chksum = hdr_chksum +     ((256*HIGH_BYTE(packet[ip_
ttlproto]))+packet[UDP_len]);
    hdrlen = swapbytes(packet[UDP_len]);
    addr = &packet[UDP_srcport];
    cksum();
    packet[UDP_cksum]= ~(hdr_chksum + ((hdr_chksum & 0xFFFF0000) >> 16));
    //echo the incoming data back to the VB program
    rc = echo_packet();
  }

  //buttons on the VB GUI are pointed towards port address 5000 decimal
  else if(packet[UDP_destport] == 0x8813)
      {
    if(packet[UDP_data] == '0')
      //received a "0" from the VB program
      LATA5=0;
    else if(packet[UDP_data] == '1')
      //received a "1" from the VB program
      LATA5=1;
    else if(packet[UDP_data] == 0x00)
      //received a 0x00 from the VB program
      LATA=0x00;
     else if(packet[UDP_data] == 0x01)
      //received a 0x01 from the VB program
      LATA1=1;
    else if(packet[UDP_data] == 0x02)
      //received a 0x02 from the VB program
      LATA2=1;
    else if(packet[UDP_data] == 0x03)
      //received a 0x03 from the VB program
      LATA3=1;
    else if(packet[UDP_data] == 0x04)
      //received a 0x04 from the VB program
          LATA4=1;

    }
}
```
****

Code Snippet 13.3: The AirDrop 802.11b driver UDP code responds to well-known Port 7 for UDP echo operations and UDP Port 5000 for the Internet Test Panel pushbuttons.

The Source port value in the UDP Header, which is displayed in Mini Sniff 13.5, hails from the hard-coded VB6 Internet Test Panel UDP port value in Sub Snippet 13.8.

Mini Sniff 13.5

```
UDP: ----- UDP Header -----
      UDP:
      UDP: Source port       = 5002
      UDP: Destination port = 7 (Echo)
      UDP: Length            = 9
      UDP: Checksum          = 1658 (correct)
      UDP: [1 byte(s) of data]
      UDP:
ADDR HEX                                                          ASCII
0000: 08 02 95 00 00 e0 98 bc 59 67 00 09 5b 6f d3 da  | ..•..à~¼Yg..[oÓÚ
0010: 00 05 3c 03 d9 2f 90 7d aa aa 03 00 00 00 08 00  | ..<.Ù/□}ªª......
0020: 45 00 00 1d 19 ee 00 00 80 11 9e ef c0 a8 00 0b  | E....î..€.žïÀ¨..
0030: c0 a8 00 97 13 8a 00 07 00 09 16 58 54           | À¨.–.Š.....XT
```

Sub Snippet 13.8

```
************************************************************
Private Sub Form_Load()
On Error Resume Next
txtip.Text = "192.168.1.150"
txtport.Text = "5000"
With udp_PC
.RemoteHost = txtip.Text
.RemotePort = Val(txtport.Text)
.Bind 5002
End With
frm_B.Show
End Sub
************************************************************
```

The UDP destination port in Mini Sniff 13.6 is the well-known Echo port. The AirDrop 802.11b driver *udp* function is coded to parse for either UDP Port 7 or UDP Port 5000. The Port 7 destination port is hard-coded in the VB6 Internet Test Panel echo application code shown in Sub Snippet 13.9. You can change the AirDrop 802.11b driver default UDP Ports to suit your application by modifying the existing AirDrop 802.11b driver UDP driver or writing your own custom AirDrop-P UDP driver/application. The UDP destination port field is a word wide. That means you can theoretically define and use up to 65,535 unique UDP ports. I say "theoretically" because you would step on some of the well-know port values along the way. If your UDP application won't be conversing with any other device that adheres to the etiquette of well-known ports, you can use any UDP port number that suits you.

Mini Sniff 13.6

```
UDP: ----- UDP Header -----
      UDP:
      UDP: Source port       = 5002
      UDP: Destination port = 7 (Echo)
      UDP: Length            = 9
      UDP: Checksum          = 1658 (correct)
      UDP: [1 byte(s) of data]
      UDP:
```

```
ADDR  HEX                                                   ASCII
0000: 08 02 95 00 00 e0 98 bc 59 67 00 09 5b 6f d3 da | ..•..à~¼Yg..[oóÚ
0010: 00 05 3c 03 d9 2f 90 7d aa aa 03 00 00 00 08 00 | ..<.Ù/□}ªª......
0020: 45 00 00 1d 19 ee 00 00 80 11 9e ef c0 a8 00 0b | E....î..€.žïÀ..
0030: c0 a8 00 97 13 8a 00 07 00 09 16 58 54           | À".—.Š...XT
```

Sub Snippet 13.9

```
************************************************************
Private Sub txtsend_Change()
On Error Resume Next
udp_PC.RemoteHost = txtip.Text
udp_PC.RemotePort = 7
udp_PC.SendData txtsend.Text
End Sub
************************************************************
```

The minimum length of a UDP packet is 8 bytes. That minimum 8 bytes includes the whole UDP header with no data. Since our UDP length is 9 bytes in Mini Sniff 13.7, there's one byte of UDP payload data. In this case, it's the "T" I sent using the Internet Test Panel.

Mini Sniff 13.7

```
UDP: ----- UDP Header -----
     UDP:
     UDP: Source port      = 5002
     UDP: Destination port = 7 (Echo)
     UDP: Length           = 9
     UDP: Checksum         = 1658 (correct)
     UDP: [1 byte(s) of data]
     UDP:
ADDR  HEX                                                   ASCII
0000: 08 02 95 00 00 e0 98 bc 59 67 00 09 5b 6f d3 da | ..•..à~¼Yg..[oóÚ
0010: 00 05 3c 03 d9 2f 90 7d aa aa 03 00 00 00 08 00 | ..<.Ù/□}ªª......
0020: 45 00 00 1d 19 ee 00 00 80 11 9e ef c0 a8 00 0b | E....î..€.žïÀ..
0030: c0 a8 00 97 13 8a 00 07 00 09 16 58 54           | À".—.Š....XT
```

The UDP checksum calculation process is a bit different than the ICMP and IP checksum procedures. You would think the UDP checksum would cover only the bytes within the UDP header and data area. That's too easy. The UDP checksum includes some choice bytes from the IP header as well. In fact, the UDP checksum is calculated using:

- The IP source address word

- The IP destination address word

- The IP protocol byte

- The UDP length word

- The UDP header

- The UDP data

And yes, the UDP length word is used twice in the calculation of the UDP checksum in Mini Sniff 13.8.

Mini Sniff 13.8

```
UDP: ----- UDP Header -----
     UDP:
     UDP: Source port      = 5002
     UDP: Destination port = 7 (Echo)
     UDP: Length           = 9
     UDP: Checksum         = 1658 (correct)
     UDP: [1 byte(s) of data]
     UDP:
ADDR HEX                                                        ASCII
0000: 08 02 95 00 00 e0 98 bc 59 67 00 09 5b 6f d3 da | ..•..à~¼Yg..[oóÚ
0010: 00 05 3c 03 d9 2f 90 7d aa aa 03 00 00 00 08 00 | ..<.Ù/□}ªª......
0020: 45 00 00 1d 19 ee 00 00 80 11 9e ef c0 a8 00 0b | E....î..€.žïÀ..
0030: c0 a8 00 97 13 8a 00 07 00 09 16 58 54           | À".−.š....XT
```

OK...We've traversed a UDP datagram encapsulated with an IP packet from the Netasyst Sniffer's point of view. Array Table 13.1 reveals how the UDP datagram is represented within the AirDrop-P packet buffer inside the PIC18LF8621's SRAM.

PACKET BUFFER LAYOUT			
Buffer Address	**Array Slot**	**DATA**	**Description**
0x06FA	[0]	0x0025	802.11b packet data length
0x06FC	[1]	0xE000	Destination MAC Address
0x06FE	[2]	0xBC98	
0x0700	[3]	0x6759	
0x0702	[4]	0x0500	Source MAC Address
0x0704	[5]	0x033C	
0x0706	[6]	0x2FD9	
0x0708	[7]	0x2500	802.3 packet data length
0x070A	[8]	0xAAAA	SNAP Header
0x070C	[9]	0x0003	UI/Protocol Identifier
0x070E	[10]	0x0000	Protocol Identifier
0x0710	[11]	0x0008	IP Protocol
0x0712	[12]	0x0045	IP version/header length/TOS
0x0714	[13]	0x1D00	IP packet length
0x0716	[14]	0xEE19	IP datagram id
0x0718	[15]	0x0000	fragment offset
0x071A	[16]	0x1180	time to live/protocol = UDP
0x071C	[17]	0xEF9E	IP header checksum
0x071E	[18]	0xA8C0	Source IP Address
0x0720	[19]	0x0B00	
0x0722	[20]	0xA8C0	Destination IP Address
0x0724	[21]	0x9700	

PACKET BUFFER LAYOUT			
Buffer Address	**Array Slot**	**DATA**	**Description**
0x0726	[22]	0x8A13	UDP source port
0x0728	[23]	0x0700	UDP destination port
0x072A	[24]	0x0900	UDP length
0x072C	[25]	0x5816	UDP checksum
0x072E	[26]	0x8254	UDP data

Array Table 13.1: Ignore the extra byte (0x82) in slot [26] as it is just some garbage holding a place in the AirDrop-P packet buffer.

As I mentioned earlier, the AirDrop 802.11b driver's *udp* function is geared to either echo the incoming UDP datagram or twiddle some of the bits of the PIC18LF8621's PORTA I/O port. The UDP destination port is set at 7. So, let's follow that thread through the AirDrop 802.11b driver's *udp* function.

The well-known Port 7 path is one of an echo operation. So, we'll use the *setipaddrs* function to redirect the source and destination IP and MAC addresses accordingly. The *setipaddrs* function doesn't do UDP datagram ports. So, we must call upon some discrete code in Sub Snippet 13.10 to perform the UDP source and destination port swap.

Sub Snippet 13.10
****

```
//************************************************************
//*    UDP Function
//*    This function uses a Visual Basic UDP program to echo the
//*    data back to the VB program and set or reset bit on PORT A
//*    under control of the VB program LCD pushbuttons.
//************************************************************
void udp()
{
    unsigned int temp;
    char rc;

    //port 7 is the well-known echo port
    if(packet[UDP_destport] == 0x0700)
    {
        //build the IP header
        setipaddrs();

        //swap the UDP source and destination ports
        temp = packet[UDP_srcport];
        packet[UDP_srcport] = packet[UDP_destport];
        packet[UDP_destport] = temp;
```

****

When setipaddrs gets done doing its thing, swapping destination and source IP and MAC addresses and recalculating the IP header checksum, the AirDrop-P packet buffer is rearranged and looks like the contents of Array Table 13.2.

PACKET BUFFER LAYOUT			
Buffer Address	**Array Slot**	**DATA**	**Description**
0x06FA	[0]	0x0025	802.11b packet data length
0x06FC	[1]	0x0500	Destination MAC Address
0x06FE	[2]	0x033C	
0x0700	[3]	0x2FD9	
0x0702	[4]	0xE000	Source MAC Address
0x0704	[5]	0xBC98	
0x0706	[6]	0x6759	
0x0708	[7]	0x2500	802.3 packet data length
0x070A	[8]	0xAAAA	SNAP Header
0x070C	[9]	0x0003	UI/Protocol Identifier
0x070E	[10]	0x0000	Protocol Identifier
0x0710	[11]	0x0008	IP Protocol
0x0712	[12]	0x0045	IP version/header length/TOS
0x0714	[13]	0x1D00	IP packet length
0x0716	[14]	0xEE19	IP datagram id
0x0718	[15]	0x0000	fragment offset
0x071A	[16]	0x1180	time to live/protocol = UDP
0x071C	[17]	0xEF9E	IP header checksum
0x071E	[18]	0xA8C0	Source IP Address
0x0720	[19]	0x9700	
0x0722	[20]	0xA8C0	Destination IP Address
0x0724	[21]	0x0B00	
0x0726	[22]	0x0700	UDP source port
0x0728	[23]	0x8A13	UDP destination port
0x072A	[24]	0x0900	UDP length
0x072C	[25]	0x5816	UDP checksum
0x072E	[26]	0x8254	UDP data

Array Table 13.2: This is simply musical chairs played by IP and MAC addresses. If you've been paying attention, you should know what every field is and why it it what it is. Again, ignore the extra byte (0x82) in slot [26] of the AirDrop-P packet buffer. Counting the bytes using the 802.11b packet data length field will tell you that the 0x82 won't get transmitted.

I ran the *UDP* function all the way up to the *echo_packet* function. So, tThe UDP checksum value is reflected in the contents of Array Table 13.2. I'll do a rewind and show you what happened.

The prerequisite clearing of the checksum-holding data spaces occurs first in Sub Snippet 13.11. The IP source address is included in the UDP checksum calculation. So, we point at that slot (slot [18]) in the AirDrop-P packet buffer. We must also include the IP destination address in the UDP checksum calculation. That's 8 bytes (hdrlen = 0x08;) we will churn through the UDP checksum mill. Once we've accumulated the UDP checksum value of the IP header address components, we can then pull out the IP protocol byte from the AirDrop-P packet buffer's slot [16] (0x11), add that to the UDP length value in slot [24] and add the result of the IP protocol byte and UDP length word to our already accumulated IP address UDP checksum value. The rest of the raw UDP checksum is accumulated from the UDP datagram itself. In this case, that's 9 bytes, which includes the UDP header and the single byte of data. Any bits that overflowed into the upper 16 bits of the *hdr_chksum* value are folded back, via mathematical addition, into the lower 16 bits of the *hdr_chksum* value. Then, just as we've always done with the accumulated checksum value, we one's complement the *hdr_chksum* value to get our UDP checksum result in AirDrop-P packet buffer slot [25].

Sub Snippet 13.11
****
```
        //calculate the UDP checksum
        packet[UDP_cksum] = 0x00;

        hdr_chksum =0;
        hdrlen = 0x08;
        addr = &packet[ip_srcaddr];
        cksum();
        hdr_chksum = hdr_chksum +    ((256*HIGH_BYTE(packet[ip_
ttlproto]))+packet[UDP_len]);
        hdrlen = swapbytes(packet[UDP_len]);
        addr = &packet[UDP_srcport];
        cksum();
        packet[UDP_cksum]= ~(hdr_chksum + ((hdr_chksum & 0xFFFF0000) >> 16));
        //echo the incoming data back to the VB program
        rc = echo_packet();
```
****

After the computed UDP checksum in placed in the appropriate AirDrop-P packet buffer slot, the UDP datagram encapsulated within an IP datagram is transmitted back to the original sending station that is running the Internet Test Panel application. The VB6 *UDP_PC_Data-Arrival* code segment in Sub Snippet 13.12 processes the incoming UDP packet, resulting in the "T" being placed in the Internet Test Panel's Echoed Data window.

Sub Snippet 13.12
****
```
Private Sub udp_PC_DataArrival(ByVal bytesTotal As Long)
On Error Resume Next
udp_PC.RemoteHost = txtip.Text
Dim strData As String
udp_PC.GetData strData
txtreceive.Text = strData
End Sub
```
****

Clicking on one of the Internet Test Panel LCD buttons sends a word associated with the selected pushbutton via UDP to the station and UDP port addressed in the Internet Test Panel's Target IP Address and Target Port windows. Sniffer Text 13.2 holds the frame information that was generated when I clicked on the *Line 2* Internet Test Panel pushbutton.

Sniffer Text 13.2

```
DLC: ----- DLC Header -----
      DLC:
      DLC: Frame 181 arrived at  11:27:42.8339; frame size is 62 (003E hex)
bytes.
      DLC: Signal level                  = 100%
      DLC: Channel                       = 1
      DLC: Data rate                     = 22 (11.0 Megabits per second)
      DLC:
      DLC: Frame Control Field #1 = 08
      DLC:                .... ..00 = 0x0 Protocol Version
      DLC:                .... 10.. = 0x2 Data Frame
      DLC:               0000 .... = 0x0 Data (Subtype)
      DLC: Frame Control Field #2 = 01
      DLC:                .... ...1 = To Distribution System
      DLC:                .... ..0. = Not from Distribution System
      DLC:                .... .0.. = Last fragment
      DLC:                .... 0... = Not retry
      DLC:                ...0 .... = Active Mode
      DLC:                ..0. .... = No more data
      DLC:                .0.. .... = Wired Equivalent Privacy is off
      DLC:                0... .... = Not ordered
      DLC: Duration                      = 213 (in microseconds)
      DLC: Basic Service Set ID          = Station Netgea6FD3DA
      DLC: Source Address                = Station Xircom03D92F
      DLC: Destination Address           = Station AbocomBC5967
      DLC: Sequence Control              = 0x02D0
      DLC: ...Sequence Number            = 0x02D (45)
      DLC: ...Fragment Number            = 0x0   (0)
      DLC:
LLC:  ----- LLC Header -----
      LLC:
      LLC:  DSAP Address = AA, DSAP IG Bit = 00 (Individual Address)
      LLC:  SSAP Address = AA, SSAP CR Bit = 00 (Command)
      LLC:  Unnumbered frame: UI
      LLC:
SNAP: ----- SNAP Header -----
      SNAP:
      SNAP: Type      = 0800 (IP)
      SNAP:
IP: ----- IP Header -----
      IP:
      IP: Version = 4, header length = 20 bytes
      IP: Type of service = 00
      IP:      000. ....    = routine
      IP:      ...0 ....    = normal delay
```

```
     IP:          .... 0... = normal throughput
     IP:          .... .0.. = normal reliability
     IP:          .... ..0. = ECT bit - transport protocol will ignore the CE bit
     IP:          .... ...0 = CE bit - no congestion
     IP: Total length    = 30 bytes
     IP: Identification  = 7371
     IP: Flags           = 0X
     IP:          .0.. .... = may fragment
     IP:          ..0. .... = last fragment
     IP: Fragment offset = 0 bytes
     IP: Time to live    = 128 seconds/hops
     IP: Protocol        = 17 (UDP)
     IP: Header checksum = 9C11 (correct)
     IP: Source address      = [192.168.0.11]
     IP: Destination address = [192.168.0.151]
     IP: No options
     IP:
UDP: ----- UDP Header -----
     UDP:
     UDP: Source port      =  5002
     UDP: Destination port =  5000 (Commplex-main)
     UDP: Length           =  10
     UDP: Checksum         =  54D5 (correct)
     UDP: [2 byte(s) of data]
     UDP:
ADDR HEX                                                   ASCII
0000: 08 01 d5 00 00 09 5b 6f d3 da 00 05 3c 03 d9 2f | ..Õ...[oóÚ..<.Ù/
0010: 00 e0 98 bc 59 67 d0 02 aa aa 03 00 00 00 08 00 | .à˜¼YgÐ.ªª......
0020: 45 00 00 1e 1c cb 00 00 80 11 9c 11 c0 a8 00 0b | E....Ë..€.œ.À¨..
0030: c0 a8 00 97 13 8a 13 88 00 0a 54 d5 02 00       | À¨.—.Š.ˆ..TÕ..
*****************************************************************
```

Sniffer Text 13.2: Note the difference in the byte order of the UDP data from the Internet Test Panel as the Netasyst Sniffer see is versus the way the AirDrop 802.11b driver sees it in Array Table 13.3.

Array Table 13.3 is a look at the AirDrop-P packet buffer after receiving the UDP frame I generated by clicking on the *Line 2* pushbutton in the Internet Test Panel.

PACKET BUFFER LAYOUT			
Buffer Address	Array Slot	DATA	Description
0x06FA	[0]	0x0026	802.11b packet data length
0x06FC	[1]	0xE000	Destination MAC Address
0x06FE	[2]	0xBC98	
0x0700	[3]	0x6759	
0x0702	[4]	0x0500	Source MAC Address
0x0704	[5]	0x033C	
0x0706	[6]	0x2FD9	
0x0708	[7]	0x2600	802.3 packet data length
0x070A	[8]	0xAAAA	SNAP Header

PACKET BUFFER LAYOUT			
Buffer Address	**Array Slot**	**DATA**	**Description**
0x070C	[9]	0x0003	UI/Protocol Identifier
0x070E	[10]	0x0000	Protocol Identifier
0x0710	[11]	0x0008	IP Protocol
0x0712	[12]	0x0045	IP version/header length/TOS
0x0714	[13]	0x1E00	IP packet length
0x0716	[14]	0xCB1C	IP datagram id
0x0718	[15]	0x0000	fragment offset
0x071A	[16]	0x1180	time to live/protocol = UDP
0x071C	[17]	0x119C	IP header checksum
0x071E	[18]	0xA8C0	Source IP Address
0x0720	[19]	0x0B00	
0x0722	[20]	0xA8C0	Destination IP Address
0x0724	[21]	0x9700	
0x0726	[22]	0x8A13	UDP source port
0x0728	[23]	0x8813	UDP destination port
0x072A	[24]	0x0A00	UDP length
0x072C	[25]	0xD554	UDP checksum
0x072E	[26]	0x0002	UDP data

Array Table 13.3: This time the 802.11b packet length is 0x0026 bytes. If you count the bytes beginning at slot [8], you'll find that the whole of slot [26] is actually UDP data.

This time, clicking an LCD pushbutton within the Internet Test Panel spawned the UDP transmission. This is no Port 7 echo operation and the destination UDP port is 5000 decimal as shown in the Internet Test Panel's Target Port window (see Screen Capture 13.1).

The AirDrop 802.11b driver's UDP application code in Sub Snippet 13.13 takes the UDP port 5000 branch and parses the UDP data in slot [26] of the AirDrop-P packet buffer. The result is that the PIC18LF8621's PORTA2 I/O pin output latch is forced to a logical high.

Sub Snippet 13.13
```
************************************************************
//buttons on the VB GUI are pointed towards port address 5000 decimal
else if(packet[UDP_destport] == 0x8813)
    {
  if(packet[UDP_data] == '0')
     //received a "0" from the VB program
     LATA5=0;
  else if(packet[UDP_data] == '1')
     //received a "1" from the VB program
     LATA5=1;
  else if(packet[UDP_data] == 0x00)
     //received a 0x00 from the VB program
     LATA=0x00;
  else if(packet[UDP_data] == 0x01)
```

```
            //received a 0x01 from the VB program
        LATA1=1;
    else if(packet[UDP_data] == 0x02)
        //received a 0x02 from the VB program
        LATA2=1;
    else if(packet[UDP_data] == 0x03)
        //received a 0x03 from the VB program
        LATA3=1;
    else if(packet[UDP_data] == 0x04)
        //received a 0x04 from the VB program
            LATA4=1;


    }
```

****

Sniffer Text 13.3 contains the capture contents of the final UDP echo frame in which all of the ASCII characters are accumulated and echoed all at once.

Sniffer Text 13.3

```
DLC: ----- DLC Header -----
      DLC:
      DLC: Frame 253 arrived at 12:17:04.2219; frame size is 96 (0060 hex) bytes.
      DLC: Signal level                = 100%
      DLC: Channel                     = 1
      DLC: Data rate                   = 22 (11.0 Megabits per second)
      DLC:
      DLC: Frame Control Field #1 = 08
      DLC:               .... ..00 = 0x0 Protocol Version
      DLC:               .... 10.. = 0x2 Data Frame
      DLC:             0000 .... = 0x0 Data (Subtype)
      DLC: Frame Control Field #2 = 01
      DLC:               .... ...1 = To Distribution System
      DLC:               .... ..0. = Not from Distribution System
      DLC:               .... .0.. = Last fragment
      DLC:               .... 0... = Not retry
      DLC:             ...0 .... = Active Mode
      DLC:             ..0. .... = No more data
      DLC:             .0.. .... = Wired Equivalent Privacy is off
      DLC:             0... .... = Not ordered
      DLC: Duration                    = 213 (in microseconds)
      DLC: Basic Service Set ID        = Station Netgea6FD3DA
      DLC: Source Address              = Station Xircom03D92F
      DLC: Destination Address         = Station AbocomBC5967
      DLC: Sequence Control            = 0x0860
      DLC: ...Sequence Number          = 0x086 (134)
      DLC: ...Fragment Number          = 0x0   (0)
      DLC:
LLC: ----- LLC Header -----
      LLC:
      LLC:  DSAP Address = AA, DSAP IG Bit = 00 (Individual Address)
      LLC:  SSAP Address = AA, SSAP CR Bit = 00 (Command)
```

```
       LLC:   Unnumbered frame: UI
       LLC:
SNAP:  ----- SNAP Header -----
       SNAP:
       SNAP: Type     = 0800 (IP)
       SNAP:
IP: ----- IP Header -----
       IP:
       IP: Version = 4, header length = 20 bytes
       IP: Type of service = 00
       IP:       000. .... = routine
       IP:       ...0 .... = normal delay
       IP:       .... 0... = normal throughput
       IP:       .... .0.. = normal reliability
       IP:       .... ..0. = ECT bit - transport protocol will ignore the CE bit
       IP:       .... ...0 = CE bit - no congestion
       IP: Total length   = 64 bytes
       IP: Identification = 7456
       IP: Flags          = 0X
       IP:       .0.. .... = may fragment
       IP:       ..0. .... = last fragment
       IP: Fragment offset = 0 bytes
       IP: Time to live    = 128 seconds/hops
       IP: Protocol        = 17 (UDP)
       IP: Header checksum = 9B9A (correct)
       IP: Source address      = [192.168.0.11]
       IP: Destination address = [192.168.0.151]
       IP: No options
       IP:
UDP: ----- UDP Header -----
       UDP:
       UDP: Source port      = 5002
       UDP: Destination port = 7 (Echo)
       UDP: Length           = 44
       UDP: Checksum         = DC98 (correct)
       UDP: [36 byte(s) of data]
       UDP:
ADDR  HEX                                                  ASCII
0000: 08 01 d5 00 00 09 5b 6f d3 da 00 05 3c 03 d9 2f | ..õ...[oóÚ..<.Ù/
0010: 00 e0 98 bc 59 67 60 08 aa aa 03 00 00 00 08 00 | .à~¼Yg`.ªª......
0020: 45 00 00 40 1d 20 00 00 80 11 9b 9a c0 a8 00 0b | E..@. ..€.>šÀ¨..
0030: c0 a8 00 97 13 8a 00 07 00 2c dc 98 54 68 69 73 | À¨.—.Š...,Ü~This
0040: 20 69 73 20 38 30 32 2e 31 31 62 20 77 69 74 68 |  is 802.11b with
0050: 20 55 44 50 20 6d 61 64 65 20 65 61 73 79 2e 20 |  UDP made easy.
```

Sniffer Text 13.2: Here's the final UDP echo frame. The Internet Test Panel application accumulated the ASCII message one byte at a time. Of course, with a little modification to the AirDrop 802.11b driver, you can fill the UDP data area with as many bytes as the PIC18LF8621's SRAM will allow and send them all at once.

I hope that you've found that even though it is considered "unreliable," UDP is a very easy way to get data from Point A to Point B with minimal hassle. In the next chapter, things get a bit more complicated as we will encounter the king of the Internet, TCP.

Notes

The band of brothers that worked the "Snakepit" at Detroit's Motown Records called themselves the Funk Brothers. The nickname was given to the band by their Rock and Roll Hall of Fame drummer William "Benny" "Papa Zita" Benjamin. If you've ever heard a Motown hit, you've heard Papa Zita's drums.

Flying Cargo with TCP/IP and the AirDrop

TCP/IP is the most complex of any of the Internet protocols used by the AirDrop we've discussed thus far. Nebulous is a good word when it comes to describing the amount of information in print that talks about various aspects of TCP/IP. I'm going to take the Jack Webb approach and only give you TCP/IP facts that you need to use TCP/IP with an AirDrop.

TCP and the AirDrop-P

The TCP/IP implementation that is effected by the AirDrop 802.11b driver is based on the TCP/IP code that was written for the series of EDTP wired Ethernet devices. The TCP/IP code I will describe was designed to conserve both microcontroller program memory and microcontroller data memory space. Thus, the AirDrop TCP/IP stack you will be introduced to here is a very minimal TCP/IP model. The AirDrop TCP/IP firmware is fashioned as a Telnet server that echoes characters it receives.

Just like all of the other internet protocols we've looked at up to this point, the TCP/IP module of the AirDrop 802.11b driver is called from the *get_frame* function. Every piece of code associated with the AirDrop 802.11b driver's TCP/IP function is listed in Code Snippet 14.1.

As you can see in Code Snippet 14.1, the Ethernet type field identifies the frame inside its data area as an IP frame and once that fact is established, the AirDrop 802.11b driver examines the IP protocol field inside the IP header. If the value in the IP protocol field of the IP header is equal to 0x06, the AirDrop 802.11b driver code routes to the TCP function's code.

Code Snippet 14.1
```
****************************

#define rxdatalength_offset 44
//***********************************************************
//*    IP Protocol Types
//***********************************************************
#define PROT_ICMP            0x01
#define PROT_TCP             0x06
#define PROT_UDP             0x11
//***********************************************************
//*    Ethernet Header Layout
//***********************************************************
unsigned int packet[761];
#define enetpacketLen11      0x00
```

```
#define enetpacketDest        0x01
#define enetpacketDest01      0x01   //destination mac address
#define enetpacketDest23      0x02
#define enetpacketDest45      0x03
#define enetpacketSrc         0x04
#define enetpacketSrc01       0x04   //source mac address
#define enetpacketSrc23       0x05
#define enetpacketSrc45       0x06
#define enetpacketLen03       0x07
#define enetpacketSnap        0x08
#define enetpacketCntrl       0x09
//                            0x0A
#define enetpacketType        0x0B   //type/length field
#define enetpacketData        0x0C   //IP data area begins here
//****************************************************************
//*     IP Header Layout
//****************************************************************
#define ip_vers_len           0x0C   //IP version and header length
//#define ip_tos                      //IP type of service
#define ip_pktlen             0x0D   //packet length
#define ip_id                 0x0E   //datagram id
#define ip_frag_offset        0x0F   //fragment offset
#define ip_ttlproto           0x10   //time to live
//#define ip_proto                    //protocol (ICMP=1, TCP=6, UDP=11)
#define ip_hdr_cksum          0x11   //header checksum
#define ip_srcaddr            0x12   //IP address of source
//                            0X13
#define ip_destaddr           0x14   //IP addess of destination
//                            0X15
#define ip_data               0x16   //IP data area
//****************************************************************
//*     TCP Header Layout
//****************************************************************
#define TCP_srcport           0x16   //TCP source port
#define TCP_destport          0x17   //TCP destination port
#define TCP_seqnum            0x18   //sequence number
//                            0x19
#define TCP_acknum            0x1A   //acknowledgment number
//                            0x1B
#define TCP_hdrflags          0x1C   //4-bit header len and flags
#define TCP_window            0x1D   //window size
#define TCP_cksum             0x1E   //TCP checksum
#define TCP_urgentptr         0x1F   //urgent pointer
#define TCP_data              0x20   //option/data
//****************************************************************
//*     TCP Flags
//*     IN flags represent incoming bits
//*     OUT flags represent outgoing bits
//****************************************************************
#define FIN_IN(packet[TCP_hdrflags] & 0x0100)
#define SYN_IN(packet[TCP_hdrflags] & 0x0200)
#define RST_IN(packet[TCP_hdrflags] & 0x0400)
```

```
#define PSH_IN          (packet[TCP_hdrflags] & 0x0800)
#define ACK_IN          (packet[TCP_hdrflags] & 0x1000)
#define URG_IN          (packet[TCP_hdrflags] & 0x2000)
#define FIN_OUT         packet[TCP_hdrflags] |= 0x0100 //0000 0001 0000 0000
#define SYN_OUT         packet[TCP_hdrflags] |= 0x0200 //0000 0010 0000 0000
#define RST_OUT         packet[TCP_hdrflags] |= 0x0400 //0000 0100 0000 0000
#define PSH_OUT         packet[TCP_hdrflags] |= 0x0800 //0000 1000 0000 0000
#define ACK_OUT         packet[TCP_hdrflags] |= 0x1000 //0001 0000 0000 0000
#define URG_OUT         packet[TCP_hdrflags] |= 0x2000 //0010 0000 0000 0000

//* COMMANDS
#define  Access_Cmd              0x0021

#define  BAP0Busy_Bit_Mask       0x8000 //busy bit in Offset0 register
#define  BAP0Err_Bit_Mask        0x4000 //Err bit in Offset0 register
#define  BAP1Busy_Bit_Mask       0x8000 //busy bit in Offset1 register
#define  BAP1Err_Bit_Mask        0x4000 //Err bit in Offset1 register

//**********************************************************
//*    PORT Definitions
//**********************************************************
#define data_in      PORTD
#define data_out     LATD
#define addr_hi      LATG
#define addr_lo      LATH
//**********************************************************
//*    PORT TRIS Definitions
//**********************************************************
#define TO_NICTRISD = 0x00
#define FROM_NIC     TRISD = 0xFF
//**********************************************************
//*    CF I/O Definitions
//**********************************************************
#define OE               0x01
#define IORD             0x02
#define WE               0x40
#define IOWR             0x80
#define set_OE           PCTL |= OE
#define clr_OE           PCTL &= ~OE
#define set_IORD         PCTL |= IORD
#define clr_IORD         PCTL &= ~IORD
#define set_IOWR         PCTL |= IOWR
#define clr_IOWR         PCTL &= ~IOWR

void wr_cf_data(char bapnum,char* buffer,unsigned int count)
{
 unsigned int addr,i;
 int *byteptr;
 char data_lo;

 if(bapnum)
  addr = Data1_Register;
```

```
else
  addr = Data0_Register;

byteptr = (int*)buffer;
for(i=0;i<(count&0xFFFE);)
  {
    wr_cf_io16(*byteptr++,addr);
    i+=2;
  }
if(count % 2)
  {
    data_lo = buffer[i++];
    wr_cf_io16(data_lo,addr);
  }
}

void wr_cf_addr(unsigned int addr)
{
 addr_hi = (make8(addr,1));
 addr_lo = addr & 0x00FF;
}

unsigned int rd_cf_io16(unsigned int addr)
{
 char data_lo,data_hi,i;
 unsigned int data16;

 wr_cf_addr(addr);
 clr_IORD;
 i=3;
 while(--i);
 data_lo=data_in;
 set_IORD;
 NOP();

 wr_cf_addr(addr+1);
 clr_IORD;
 i=3;
 while(--i);
 data_hi=data_in;
 set_IORD;
 NOP();
 data16 = make16(data_hi,data_lo);
 return(data16);
}

void wr_cf_io16(unsigned int data16,unsigned int addr)
{
 char i;

 wr_cf_addr(addr);
 data_out = LOW_BYTE(data16);
```

```
 TO_NIC;
 clr_IOWR;
 i=1;
 while(--i);
 set_IOWR;
 NOP();
 data_out = 0xFF;

 wr_cf_addr(addr+1);
 data_out = HIGH_BYTE(data16);
 clr_IOWR;
 i=1;
 while(--i);
 set_IOWR;
 NOP();
 data_out = 0xFF;
 FROM_NIC;
}

//********************************************************
//* get a free bap
//********************************************************
char get_free_bap(void)
{
 unsigned int temp;
 char rc;

 rc = 2;
 temp = rd_cf_io16(Offset1_Register);
 if(!(temp & BAP1Busy_Bit_Mask))
  rc = 1;
 temp = rd_cf_io16(Offset0_Register);
 if(!(temp & BAP0Busy_Bit_Mask))
  rc = 0;
 return(rc);
}
***************************

//********************************************************
//*    Get A Frame From the CF Receive Buffer
//*    This routine removes an Ethernet frame from the receive buffer.
//********************************************************
void get_frame()
{
 unsigned int temp,i;
 char rc,bapnum;

 rc = 0;
 RxFID = rd_cf_io16(RxFID_Register);     // Get the FID of the allocated buf-
fer
 bapnum = get_free_bap();
 switch(bapnum)
```

```
   {
    case 0:
     wr_cf_io16(RxFID,Select0_Register);
     wr_cf_io16(rxdatalength_offset,Offset0_Register);
     do{
           temp = rd_cf_io16(Offset0_Register);
          }while(temp & BAP0Busy_Bit_Mask);
     packet[enetpacketLen11] = rd_cf_io16(Data0_Register);//get data length
     if(packet[enetpacketLen11] > 0x05DC)
      packet[enetpacketLen11] = 0x05DC;
     for(i=0;i<3;++i)
      packet[enetpacketDest+i] = rd_cf_io16(Data0_Register);
     for(i=0;i<3;++i)
      packet[enetpacketSrc+i] = rd_cf_io16(Data0_Register);
     for(i=0;i<packet[enetpacketLen11];++i)
      packet[enetpacketLen03+i] = rd_cf_io16(Data0_Register);
     // Acknowledge that we now own the FID
     wr_cf_io16(EvStat_Rx_Bit_Mask,EvAck_Register );
     break;

    case 1:
     wr_cf_io16(RxFID,Select1_Register);
     wr_cf_io16(rxdatalength_offset,Offset1_Register);
     do{
           temp = rd_cf_io16(Offset1_Register);
          }while(temp & BAP0Busy_Bit_Mask);
     packet[enetpacketLen11] = rd_cf_io16(Data1_Register);//get data length
     if(packet[enetpacketLen11] > 0x05DC)
      packet[enetpacketLen11] = 0x05DC;
     for(i=0;i<3;++i)
      packet[enetpacketDest+i] = rd_cf_io16(Data1_Register);
     for(i=0;i<3;++i)
      packet[enetpacketSrc+i] = rd_cf_io16(Data1_Register);
     for(i=0;i<packet[enetpacketLen11];++i)   //removed +1
      packet[enetpacketLen03+i] = rd_cf_io16(Data1_Register);
     // Acknowledge that we now own the FID
     wr_cf_io16(EvStat_Rx_Bit_Mask,EvAck_Register );
     break;
   default:
     rc = 1;
     break;
   }
  if(!rc)
  {
   //process an ARP packet
   if(packet[enetpacketType] == 0x0608)
   {
    if(packet[arp_hwtype] == 0x0100 &&
     packet[arp_prtype] == 0x0008 &&
    packet[arp_hwprlen] == 0x0406 &&
    packet[arp_op] == 0x0100 &&
    ipaddri[0] == packet[arp_tipaddr] &&
```

```
      ipaddri[1] == packet[arp_tipaddr+1])
      arp();
  }

  //process an IP packet
  else if(packet[enetpacketType] == 0x0008
  && packet[ip_destaddr] == ipaddri[0]
  && packet[ip_destaddr+1] == ipaddri[1])
  {
    if(HIGH_BYTE(packet[ip_ttlproto]) == PROT_ICMP)
      icmp();
    else if(HIGH_BYTE(packet[ip_ttlproto]) == PROT_UDP)
      udp();
    else if(HIGH_BYTE(packet[ip_ttlproto]) == PROT_TCP)
      tcp();
  }
}
else
  do_event_housekeeping();
}
*****************************
```

Code Snippet 14.1: One small step in AirDrop 802.11b driver code, one giant leap in your 802.11b knowledge.

Earlier, you saw how the process of encapsulation ties IP and UDP together in the IP data area. TCP (Transmission Control Protocol) works with IP in the same manner, as shown in the highlighted area of Sub Snippet 14.1. The concept of encapsulation prevails in TCP's relationship to IP. In an IP datagram that contains a TCP payload, the TCP payload lies within the IP data area just as in a IP datagram carrying a UDP message where the UDP package lies inside of the IP data area.

Sub Snippet 14.1

```
************************************************************

//*********************************************************
//*    IP Header Layout
//*********************************************************
#define ip_vers_len        0x0C   //IP version and header length
//#define ip_tos                  //IP type of service
#define ip_pktlen          0x0D   //packet length
#define ip_id              0x0E   //datagram id
#define ip_frag_offset     0x0F   //fragment offset
#define ip_ttlproto        0x10   //time to live
//#define ip_proto                //protocol (ICMP=1, TCP=6, UDP=11)
#define ip_hdr_cksum       0x11   //header checksum
#define ip_srcaddr         0x12   //IP address of source
//                         0X13
#define ip_destaddr        0x14   //IP addess of destination
//                         0X15
#define ip_data            0x16   //IP data area
```

```
//*************************************************************
//*     TCP Header Layout
//*************************************************************
#define TCP_srcport        0x16   //TCP source port
#define TCP_destport       0x17   //TCP destination port
#define TCP_seqnum         0x18   //sequence number
//                         0x19
#define TCP_acknum         0x1A   //acknowledgment number
//                         0x1B
#define TCP_hdrflags       0x1C   //4-bit header len and flags
#define TCP_window         0x1D   //window size
#define TCP_cksum          0x1E   //TCP checksum
#define TCP_urgentptr      0x1F   //urgent pointer
#define TCP_data           0x20   //option/data

*************************************************************
```

TCP/IP was designed from the ground up to be platform independent. That's why we can run a flavor of TCP/IP on a relatively tiny microcontroller-based system or on a mainframe complex the size of a football field. TCP/IP is a collection of protocols that are standardized across the world of networking. TCP/IP can't be used for every situation. However, while TCP/IP is not the total solution to all networking problems, TCP/IP stands as a pretty good model of what all of internetworking should be. Normally, to implement TCP/IP one must employ the use of what is termed a "TCP/IP stack." Complete TCP/IP stacks can be very large and usually aren't easily ported fully to smaller platforms like our AirDrop.

Telnet is a protocol that is common to most users of the Internet. The purpose of the Telnet Protocol is to provide a general-purpose means of interfacing terminal devices and terminal-oriented applications to each other.

Just as you've seen with UDP network devices, every machine or device on a TCP/IP-based network no matter how big or how small is called a host. That includes clients as well as servers. It used to be that a server was always the big machine on the wire and the client was a workstation on someone's desk in accounting. Today, servers sit on desks and clients can be worn on your wrist or vice versa. Regardless of the host's size, the idea is that all hosts can communicate with each other. This implies that all hosts on all networks can communicate host to host across differing networks. This may sound impractical, but that's how the Internet works. Today hosts all over the world communicate by passing messages, accessing data and transferring files between themselves sometimes using dissimilar networks. A perfect example is our AirDrop module. The AirDrop is a wireless device that when incorporated with an AP, seamlessly communicates with other wired and wireless devices on the network.

In the TCP/IP world messages are generally short packets of data. Just like UDP, each TCP data packet is addressed to reach a particular host on a particular network. There is no difference in the addressing scheme used for TCP and UDP. The protocol or IP address for both protocols is the familiar four bytes divided by dots address scheme (192.168.0.151). Physical addresses or MAC addresses apply to both UDP and TCP and are used in the identical manner by both protocols.

You're probably used to hearing the term "TCP/IP" and thinking about it as one protocol. However, you know from our UDP experiences that IP is indeed its own protocol. The combination of Transmission Control Protocol and Internet Protocol was not done by accident but by design. Just as it is in the UDP world, IP is the unreliable component of the TCP/IP pair. You will remember that IP is termed unreliable because there is no way that IP itself can guarantee that a data packet will actually be delivered to its destination. All of the hosts in the Internet or on a local LAN give their best effort to deliver an IP data packet. Despite all of the good intentions, the problem is that nobody cares how it looks when it gets where it's going. If you've ever sat and watched baggage handlers load baggage onto an aircraft, you already have a pretty good concept of how IP works. For example, relating to the IP datagram, an IP data packet is just like a piece of freight in the hands of the guy or gal loading baggage onto an airplane. He or she gets the bag (our IP datagram) from one of those mobile baggage carts they drive around (the host) and slings it towards the moving-belt ramp (LAN, Internet). Sometimes the handlers throw a bag and it misses the conveyor. One of the other baggage handlers (other hosts) may pick up the dropped bag (corrupted IP datagram) and either sling it back on the ramp (passed it along to the next host for inspection) or put it directly on the plane (delivered the damaged IP datagram to the receiving host). The plane (receiving host), although being loaded with a bag full of broken goods, never reported back to the cart (sending host) that the bag (IP datagram) that was finally loaded had its contents damaged in transit. My little story implies that each bag or each IP datagram is independent of any other IP datagram and nobody really cares what's inside the bag while it is in transit.

The TCP/IP protocol stack is composed of five layers as shown in Figure 14.1. The Physical layer is the simplest layer with the Application layer potentially being the most complex of the five layers. It's rather obvious that even though the layers depend upon each other, each layer of the stack has a distinct job to do.

The confusion that could exist between the layers is eliminated by the use of unique headers in conjunction with datagram encapsulation. Each layer passes only properly formatted output to the next layer for processing. Encapsulation also allows each layer to treat the data in the way it prefers without affecting the way the data is treated in other layers. For example, the Transport Layer likes to pretend that data is entering in a constant stream while the Internet or IP layer sees data as separate connectionless datagrams. To write a successful embedded TCP/IP application it is necessary to understand the functions of each protocol layer. Let's take a look at each of them from the bottom of the stack to the top of the stack.

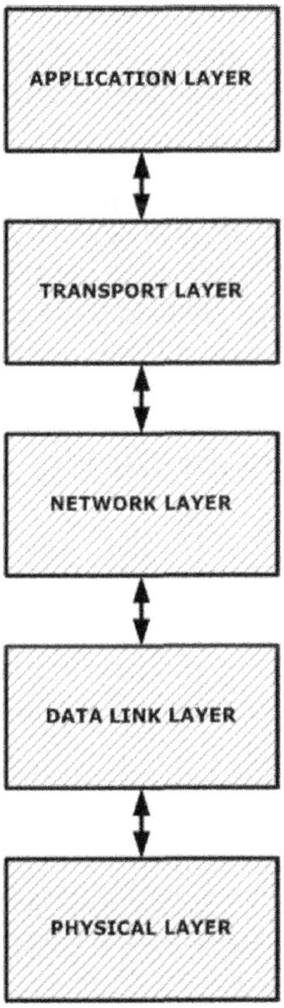

APPLICATION LAYER

TRANSPORT LAYER

NETWORK LAYER

DATA LINK LAYER

PHYSICAL LAYER

Figure 14.1: Data can flow from the Application Layer down or from the Physical Layer up.

The TCP/IP Stack's Physical Layer

Physical layer is another way of saying hardware layer. This is the wire, cable and electronics that connect the devices and networks to each other. Physical also implies "touchable" or real. For the AirDrop module, the physical layer consists of RF waves. RF isn't "touchable", but it exists and thus, is real. On a wired network, the physical network consists of an Ethernet hub/router, possibly an AP (most of the APs aimed at home wireless networks have a wired switch functionality as well as wireless capability) and any other cabling or electronic devices that tie the physical network together using wire. The bottom line is that the physical layer always sees the entire packet whether it is receiving it or transmitting it and never adds or subtracts to a packet's contents.

The TCP/IP Stack's Data Link Layer

The Data Link Layer is only responsible for transferring a datagram from one host over a single physical link to another host. Most if not all of the Data Link Layer functionality resides in the AirDrop's MAC engine. The AirDrop MAC engine accepts data and wraps it into an 802.11b frame that can be received and "unwrapped" by the station the data was addressed to. The AirDrop MAC engine also automatically generates the preamble and CRC when an 802.11b frame is transmitted. In addition, the AirDrop's MAC engine also makes sure the RF medium is clear before attempting to send a message and automatically retries failed send operations. The AirDrop MAC engine is also responsible for checking the incoming frame's hardware address to determine if the incoming frame belongs to it or another station on the network.

The TCP/IP Stack's Network Layer

The Network Layer encapsulates messages passed from the Transport Layer and produces datagrams. This is where IP lives. The Network Layer encapsulates a UDP datagram or TCP segment inside an IP datagram. An IP header is added, which calls out the handling of the IP datagram, and the datagram is then passed along to the Data Link Layer for transmission.

The TCP/IP Stack's Transport Layer

Between the Application Layer and the Network Layer lies the Transport Layer. The job of the Transport layer is to pass data between the Application Layer and the Transport Layer using TCP or UDP protocols.

UDP lives here but TCP is what makes the Transport Layer famous. TCP unlike UDP uses a virtual connection to make sure that the data arrives at its destination intact and in order. TCP accomplishes this "connection" via handshaking and special codes in each TCP data segment.

TCP and UDP receive data from the application and form segments or packets (datagrams) respectively. A destination address is added before passing the packet or segment to the Network Layer.

The TCP/IP Stack's Application Layer

The final and topmost layer is the Application Layer. This is where the programmer reigns. There are more protocols used in this layer than I care to mention. Some familiar and some home grown. In simplest terms, data flows from the Application layer of the originating host down through the TCP/IP stack and out the Physical layer across to the Physical layer of the destination host. Once the data enters the destination host's Physical layer, the process is reversed and the data flows up through the TCP/IP stack to the Application layer where it is processed by the application code.

A host could be running multiple UDP or TCP/IP applications at once. With all of that data flying around, one might ask how does the TCP/IP stack know which messages belong to it and where to route the messages when it receives them. Just like UDP, TCP/IP handles these situations by assigning each network connection its own protocol port. A protocol port is actually an internal TCP or UDP address. This address is passed down the stack in the header of each packet of data. IP sends logical host addresses (192.168.0.151) that we humans read in dotted decimal format and TCP and UDP send hexadecimal protocol port addresses (7, 23, 5000,etc.) that we interpret in decimal. You were exposed to IP and port addressing in the UDP chapter. TCP uses port and IP addressing in a similar manner.

TCP/IP – The Big Ugly

The TCP/IP and supporting code contained within Code Snippet 14.2 is intimidating. Rather than tremble at it's feet, let's throw some bits at this big hunk of code and see what it will do.

Code Snippet 14.2
****

```
//**********************************************************
//*    TELNET SERVER BANNER STATEMENT CONSTANT
//**********************************************************
char const telnet_banner[] = "\r\nAirDrop-P>";
//**********************************************************
//*    Port Definitions
//*    This address is used by TCP and the Telnet function.
//*    This can be changed to any valid port number as long as
//*    you modify your code to recognize the new port number.
//**********************************************************
#define  MY_PORT_ADDRESS      0x981F  // 8088 DECIMAL
//**********************************************************
//*    FLAGS
//**********************************************************
char flags;
#define synflag 0x01
#define finflag 0x02
#define bsynflag        flags & synflag
#define bfinflag        flags & finflag
```

```
#define clr_synflag     flags &= ~synflag
#define set_synflag     flags |= synflag
#define clr_finflag     flags &= ~finflag
#define set_finflag     flags |= finflag
//*************************************************************
//*     IP Header Layout
//*************************************************************
#define ip_vers_len         0x0C   //IP version and header length
//#define ip_tos                   //IP type of service
#define ip_pktlen           0x0D   //packet length
#define ip_id               0x0E   //datagram id
#define ip_frag_offset      0x0F   //fragment offset
#define ip_ttlproto         0x10   //time to live
//#define ip_proto                 //protocol (ICMP=1, TCP=6, UDP=11)
#define ip_hdr_cksum        0x11   //header checksum
#define ip_srcaddr          0x12   //IP address of source
//                          0X13
#define ip_destaddr         0x14   //IP addess of destination
//                          0X15
#define ip_data             0x16   //IP data area
//*************************************************************
//*     TCP Header Layout
//*************************************************************
#define TCP_srcport         0x16   //TCP source port
#define TCP_destport        0x17   //TCP destination port
#define TCP_seqnum          0x18   //sequence number
//                          0x19
#define TCP_acknum          0x1A   //acknowledgment number
//                          0x1B
#define TCP_hdrflags        0x1C   //4-bit header len and flags
#define TCP_window          0x1D   //window size
#define TCP_cksum           0x1E   //TCP checksum
#define TCP_urgentptr       0x1F   //urgent pointer
#define TCP_data            0x20   //option/data
//*************************************************************
//*     TCP Flags
//*     IN flags represent incoming bits
//*     OUT flags represent outgoing bits
//*************************************************************
#define FIN_IN       (packet[TCP_hdrflags] & 0x0100)
#define SYN_IN       (packet[TCP_hdrflags] & 0x0200)
#define RST_IN       (packet[TCP_hdrflags] & 0x0400)
#define PSH_IN       (packet[TCP_hdrflags] & 0x0800)
#define ACK_IN       (packet[TCP_hdrflags] & 0x1000)
#define URG_IN       (packet[TCP_hdrflags] & 0x2000)
#define FIN_OUT      packet[TCP_hdrflags] |= 0x0100 //0000 0001 0000 0000
#define SYN_OUT      packet[TCP_hdrflags] |= 0x0200 //0000 0010 0000 0000
#define RST_OUT      packet[TCP_hdrflags] |= 0x0400 //0000 0100 0000 0000
#define PSH_OUT      packet[TCP_hdrflags] |= 0x0800 //0000 1000 0000 0000
#define ACK_OUT      packet[TCP_hdrflags] |= 0x1000 //0001 0000 0000 0000
#define URG_OUT      packet[TCP_hdrflags] |= 0x2000 //0010 0000 0000 0000
```

```
//**********************************************************
//*    Macros
//**********************************************************
#define bitset(var, bitno) ((var) |= 1 << (bitno))
#define bitclr(var, bitno) ((var) &= ~(1 << (bitno)))
#define make8(var,offset)   ((unsigned int)var >> (offset * 8)) & 0x00FF
#define make16(varhigh,varlow)      (((unsigned int)varhigh & 0xFF)* 0x100) +
((unsigned int)varlow & 0x00FF)
#define make32(var1,var2,var3,var4) \
            ((unsigned long)var1<<24)+((unsigned long)var2<<16)+ \
            ((unsigned long)var3<<8)+((unsigned long)var4)
#define make32i(var1,var2) ((unsigned long)var1<<16)+((unsigned long)var2)


//**********************************************************
//*    Send TCP Packet
//*    This routine assembles and sends a complete TCP/IP packet.
//*    40 bytes of IP and TCP header data is assumed.
//**********************************************************
void send_tcp_packet()
{
        unsigned int temp;

    //count IP and TCP header bytes.. Total = 40 bytes
    ip_packet_len = 40 + tcpdatalen_out;
    packet[ip_pktlen] = swapbytes(ip_packet_len);
    temp = packet[enetpacketLen11];
    temp += tcpdatalen_out;
    packet[enetpacketLen11] = temp;
    packet[enetpacketLen03] = swapbytes(temp);

    setipaddrs();

        //swap the TCP source and destination ports
        temp = packet[TCP_srcport];
        packet[TCP_srcport] = packet[TCP_destport];
        packet[TCP_destport] = temp;

    //assemble the acknowledgment
    client_seqnum=make32i(packet[TCP_seqnum],packet[TCP_seqnum+1]);
    client_seqnum = client_seqnum + swapbytes(tcpdatalen_in);
    packet[TCP_acknum] = (client_seqnum & 0xFFFF0000) >> 16;
    packet[TCP_acknum+1] = client_seqnum & 0x0000FFFF;
    packet[TCP_seqnum+1] = my_seqnum & 0x0000FFFF;
    packet[TCP_seqnum] = (my_seqnum & 0xFFFF0000) >> 16;

    packet[TCP_hdrflags] &= 0xC0FF;
    ACK_OUT;
    if(bfinflag)
    {
        FIN_OUT;
        clr_finflag;
    }
```

```
   packet[TCP_cksum] = 0x00;

   hdr_chksum =0;
   hdrlen = 0x08;
   addr = &packet[ip_srcaddr];
   cksum();
   hdr_chksum = hdr_chksum + 256*HIGH_BYTE(packet[ip_ttlproto]);
   tcplen = swapbytes(packet[ip_pktlen]) - (LOW_BYTE(packet[ip_vers_len]) &
0x0F) *4;
   hdr_chksum = hdr_chksum + swapbytes(tcplen);
   hdrlen = tcplen;
   addr = &packet[TCP_srcport];
   cksum();
   packet[TCP_cksum]= ~(hdr_chksum + ((hdr_chksum & 0xFFFF0000) >> 16));

      echo_packet();
}
//*************************************************************
//*    SETIPADDRS
//*    This function builds the IP header.
//*************************************************************
void setipaddrs(void)
{
   //move IP source address to destination address
   packet[ip_destaddr]=packet[ip_srcaddr];
   packet[ip_destaddr+1]=packet[ip_srcaddr+1];
   //make ethernet module IP address source address
   packet[ip_srcaddr]=ipaddri[0];
   packet[ip_srcaddr+1]=ipaddri[1];
   //move hardware source address to destinatin address
   packet[enetpacketDest01]=packet[enetpacketSrc01];
   packet[enetpacketDest23]=packet[enetpacketSrc23];
   packet[enetpacketDest45]=packet[enetpacketSrc45];
   //make ethernet module mac address the source address
   packet[enetpacketSrc01]=macaddri[0];
   packet[enetpacketSrc23]=macaddri[1];
   packet[enetpacketSrc45]=macaddri[2];

   //calculate the IP header checksum
   packet[ip_hdr_cksum]=0x00;

   hdr_chksum =0;
   hdrlen = (LOW_BYTE(packet[ip_vers_len]) & 0x0F) *4;
   addr = &packet[ip_vers_len];
   cksum();
   packet[ip_hdr_cksum]= ~(hdr_chksum + ((hdr_chksum & 0xFFFF0000) >> 16));
 }
//*************************************************************
//*    CHECKSUM CALCULATION ROUTINE
//*************************************************************
void cksum()
{
```

```
        //hdr_chksum = 0;
     while(hdrlen > 1)
     {
          hdr_chksum = hdr_chksum + *addr++;
        hdrlen -=2;
        }
     if(hdrlen > 0)
        {
        hdr_chksum = hdr_chksum + (*addr & 0x00FF);
     --hdrlen;
        }
}

unsigned int swapbytes(unsigned int val)
{
     char temphi,templo;

     temphi = make8(val,1);
     templo = make8(val,0);
     return(make16(templo,temphi));
}


//***********************************************************
//*    Echo Packet Function
//*    This routine does not modify the incoming packet size and
//*    thus echoes the original packet structure.
//***********************************************************
char echo_packet()
{
     char rc;
     unsigned int TxFID;

     TxFID = get_free_TxFID();
     if(TxFID == 0)
      rc = 1;
     else
      {
   rc = bap_write(TxFID,rxdatalength_offset,(char*)packet,packet[enetpacke
tLen11]+16);
        if(rc=send_command(TransmitReclaim_Cmd,TxFID))
             printf("\r\nTransmit failed");

   do_event_housekeeping();
      }
     return(rc);
}

unsigned int swapbytes(unsigned int val)
{
     char temphi,templo;

     temphi = make8(val,1);
```

```
        templo = make8(val,0);
        return(make16(templo,temphi));
}


//********************************************************
//*    TCP Function
//*    This function uses TCP protocol to act as a Telnet server on
//*    port 8088 decimal.  The application function is called with
//*    every incoming character.
//********************************************************
void tcp()
{
        char i,j,k;
        unsigned int temp;

    //ASSEMBLE THE DESTINATION PORT ADDRESS FROM THE INCOMING PACKET
    portaddr = packet[TCP_destport];

    //CALCULATE THE LENGTH OF THE DATA COMING IN WITH THE PACKET
    tcpdatalen_in = (swapbytes(packet[ip_pktlen]))-((packet[ip_vers_len] &
0x0F) * 4)-(((packet[TCP_hdrflags] & 0x00FF) >> 4) * 4);

    //IF AN ACK IS RECEIVED AND THE DESTINATION PORT ADDRESS IS VALID AND NO
DATA IS IN THE PACKET
    if(ACK_IN && portaddr == MY_PORT_ADDRESS && tcpdatalen_in == 0x00)
    {
        //ASSEMBLE THE ACKNOWLEDGMENT NUMBER FROM THE INCOMING PACKET
        incoming_ack =make32i(packet[TCP_acknum],packet[TCP_acknum+1]);

        //IF THE INCOMING PACKET IS A RESULT OF SESSION ESTABLISHMENT
        if(bsynflag)
        {
            //CLEAR THE SYN FLAG
            clr_synflag;

            //THE INCOMING ACKNOWLEDGMENT IS MY NEW SEQUENCE NUMBER
            my_seqnum = incoming_ack;

            //SEND THE TELNET SERVER BANNER
            //LIMIT THE CHARACTER COUNT TO 40 DECIMAL
            j = sizeof(telnet_banner);
            //LENGTH OF THE BANNER MESSAGE
            tcpdatalen_out = j;
            i = 0;
            k=0;
            do{
              packet[TCP_data+k] = (telnet_banner[i+1] << 8) | telnet_banner[i];
              i+=2;
              ++k;
            }while(j-- > 1);
            do{
              packet[TCP_data+k] = telnet_banner[i];
```

```
        }while(j-- > 0);

        //EXPECT TO GET AN ACKNOWLEDGMENT OF THE BANNER MESSAGE
        expected_ack = my_seqnum + swapbytes(tcpdatalen_out);

        //SEND THE TCP/IP PACKET
        send_tcp_packet();
    }
}

        //EXPECT TO GET AN ACKNOWLEDGMENT OF THE BANNER MESSAGE
        expected_ack = my_seqnum + swapbytes(tcpdatalen_out);

        //SEND THE TCP/IP PACKET
        send_tcp_packet();
    }
}

    //IF AN ACK IS RECEIVED AND THE PORT ADDRESS IS VALID AND THERE IS DATA
IN THE INCOMING PACKET
    if(ACK_IN && portaddr == MY_PORT_ADDRESS && tcpdatalen_in)
    {
        //RECEIVE THE DATA AND PUT IT INTO THE INCOMING DATA BUFFER
        temp = tcpdatalen_in;
        i=0;
    while(temp > 1)
        {
            aux_data[i] = LOW_BYTE(packet[TCP_data+i]);
            aux_data[i+1] = HIGH_BYTE(packet[TCP_data+i]);
            i+=2;
            temp-=2;
        }
        while(temp > 0)
        {
            aux_data[i] = LOW_BYTE(packet[TCP_data+i]);
            --temp;
        }
        //RUN THE TCP APPLICATION
        application_code();

    //ASSEMBLE THE ACKNOWLEDGMENT NUMBER FROM THE INCOMING PACKET
    incoming_ack =make32i(packet[TCP_acknum],packet[TCP_acknum+1]);

    //CHECK FOR THE NUMBER OF BYTES ACKNOWLEDGED
    //DETERMINE HOW MANY BYTES ARE OUTSTANDING AND ADJUST THE OUTGOING SE-
QUENCE NUMBER ACCORDINGLY
        if(incoming_ack <= expected_ack)
            my_seqnum = expected_ack - (expected_ack - incoming_ack);

    //MY EXPECTED ACKNOWLEDGMENT NUMBER
    expected_ack = my_seqnum + swapbytes(tcpdatalen_out);
```

```
        send_tcp_packet();
    }

    //this code segment processes the incoming SYN from the Telnet client
    //and sends back the initial sequence number (ISN) and acknowledges
    //the incoming SYN packet
    if(SYN_IN && portaddr == MY_PORT_ADDRESS)
    {
        tcpdatalen_in = 0x01;
        set_synflag;

        setipaddrs();

        //SWAP THE TCP SOURCE AND DESTINATION PORTS
        temp = packet[TCP_srcport];
        packet[TCP_srcport] = packet[TCP_destport];
        packet[TCP_destport] = temp;

        //ASSEMBLE THE ACKNOWLEDGMENT
        client_seqnum=make32i(packet[TCP_seqnum],packet[TCP_seqnum+1]);
        client_seqnum = client_seqnum + swapbytes(tcpdatalen_in);
        packet[TCP_acknum] = (client_seqnum & 0xFFFF0000) >> 16;
        packet[TCP_acknum+1] = client_seqnum & 0x0000FFFF;

        if(++ISN == 0x0000 || ++ISN == 0xFFFF)
            ISN = 0x1234;
        my_seqnum = 0xFFFF3412;

        //set_packet32(TCP_seqnum,my_seqnum);
            packet[TCP_seqnum] = my_seqnum & 0x0000FFFF;
            packet[TCP_seqnum+1] = (my_seqnum & 0xFFFF0000) >> 16;
        packet[TCP_hdrflags] &= 0xC0FF;
        SYN_OUT;
        ACK_OUT;

        packet[TCP_cksum] = 0x00;
        //packet[TCP_cksum+1] = 0x00;

        hdr_chksum =0;
        hdrlen = 0x08;
        addr = &packet[ip_srcaddr];
        cksum();
        hdr_chksum = hdr_chksum + 256*HIGH_BYTE(packet[ip_ttlproto]);
        tcplen = swapbytes(packet[ip_pktlen]) - (LOW_BYTE(packet[ip_vers_len])
& 0x0F) *4;
        hdr_chksum = hdr_chksum + swapbytes(tcplen);
        hdrlen = tcplen;
        addr = &packet[TCP_srcport];
        cksum();
        packet[TCP_cksum]= ~(hdr_chksum + ((hdr_chksum & 0xFFFF0000) >> 16));
        echo_packet();
    }
```

```
//this code segment processes a FIN from the Telnet client
//and acknowledges the FIN and any incoming data.
if(FIN_IN && portaddr == MY_PORT_ADDRESS)
{
    if(tcpdatalen_in)
    {
        for(i=0;i<tcpdatalen_in;++i)
        {
            aux_data[i] = packet[TCP_data+i];
            application_code();
        }
    }

    set_finflag;

    ++tcpdatalen_in;

    incoming_ack =make32i(packet[TCP_acknum],packet[TCP_acknum+1]);
    if(incoming_ack <= expected_ack)
        my_seqnum = expected_ack - (expected_ack - incoming_ack);

    expected_ack = my_seqnum + swapbytes(tcpdatalen_out);
    send_tcp_packet();

}
}
```
★★★★★★★★★★★★★★★★★★★★★★★★★★★★★

Code Snippet 14.2: This is one big and seemingly confusing jumble of code. However, just like anything else that appears to be difficult, the mess of code is better understood when broken down into smaller more manageable chunks.

I opened up a Telnet session on SNOOPER with the following:

Microsoft Telnet> open 192.168.0.151 8088

Screen Capture 14.1: The Telnet application you see running in the shot is part of Microsoft Windows XP Professional, which is running on SNOOPER.

After pressing SNOOPER's ENTER key, I was welcomed with the following screen.

Screen Capture 14.2: This is the AirDrop module's response to the initiation of a TCP/IP session by the SNOOPER laptop. The banner AirDrop-P was generated by the AirDrop-P module's TCP/IP application code contained within the AirDrop 802.11b driver.

As I typed, each character I entered was echoed by the AirDrop-P module.

Screen Capture 14.3: The echoed message is "edtp".

So much for the human Telnet application play-by-play. Let's look at what the Netasyst Sniffer and the PIC18LF8621 saw.

Sniffer Text 14.1
```
DLC: ----- DLC Header -----
      DLC:
      DLC: Frame 43 arrived at 10:15:03.0965; frame size is 80 (0050 hex) bytes.
      DLC: Signal level               = 100%
      DLC: Channel                    = 1
      DLC: Data rate                  = 22 (11.0 Megabits per second)
      DLC:
```

```
      DLC: Frame Control Field #1 = 08
      DLC:                 .... ..00 = 0x0 Protocol Version
      DLC:                 .... 10.. = 0x2 Data Frame
      DLC:              0000 .... = 0x0 Data (Subtype)
      DLC: Frame Control Field #2 = 02
      DLC:                 .... ...0 = Not to Distribution System
      DLC:                 .... ..1. = From Distribution System
      DLC:                 .... .0.. = Last fragment
      DLC:                 .... 0... = Not retry
      DLC:              ...0 .... = Active Mode
      DLC:              ..0. .... = No more data
      DLC:              .0.. .... = Wired Equivalent Privacy is off
      DLC:              0... .... = Not ordered
      DLC: Duration                    = 149 (in microseconds)
      DLC: Destination Address         = Station AbocomBC5967
      DLC: Basic Service Set ID        = Station Netgea6FD3DA
      DLC: Source address              = Station Xircom03D92F
      DLC: Sequence Control            = 0xCB60
      DLC: ...Sequence Number          = 0xCB6 (3254)
      DLC: ...Fragment Number          = 0x0   (0)
      DLC:
LLC:  ----- LLC Header -----
      LLC:
      LLC:  DSAP Address = AA, DSAP IG Bit = 00 (Individual Address)
      LLC:  SSAP Address = AA, SSAP CR Bit = 00 (Command)
      LLC:  Unnumbered frame: UI
      LLC:
SNAP: ----- SNAP Header -----
      SNAP:
      SNAP: Type     = 0800 (IP)
      SNAP:
IP: ----- IP Header -----
      IP:
      IP: Version = 4, header length = 20 bytes
      IP: Type of service = 00
      IP:        000. .... = routine
      IP:        ...0 .... = normal delay
      IP:        .... 0... = normal throughput
      IP:        .... .0.. = normal reliability
      IP:        .... ..0. = ECT bit - transport protocol will ignore the CE bit
      IP:        .... ...0 = CE bit - no congestion
      IP: Total length   = 48 bytes
      IP: Identification = 49951
      IP: Flags          = 4X
      IP:       .1.. .... = don't fragment
      IP:       ..0. .... = last fragment
      IP: Fragment offset = 0 bytes
      IP: Time to live    = 128 seconds/hops
      IP: Protocol        = 6 (TCP)
      IP: Header checksum = B5B5 (correct)
      IP: Source address      = [192.168.0.11]
      IP: Destination address = [192.168.0.151]
```

```
          IP: No options
          IP:
TCP: ----- TCP header -----
          TCP:
          TCP: Source port              =  1859
          TCP: Destination port         =  8088
          TCP: Initial sequence number = 3238359086
          TCP: Next expected Seq number= 3238359087
          TCP: Data offset              = 28 bytes (4 bits)
          TCP: Reserved Bits: Reserved for Future Use (6 bits)
          TCP: Flags                     = 02
          TCP:                  ..0. .... = (No urgent pointer)
          TCP:                  ...0 .... = (No acknowledgment)
          TCP:                  .... 0... = (No push)
          TCP:                  .... .0.. = (No reset)
          TCP:                  .... ..1. = SYN
          TCP:                  .... ...0 = (No FIN)
          TCP: Window                    = 16384
          TCP: Checksum                  = 691D (correct)
          TCP: Urgent pointer            = 0
          TCP:
          TCP: Options follow
          TCP: Maximum segment size = 1460
          TCP: No-Operation
          TCP: No-Operation
          TCP: SACK-Permitted Option
          TCP:
ADDR  HEX                                                        ASCII
0000: 08 02 95 00 00 e0 98 bc 59 67 00 09 5b 6f d3 da | ..•...à˜¼Yg..[oÓÚ
0010: 00 05 3c 03 d9 2f 60 cb aa aa 03 00 00 00 08 00 | ..<.Ù/`Ëªª......
0020: 45 00 00 30 c3 1f 40 00 80 06 b5 b5 c0 a8 00 0b | E..0Ã.@.€.µµÀ¨..
0030: c0 a8 00 97 07 43 1f 98 c1 05 70 2e 00 00 00 00 | À¨.—.C.˜Á.p.....
0040: 70 02 40 00 69 1d 00 00 02 04 05 b4 01 01 04 02 | p.@.i......´....
```

Sniffer Text 14.1: I'm sure by now that you are an expert IP header marksman. However, there's a new internet protocol sheriff in town calling himself "TCP" and he, along with the good doctor, "IP", intend to keep the townsfolk orderly.

Although TCP/IP looks to be a buggerbear, it is treated like all of the other internet protocols in the DLC Header area. In the TCP/IP discussions that will follow, I'll show all of the Netasyst Sniffer captures as they leave the AP (From Distribution System). As you can see in Mini Sniff 14.1, the first TCP/IP frame fired with intention originates from the SNOOPER laptop and is aimed at the AirDrop-P module.

Mini Sniff 14.1

```
DLC: ----- DLC Header -----
          DLC:
          DLC: Frame 43 arrived at 10:15:03.0965; frame size is 80 (0050 hex) bytes.
          DLC: Signal level             = 100%
          DLC: Channel                  = 1
```

```
       DLC: Data rate                           = 22 (11.0 Megabits per second)
       DLC:
       DLC: Frame Control Field #1 = 08
       DLC:                  .... ..00 = 0x0 Protocol Version
       DLC:                  .... 10.. = 0x2 Data Frame
       DLC:                  0000 .... = 0x0 Data (Subtype)
       DLC: Frame Control Field #2 = 02
       DLC:                  .... ...0 = Not to Distribution System
       DLC:                  .... ..1. = From Distribution System
       DLC:                  .... .0.. = Last fragment
       DLC:                  .... 0... = Not retry
       DLC:                  ...0 .... = Active Mode
       DLC:                  ..0. .... = No more data
       DLC:                  .0.. .... = Wired Equivalent Privacy is off
       DLC:                  0... .... = Not ordered
       DLC: Duration                            = 149 (in microseconds)
       DLC: Destination Address                 = Station AbocomBC5967
       DLC: Basic Service Set ID                = Station Netgea6FD3DA
       DLC: Source address                      = Station Xircom03D92F
       DLC: Sequence Control                    = 0xCB60
       DLC: ...Sequence Number                  = 0xCB6 (3254)
       DLC: ...Fragment Number                  = 0x0   (0)
ADDR   HEX                                                    ASCII
0000:  08 02 95 00 00 e0 98 bc 59 67 00 09 5b 6f d3 da | ..•..à˜¼Yg..[oÓÚ
0010:  00 05 3c 03 d9 2f 60 cb aa aa 03 00 00 00 08 00 | ..<.Ù/`Ëªª......
0020:  45 00 00 30 c3 1f 40 00 80 06 b5 b5 c0 a8 00 0b | E..0Ã.@.€.µµÀ"..
0030:  c0 a8 00 97 07 43 1f 98 c1 05 70 2e 00 00 00 00 | À"–.C.˜Á.p.....
0040:  70 02 40 00 69 1d 00 00 02 04 05 b4 01 01 04 02 | p.@.i......´....
```

There's nothing new in Mini Sniff 14.2 except the IP Protocol field, which is telling us that the IP payload is a TCP segment.

Mini Sniff 14.2

```
LLC:    ----- LLC Header -----
       LLC:
       LLC:  DSAP Address = AA, DSAP IG Bit = 00 (Individual Address)
       LLC:  SSAP Address = AA, SSAP CR Bit = 00 (Command)
       LLC:  Unnumbered frame: UI
       LLC:
SNAP:   ----- SNAP Header -----
       SNAP:
       SNAP: Type       = 0800 (IP)
       SNAP:
IP: ----- IP Header -----
       IP:
       IP: Version = 4, header length = 20 bytes
       IP: Type of service = 00
       IP:         000. ....    = routine
       IP:         ...0 ....    = normal delay
       IP:         .... 0...    = normal throughput
       IP:         .... .0..    = normal reliability
```

```
IP:           .... ..0. = ECT bit - transport protocol will ignore the CE bit
IP:           .... ...0 = CE bit - no congestion
IP: Total length      = 48 bytes
IP: Identification    = 49951
IP: Flags             = 4X
IP:          .1.. .... = don't fragment
IP:          ..0. .... = last fragment
IP: Fragment offset   = 0 bytes
IP: Time to live      = 128 seconds/hops
IP: Protocol          = 6 (TCP)
IP: Header checksum   = B5B5 (correct)
IP: Source address      = [192.168.0.11]
IP: Destination address = [192.168.0.151]
IP: No options
ADDR  HEX                                                   ASCII
0000: 08 02 95 00 00 e0 98 bc 59 67 00 09 5b 6f d3 da | ..•...à~¼Yg..[oÓÚ
0010: 00 05 3c 03 d9 2f 60 cb aa aa 03 00 00 00 08 00 | ..<.Ù/`Ëªª......
0020: 45 00 00 30 c3 1f 40 00 80 06 b5 b5 c0 a8 00 0b | E..0Ã.@.€.µµÀ¨..
0030: c0 a8 00 97 07 43 1f 98 c1 05 70 2e 00 00 00 00 | À¨.—.C.˜Á.p.....
0040: 70 02 40 00 69 1d 00 00 02 04 05 b4 01 01 04 02 | p.@.i......´....
```

There's lots of brand new stuff here. So, before I take you through each of the fields in the TCP header, I'll put up a TCP header graphic and the AirDrop 802.11b driver TCP header layout so you can match up the Netasyst Sniffer fields with the AirDrop 802.11b driver TCP header layout.

Before we start our indepth look at the TCP header fields, which are shown from the AirDrop 802.11b driver perspective in Sub Snippet 14.3, there are some things I would like to point out to you. The Reserved and ECN fields are combined into a single 6-bit field by the Netasyst Sniffer. If you read the contents of RFC 3540, you'll find that the ECN (Explicit Congestion Notification) field is 3 bits in length. That leaves a Reserved area that is only 3 bits long. The ECN field is optional and is intended to be used to protect against accidental or malicious concealment of marked packets from the TCP sender. What that really means is that ECN is a methodology that can help prevent intentionally or accidentally greedy stations from hogging too much bandwidth. The reason the Netasyst Sniffer doesn't tag the ECN field is because ECN isn't widely deployed. And, if you're trying to match up the Netasyst Sniffer's *Next expected Seq number* field, forget about it. That field is a feature of the Netasyst Sniffer. What it's telling you is that the next incoming TCP segment should have the sequence number contained within the *Next expected Seq number* field. Finally, note the absence of the Acknowledgment Number field in the Netasyst Sniffer capture. That's because the *Acknowledgement Flag* is not set. The Acknowledgment field is where it's supposed to be in the header and it is filled with zeroes. I've highlighted the "missing" Acknowledgment field in the Mini Sniff 14.3 capture dump area.

Mini Sniff 14.3

TCP HEADER LAYOUT					
Source Port				Destination Port	
Sequence Number					
Acknowledgment Number					
Data Offset	Reserved	ECN	Flags	Window	
Checksum				Urgent Pointer	
Options and Padding					
Data					

```
TCP: ----- TCP header -----
     TCP:
     TCP: Source port             = 1859
     TCP: Destination port        = 8088
     TCP: Initial sequence number = 3238359086
     TCP: Next expected Seq number= 3238359087
     TCP: Data offset             = 28 bytes (4 bits)
     TCP: Reserved Bits: Reserved for Future Use (6 bits)
     TCP: Flags                   = 02
     TCP:                ..0. .... = (No urgent pointer)
     TCP:                ...0 .... = (No acknowledgment)
     TCP:                .... 0... = (No push)
     TCP:                .... .0.. = (No reset)
     TCP:                .... ..1. = SYN
     TCP:                .... ...0 = (No FIN)
     TCP: Window                  = 16384
     TCP: Checksum                = 691D (correct)
     TCP: Urgent pointer          = 0
     TCP:
     TCP: Options follow
     TCP: Maximum segment size = 1460
     TCP: No-Operation
     TCP: No-Operation
     TCP: SACK-Permitted Option
ADDR  HEX                                               ASCII
0000: 08 02 95 00 00 e0 98 bc 59 67 00 09 5b 6f d3 da | ..•..à~¼Yg..[oÓÚ
0010: 00 05 3c 03 d9 2f 60 cb aa aa 03 00 00 00 08 00 | ..<.Ù/`Ëªª......
0020: 45 00 00 30 c3 1f 40 00 80 06 b5 b5 c0 a8 00 0b | E..0Ã.@.€.µµÀ¨..
0030: c0 a8 00 97 07 43 1f 98 c1 05 70 2e 00 00 00 00 | À¨.—.C.˜Á.p.....
0040: 70 02 40 00 69 1d 00 00 02 04 05 b4 01 01 04 02 | p.@.i......´....
```

Sub Snippet 14.3

```
**************************************************************

//*************************************************************
//*    TCP Header Layout
//*************************************************************
#define TCP_srcport       0x16    //TCP source port
#define TCP_destport      0x17    //TCP destination port
#define TCP_seqnum        0x18    //sequence number
//                        0x19
#define TCP_acknum        0x1A    //acknowledgment number
//                        0x1B
#define TCP_hdrflags      0x1C    //4-bit header len and flags
#define TCP_window        0x1D    //window size
#define TCP_cksum         0x1E    //TCP checksum
#define TCP_urgentptr     0x1F    //urgent pointer
#define TCP_data          0x20    //option/data


**************************************************************
```

The *Source port* value in Mini Sniff 14.4 represents the SNOOPER laptop's TCP port. Since SNOOPER is running XP Professional, we must assume that the Windows Telnet application chose this 16-bit port number.

Mini Sniff 14.4

```
TCP: ----- TCP header -----
     TCP:
     TCP: Source port              = 1859
     TCP: Destination port         = 8088
     TCP: Initial sequence number = 3238359086
     TCP: Next expected Seq number= 3238359087
     TCP: Data offset              = 28 bytes (4 bits)
     TCP: Reserved Bits: Reserved for Future Use (6 bits)
     TCP: Flags                    = 02
     TCP:                   ..0. .... = (No urgent pointer)
     TCP:                   ...0 .... = (No acknowledgment)
     TCP:                   .... 0... = (No push)
     TCP:                   .... .0.. = (No reset)
     TCP:                   .... ..1. = SYN
     TCP:                   .... ...0 = (No FIN)
     TCP: Window                   = 16384
     TCP: Checksum                 = 691D (correct)
     TCP: Urgent pointer           = 0
     TCP:
     TCP: Options follow
     TCP: Maximum segment size = 1460
     TCP: No-Operation
     TCP: No-Operation
     TCP: SACK-Permitted Option
     TCP:
```

```
ADDR   HEX                                                       ASCII
0000:  08 02 95 00 00 e0 98 bc 59 67 00 09 5b 6f d3 da  | ..•..à~¼Yg..[oÓÚ
0010:  00 05 3c 03 d9 2f 60 cb aa aa 03 00 00 00 08 00  | ..<.Ù/`Ëªª......
0020:  45 00 00 30 c3 1f 40 00 80 06 b5 b5 c0 a8 00 0b  | E..0Ã.@.€.µµÀ"..
0030:  c0 a8 00 97 07 43 1f 98 c1 05 70 2e 00 00 00 00  | À".−.C.~Á.p.....
0040:  70 02 40 00 69 1d 00 00 02 04 05 b4 01 01 04 02  | p.@.i......´....
```

There's no doubt about where the 16-bit *Destination port* value originated. We hard-coded the TCP port value of 8088 decimal (0x1F98) highlighted in Mini Sniff 14.5 in the AirDrop 802.11b driver using the AirDrop 802.11b driver code in Sub Snippet 14.4.

Mini Sniff 14.5

```
TCP: ----- TCP header -----
TCP:
TCP: Source port            =  1859
TCP: Destination port       =  8088
TCP: Initial sequence number = 3238359086
TCP: Next expected Seq number= 3238359087
TCP: Data offset            = 28 bytes (4 bits)
TCP: Reserved Bits: Reserved for Future Use (6 bits)
TCP: Flags                  = 02
TCP:                  ..0. .... = (No urgent pointer)
TCP:                  ...0 .... = (No acknowledgment)
TCP:                  .... 0... = (No push)
TCP:                  .... .0.. = (No reset)
TCP:                  .... ..1. = SYN
TCP:                  .... ...0 = (No FIN)
TCP: Window                 = 16384
TCP: Checksum               = 691D (correct)
TCP: Urgent pointer         = 0
TCP:
TCP: Options follow
TCP: Maximum segment size = 1460
TCP: No-Operation
TCP: No-Operation
TCP: SACK-Permitted Option
TCP:
ADDR   HEX                                                       ASCII
0000:  08 02 95 00 00 e0 98 bc 59 67 00 09 5b 6f d3 da  | ..•..à~¼Yg..[oÓÚ
0010:  00 05 3c 03 d9 2f 60 cb aa aa 03 00 00 00 08 00  | ..<.Ù/`Ëªª......
0020:  45 00 00 30 c3 1f 40 00 80 06 b5 b5 c0 a8 00 0b  | E..0Ã.@.€.µµÀ"..
0030:  c0 a8 00 97 07 43 1f 98 c1 05 70 2e 00 00 00 00  | À".−.C.~Á.p.....
0040:  70 02 40 00 69 1d 00 00 02 04 05 b4 01 01 04 02  | p.@.i......´....
```

Sub Snippet 14.4

```
*********************************************************
```

```
//*********************************************************
//*     Port Definitions
//*     This address is used by TCP and the Telnet function.
//*     This can be changed to any valid port number as long as
//*     you modify your code to recognize the new port number.
//*********************************************************
#define  MY_PORT_ADDRESS      0x981F  // 8088 DECIMAL
```

```
*********************************************************
```

The *Initial sequence number* shown in Mini Sniff 14.6 is a randomly generated 32-bit number. In this case, SNOOPER is attempting to establish a TCP/IP connection with the Air-Drop-P module. This is SNOOPER's up-front way of communicating its sequence number to the AirDrop-P module.

Mini Sniff 14.6

```
TCP: ----- TCP header -----
     TCP:
     TCP: Source port            =  1859
     TCP: Destination port       =  8088
     TCP: Initial sequence number = 3238359086
     TCP: Next expected Seq number= 3238359087
     TCP: Data offset             = 28 bytes (4 bits)
     TCP: Reserved Bits: Reserved for Future Use (6 bits)
     TCP: Flags                   = 02
     TCP:                 ..0. .... = (No urgent pointer)
     TCP:                 ...0 .... = (No acknowledgment)
     TCP:                 .... 0... = (No push)
     TCP:                 .... .0.. = (No reset)
     TCP:                 .... ..1. = SYN
     TCP:                 .... ...0 = (No FIN)
     TCP: Window                  = 16384
     TCP: Checksum                = 691D (correct)
     TCP: Urgent pointer          = 0
     TCP:
     TCP: Options follow
     TCP: Maximum segment size = 1460
     TCP: No-Operation
     TCP: No-Operation
     TCP: SACK-Permitted Option
     TCP:
ADDR  HEX                                                ASCII
0000: 08 02 95 00 00 e0 98 bc 59 67 00 09 5b 6f d3 da | ...•...à~¼Yg..[oÓÚ
0010: 00 05 3c 03 d9 2f 60 cb aa aa 03 00 00 00 08 00 | ..<.Ù/`Ëªª......
0020: 45 00 00 30 c3 1f 40 00 80 06 b5 b5 c0 a8 00 0b | E..0Ã.@.€.µµÀ¨..
0030: c0 a8 00 97 07 43 1f 98 c1 05 70 2e 00 00 00 00 | À¨.—.C.~Á.p.....
0040: 70 02 40 00 69 1d 00 00 02 04 05 b4 01 01 04 02 | p.@.i......´....
```

Without any added complications, the TCP header is interesting enough on its own. Many of the TCP header fields play on other TCP header fields. The relationships between dependent TCP header fields are very important and we must pay particular attention to them in the AirDrop 802.11b driver. One of those "interesting" TCP header fields is the *Data offset* field, which I've illuminated in Mini Sniff 14.7. The Data offset field is 4 bits long and tells us where the data begins within the TCP header structure in 32-bit (4-byte) chunks. The current *Data offset* value is 7. That should equate to seven 4-byte slices (28 bytes) sitting in front of the TCP header data field. Let's check the Netasyst Sniffer's work:

Mini Sniff 14.7

TCP Header Data Offset Calculation	
TCP Header Field	**Field Length (Bytes)**
Source Port	2
Destination Port	2
Initial Sequence Number	4
Acknowledgment Number	4
Data Offset/Reserved/ECN/Flags	2
Window	2
Checksum	2
Urgent Pointer	2
Options and Padding	8
Data Starts Here	**28**

```
TCP: ----- TCP header -----
     TCP:
     TCP: Source port          =  1859
     TCP: Destination port     =  8088
     TCP: Initial sequence number = 3238359086
     TCP: Next expected Seq number= 3238359087
     TCP: Data offset           = 28 bytes (4 bits)
     TCP: Reserved Bits: Reserved for Future Use (6 bits)
     TCP: Flags                 = 02
     TCP:                ..0. .... = (No urgent pointer)
     TCP:                ...0 .... = (No acknowledgment)
     TCP:                .... 0... = (No push)
     TCP:                .... .0.. = (No reset)
     TCP:                .... ..1. = SYN
     TCP:                .... ...0 = (No FIN)
     TCP: Window                = 16384
     TCP: Checksum              = 691D (correct)
     TCP: Urgent pointer        = 0
     TCP:
     TCP: Options follow
     TCP: Maximum segment size = 1460
     TCP: No-Operation
     TCP: No-Operation
```

```
      TCP: SACK-Permitted Option
      TCP:
ADDR  HEX                                                                    ASCII
0000: 08 02 95 00 00 e0 98 bc 59 67 00 09 5b 6f d3 da | ..•..à˜¼Yg..[oóÚ
0010: 00 05 3c 03 d9 2f 60 cb aa aa 03 00 00 00 08 00 | ..<.Ù/`Ëªª......
0020: 45 00 00 30 c3 1f 40 00 80 06 b5 b5 c0 a8 00 0b | E..0Ã.@.€.µµÀ"..
0030: c0 a8 00 97 07 43 1f 98 c1 05 70 2e 00 00 00 00 | À".—.C.˜Á.p.....
0040: 70 02 40 00 69 1d 00 00 02 04 05 b4 01 01 04 02 | p.@.i......´....
```

Just when you thought you had a handle on what the TCP header is all about, it gets more interesting. We just manually counted the number of bytes to the data area of the TCP header to double check a machine. The funny thing is that there is not data in the data area of our TCP header. In that this is a TCP initialization segment, the data lies in the *Flags* field. The SYN (Synchronize sequence numbers) flag is set in Mini Sniff 14.8 and is considered data in this mode. The AirDrop 802.11b driver must acknowledge any data that arrives in the TCP segment. In this case, that data is one byte of SYN flag. The AirDrop-P had better acknowledge that one byte in its response. That's why the Netasyst Sniffer set the *Next expected Seq number* field one byte beyond the *Initial sequence number field*. The TCP header *Flags* are integral to the proper operation of the TCP/IP mechanism and the AirDrop 802.11b driver must keep a close watch on them.

Mini Sniff 14.8

```
TCP: ----- TCP header -----
     TCP:
     TCP: Source port           =  1859
     TCP: Destination port      =  8088
     TCP: Initial sequence number = 3238359086
     TCP: Next expected Seq number= 3238359087
     TCP: Data offset            = 28 bytes (4 bits)
     TCP: Reserved Bits: Reserved for Future Use (6 bits)
     TCP: Flags                  = 02
     TCP:                ..0. .... = (No urgent pointer)
     TCP:                ...0 .... = (No acknowledgment)
     TCP:                .... 0... = (No push)
     TCP:                .... .0.. = (No reset)
     TCP:                .... ..1. = SYN
     TCP:                .... ...0 = (No FIN)
     TCP: Window                 = 16384
     TCP: Checksum               = 691D (correct)
     TCP: Urgent pointer         = 0
     TCP:
     TCP: Options follow
     TCP: Maximum segment size = 1460
     TCP: No-Operation
     TCP: No-Operation
     TCP: SACK-Permitted Option
     TCP:
```

```
ADDR   HEX                                                          ASCII
0000:  08 02 95 00 00 e0 98 bc 59 67 00 09 5b 6f d3 da  | ..•..à˜¼Yg..[oÓÚ
0010:  00 05 3c 03 d9 2f 60 cb aa aa 03 00 00 00 08 00  | ..<.Ù/`Ëªª......
0020:  45 00 00 30 c3 1f 40 00 80 06 b5 b5 c0 a8 00 0b  | E..0Ã.@.€.µµÀ"..
0030:  c0 a8 00 97 07 43 1f 98 c1 05 70 2e 00 00 00 00  | À".—.C.˜Á.p.....
0040:  70 02 40 00 69 1d 00 00 02 04 05 b4 01 01 04 02  | p.@.i......´....
```

Don't get real excited about the rest of the TCP header you see in Mini Sniff 14.9. The *Window* number tells the AirDrop 802.11b driver how many bytes the sender of this TCP segment is willing to accept. That's nice. The AirDrop 802.11b driver ignores this field.

Mini Sniff 14.9

```
TCP: ----- TCP header -----
     TCP:
     TCP: Source port            =  1859
     TCP: Destination port       =  8088
     TCP: Initial sequence number = 3238359086
     TCP: Next expected Seq number= 3238359087
     TCP: Data offset            = 28 bytes (4 bits)
     TCP: Reserved Bits: Reserved for Future Use (6 bits)
     TCP: Flags                  = 02
     TCP:                ..0. .... = (No urgent pointer)
     TCP:                ...0 .... = (No acknowledgment)
     TCP:                .... 0... = (No push)
     TCP:                .... .0.. = (No reset)
     TCP:                .... ..1. = SYN
     TCP:                .... ...0 = (No FIN)
     TCP: Window                 = 16384
     TCP: Checksum               = 691D (correct)
     TCP: Urgent pointer         = 0
     TCP:
     TCP: Options follow
     TCP: Maximum segment size = 1460
     TCP: No-Operation
     TCP: No-Operation
     TCP: SACK-Permitted Option
     TCP:
ADDR   HEX                                                          ASCII
0000:  08 02 95 00 00 e0 98 bc 59 67 00 09 5b 6f d3 da  | ..•..à˜¼Yg..[oÓÚ
0010:  00 05 3c 03 d9 2f 60 cb aa aa 03 00 00 00 08 00  | ..<.Ù/`Ëªª......
0020:  45 00 00 30 c3 1f 40 00 80 06 b5 b5 c0 a8 00 0b  | E..0Ã.@.€.µµÀ"..
0030:  c0 a8 00 97 07 43 1f 98 c1 05 70 2e 00 00 00 00  | À".—.C.˜Á.p.....
0040:  70 02 40 00 69 1d 00 00 02 04 05 b4 01 01 04 02  | p.@.i......´....
```

What would we do if we didn't have to compute or verify a checksum. Rest assured that we will be computing a value for this checksum field in our outgoing TCP segments just as it was done for the incoming TCP segment in Mini Sniff 14.10.

Mini Sniff 14.10

```
TCP: ----- TCP header -----
       TCP:
       TCP: Source port          =  1859
       TCP: Destination port     =  8088
       TCP: Initial sequence number = 3238359086
       TCP: Next expected Seq number= 3238359087
       TCP: Data offset          = 28 bytes (4 bits)
       TCP: Reserved Bits: Reserved for Future Use (6 bits)
       TCP: Flags               = 02
       TCP:              ..0. .... = (No urgent pointer)
       TCP:              ...0 .... = (No acknowledgment)
       TCP:              .... 0... = (No push)
       TCP:              .... .0.. = (No reset)
       TCP:              .... ..1. = SYN
       TCP:              .... ...0 = (No FIN)
       TCP: Window              = 16384
       TCP: Checksum            = 691D (correct)
       TCP: Urgent pointer      = 0
       TCP:
       TCP: Options follow
       TCP: Maximum segment size = 1460
       TCP: No-Operation
       TCP: No-Operation
       TCP: SACK-Permitted Option
       TCP:
ADDR   HEX                                                    ASCII
0000:  08 02 95 00 00 e0 98 bc 59 67 00 09 5b 6f d3 da  |  ..•..à˜¼Yg..[oÓÚ
0010:  00 05 3c 03 d9 2f 60 cb aa aa 03 00 00 00 08 00  |  ..<.Ù/`Ëªª......
0020:  45 00 00 30 c3 1f 40 00 80 06 b5 b5 c0 a8 00 0b  |  E..0Ã.@.€.µµÀ¨..
0030:  c0 a8 00 97 07 43 1f 98 c1 05 70 2e 00 00 00 00  |  À¨.—.C.˜Á.p.....
0040:  70 02 40 00 69 1d 00 00 02 04 05 b4 01 01 04 02  |  p.@.i.....´....
```

The *Urgent pointer* field in Mini Sniff 14.11 only comes into play if the *URG* bit is set in the *Flags* bit field. It's a TCP family thing. Again, that's nice. The AirDrop 802.11b driver ignores this field too.

Mini Sniff 14.11

```
TCP: ----- TCP header -----
       TCP:
       TCP: Source port          =  1859
       TCP: Destination port     =  8088
       TCP: Initial sequence number = 3238359086
       TCP: Next expected Seq number= 3238359087
       TCP: Data offset          = 28 bytes (4 bits)
       TCP: Reserved Bits: Reserved for Future Use (6 bits)
       TCP: Flags               = 02
       TCP:              ..0. .... = (No urgent pointer)
       TCP:              ...0 .... = (No acknowledgment)
       TCP:              .... 0... = (No push)
       TCP:              .... .0.. = (No reset)
```

```
TCP:                       .... ..1. = SYN
TCP:                       .... ...0 = (No FIN)
TCP: Window                 = 16384
TCP: Checksum               = 691D (correct)
TCP: Urgent pointer         = 0
TCP:
TCP: Options follow
TCP: Maximum segment size = 1460
TCP: No-Operation
TCP: No-Operation
TCP: SACK-Permitted Option
TCP:
```

```
ADDR  HEX                                              ASCII
0000: 08 02 95 00 00 e0 98 bc 59 67 00 09 5b 6f d3 da | ..•..à~¼Yg..[oÓÚ
0010: 00 05 3c 03 d9 2f 60 cb aa aa 03 00 00 00 08 00 | ..<.Ù/`Ëªª......
0020: 45 00 00 30 c3 1f 40 00 80 06 b5 b5 c0 a8 00 0b | E..0Ã.@.€.µµÀ"..
0030: c0 a8 00 97 07 43 1f 98 c1 05 70 2e 00 00 00 00 | À".—.C.~Á.p.....
0040: 70 02 40 00 69 1d 00 00 02 04 05 b4 01 01 04 02 | p.@.i.....´....
```

I considered talking about the Options area in Mini Sniff 14.12. However, why waste your time with something that the AirDrop 802.11b driver ignores anyway. The AirDrop TCP/IP stack is only intended to provide basic functionality. And, we're not here to learn to write full-blown TCP/IP stacks. We're here to get microcontrollers on wireless LANs using 802.11b technology. The key thing to note here is that by design there is no official data area in this TCP segment as called out in the TCP header layout.

Mini Sniff 14.12

```
TCP: ----- TCP header -----
TCP:
TCP: Source port            = 1859
TCP: Destination port       = 8088
TCP: Initial sequence number = 3238359086
TCP: Next expected Seq number= 3238359087
TCP: Data offset            = 28 bytes (4 bits)
TCP: Reserved Bits: Reserved for Future Use (6 bits)
TCP: Flags                  = 02
TCP:                       ..0. .... = (No urgent pointer)
TCP:                       ...0 .... = (No acknowledgment)
TCP:                       .... 0... = (No push)
TCP:                       .... .0.. = (No reset)
TCP:                       .... ..1. = SYN
TCP:                       .... ...0 = (No FIN)
TCP: Window                 = 16384
TCP: Checksum               = 691D (correct)
TCP: Urgent pointer         = 0
TCP:
TCP: Options follow
TCP: Maximum segment size = 1460
TCP: No-Operation
TCP: No-Operation
TCP: SACK-Permitted Option
TCP:
```

```
ADDR  HEX                                              ASCII
0000: 08 02 95 00 00 e0 98 bc 59 67 00 09 5b 6f d3 da | ..•..à˜¼Yg..[oÓÚ
0010: 00 05 3c 03 d9 2f 60 cb aa aa 03 00 00 00 08 00 | ..<.Ù/`Ëªª......
0020: 45 00 00 30 c3 1f 40 00 80 06 b5 b5 c0 a8 00 0b | E..0Ã.@.€.µµÀ˝..
0030: c0 a8 00 97 07 43 1f 98 c1 05 70 2e 00 00 00 00 | À˝.–.C.˜Á.p.....
0040: 70 02 40 00 69 1d 00 00 02 04 05 b4 01 01 04 02 | p.@.i.........´....
```

OK...Now you've got a good idea of what's important and what's not important to us in the TCP header. The Netasyst Sniffer has given us its story. So, let's see what the AirDrop 802.11b driver code has to say about this.

There are some things that we already know about TCP. To establish a TCP/IP session, the originating station starts a process called the 3-way handshake. We know from our study of the fields of the Netasyst Sniffer capture that the first message sent in the 3-way handshake process is a data-less TCP segment with the SYN bit set in the TCP header flags field.

The AirDrop 802.11b driver must always determine if the destination TCP port address is its own. Also, just in case the incoming destination TCP port address is valid for the AirDrop module, the AirDrop 802.11b driver must know how much, if any, data is being carried in the incoming TCP segment. The variable *portaddr* in Sub Snippet 14.5 is loaded with the incoming destination TCP port value. To get what we need to perform our length calculations we must turn to fields in the IP header and the TCP header. The TCP data length is calculated by subtracting the IP header length and TCP header length from the total IP packet length. We already have the IP datagram length in the *ip_pktlen* field of the IP header. The IP header length is calculated using the *ip_vers_len* field of the IP header multiplied times four. In a similar fashion, the high nibble of the *TCP_hdrflags* field in the TCP header is multiplied by four to obtain the TCP header length.

Sub Snippet 14.5
```
*************************************************************

//***********************************************************
//*    TCP Function
//*    This function uses TCP protocol to act as a Telnet server on
//*    port 8088 decimal.  The application function is called with
//*    every incoming character.
//***********************************************************
void tcp()
{
        char i,j,k;
        unsigned int temp;

    //assemble the destination port address from the incoming packet
    portaddr = packet[TCP_destport];

    //calculate the length of the data coming in with the packet
    tcpdatalen_in = (swapbytes(packet[ip_pktlen]))-((packet[ip_vers_len] &
0x0F) * 4)-(((packet[TCP_hdrflags] & 0x00FF) >> 4) * 4);
*************************************************************
```

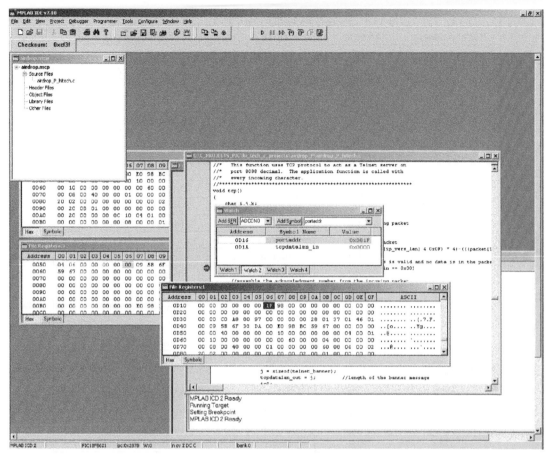

*Screen Capture 14.4: I executed the tcp function code we just discussed and here's what the MPLAB IDE Watch window showed for the variables **portaddr** and **tcpdatalen_in**.*

Part one of the TCP 3-way handshake from the PIC18LF8621's point of view is shown in Array Table 14.1. Let's use the contents of Array Table 14.1 to see how the AirDrop 802.11b driver determined the value of *tcpdatalen_in*.

tcpdatalen_in = 0x0030 – 0x14 – 0x1C = 48 – 20 - 28 = 0

First we apply the *swapbytes* function to the *ip_pktlen* value in AirDrop-P packet buffer slot [13], which results in 0x0030. The *ip_vers_len* contained in AirDrop 802.11b driver slot [12] logically ANDED with 0x0F gives us 0x05 and the result of the logical AND multiplied by 4 yields 20 decimal, or hexadecimal 0x14. Logically ANDING the *Data Offset* value, which is part of the TCP header Flags in AirDrop 802.11b driver slot [28], results in 0x70. Shifting 0x70 right four times results in 0x07. Multiplying the shifted value of 0x07 by four gives us 28 decimal, or 0x1C.

PACKET BUFFER LAYOUT			
Buffer Address	**Array Slot**	**DATA**	**Description**
0x06FA	[0]	0x0038	802.11b packet data length
0x06FC	[1]	0xE000	Destination MAC Address
0x06FE	[2]	0xBC98	
0x0700	[3]	0x6759	
0x0702	[4]	0x0500	Source MAC Address
0x0704	[5]	0x033C	
0x0706	[6]	0x2FD9	
0x0708	[7]	0x3800	802.3 packet data length
0x070A	[8]	0xAAAA	SNAP Header
0x070C	[9]	0x0003	UI/Protocol Identifier
0x070E	[10]	0x0000	Protocol Identifier
0x0710	[11]	0x0008	IP Protocol
0x0712	[12]	0x0045	IP version/header length/TOS
0x0714	[13]	0x3000	IP packet length
0x0716	[14]	0x1FC3	IP datagram id
0x0718	[15]	0x0040	fragment offset
0x071A	[16]	0x0680	time to live/protocol = TCP
0x071C	[17]	0xB5B5	IP header checksum
0x071E	[18]	0xA8C0	Source IP Address
0x0720	[19]	0x0B00	
0x0722	[20]	0xA8C0	Destination IP Address
0x0724	[21]	0x9700	
0x0726	[22]	0x4307	TCP source port
0x0728	[23]	0x981F	TCP destination port
0x072A	[24]	0x05C1	Initial Sequence Number
0x072C	[25]	0x2E70	
0x072E	[26]	0x0000	Acknowledgment Number
0x0730	[27]	0x0000	
0x0732	[28]	0x0270	Data Offset/Reserved/Flags
0x0734	[29]	0x0040	Window
0x0736	[30]	0x1D69	Checksum
0x0738	[31]	0x0000	Urgent Pointer
0x073A	[32]	0x0402	Options
0x073C	[33]	0xB405	
0x073E	[34]	0x0101	
0x0740	[35]	0x0204	

Array Table 14.1: This is a microcontroller view of the first of the 3-way handshake frames. Note that there is no physical data area and that the Acknowledgment double word is "invisible" to the TCP/IP session initialization process at this point. The Acknowledge double word is only valid if the Acknowledge Flag bit is set.

There are four major branch points in the *tcp* function. Three of those branch points look for an active *Acknowledgment Flag* (ACK) or active *FIN Flag* (FIN) bit. Only one of the four *tcp* function branch points check for an active *SYN Flag* bit. I've culled the SYN AirDrop 802.11b driver code into Code Snippet 14.3 so we can look at it line by line.

Code Snippet 14.1
****

```
//this code segment processes the incoming SYN from the Telnet client
//and sends back the initial sequence number (ISN) and acknowledges
//the incoming SYN packet
if(SYN_IN && portaddr == MY_PORT_ADDRESS)
{
    tcpdatalen_in = 0x01;
    set_synflag;

    setipaddrs();

    //swap the TCP source and destination ports
    temp = packet[TCP_srcport];
    packet[TCP_srcport] = packet[TCP_destport];
    packet[TCP_destport] = temp;

    //ASSEMBLE THE ACKNOWLEDGMENT
    client_seqnum=make32i(packet[TCP_seqnum],packet[TCP_seqnum+1]);
    client_seqnum = client_seqnum + swapbytes(tcpdatalen_in);
    packet[TCP_acknum] = (client_seqnum & 0xFFFF0000) >> 16;
    packet[TCP_acknum+1] = client_seqnum & 0x0000FFFF;

    if(++ISN == 0x0000 || ++ISN == 0xFFFF)
        ISN = 0x1234;
    my_seqnum = 0xFFFF3412;

    //set_packet32(TCP_seqnum,my_seqnum);
        packet[TCP_seqnum] = my_seqnum & 0x0000FFFF;
        packet[TCP_seqnum+1] = (my_seqnum & 0xFFFF0000) >> 16;
    packet[TCP_hdrflags] &= 0xC0FF;
    SYN_OUT;
    ACK_OUT;

    packet[TCP_cksum] = 0x00;

    hdr_chksum =0;
    hdrlen = 0x08;
    addr = &packet[ip_srcaddr];
    cksum();
    hdr_chksum = hdr_chksum + 256*HIGH_BYTE(packet[ip_ttlproto]);
    tcplen = swapbytes(packet[ip_pktlen]) - (LOW_BYTE(packet[ip_vers_len])
& 0x0F) *4;
    hdr_chksum = hdr_chksum + swapbytes(tcplen);
    hdrlen = tcplen;
```

```
    addr = &packet[TCP_srcport];
    cksum();
    packet[TCP_cksum]= ~(hdr_chksum + ((hdr_chksum & 0xFFFF0000) >> 16));
    echo_packet();
}
```

Code Snippet 14.1: This part of the AirDrop 802.11b driver processes a TCP segment that arrives with the SYN bit set.

Nothing will occur if the destination TCP port address we gleaned in the beginning of the *tcp* function does not match the AirDrop-P's hard-coded TCP port value of 8088 decimal. Once the TCP destination port match is verified and the SYN flag is found to be set, the AirDrop module prepares for part 2 of the 3-way handshake by telling itself that the incoming sequence number from SNOOPER should be incremented in the outgoing message in the AirDrop's outgoing acknowledgment number (*tcpdatalen_in = 0x01*). A software flag (*synflag*) is set to signal to the AirDrop 802.11b driver's *TCP* function that the first part of the 3-way handshake has already taken place.

The AirDrop in this scenario is the server and SNOOPER, the laptop, is the client. The SYN flag in the incoming frame of SNOOPER's first TCP handshake segment must be treated as a sequence number in the AirDrop's response, which is the second part of the 3-way handshake. This is clearly illustrated by the Netasyst Sniffer feature that suggests the next sequence number that should be sent. The value of the randomly generated *Initial sequence number* field in Sniffer Text 14.1 is 3238359086 decimal. Note that SNOOPER s expecting the value of the *Initial sequence number* field to be incremented by one as the AirDrop's *Next expected Seq number* field is set at 3238359087. The SYN flag bit is set. So, the incoming data length in Sub Snippet 14.6 is defined as one byte even though there is no physical data in the TCP segment.

Sub Snippet 14.6
```
*********************************************************
char flags;
#define synflag 0x01
#define set_synflag    flags |= synflag
#define SYN_IN (packet[TCP_hdrflags] & 0x0200)


****************************

    if(SYN_IN && portaddr == MY_PORT_ADDRESS)
    {
        tcpdatalen_in = 0x01;
        set_synflag;


*********************************************************
```

You have seen the *setipaddrs* function put to use in various functions throughout the AirDrop 802.11b driver. The *setipaddrs* function's mission has not changed. The AirDrop 802.11b driver's *tcp* function puts the *setipaddrs* function to work in Sub Snippet 14.7 getting the frame ready to return from whence it came by turning around the source and destination IP and MAC addresses and recalculating the IP header checksum.

The reversal of source and destination addresses has to be done for the source and destination TCP port numbers as well. A scratch register holds the original TCP source port value for the *packet[TCP_srcport]* memory location while the TCP source and destination port values get turned around in the AirDrop-P packet buffer.

Sub Snippet 14.7

```
//************************************************************

//************************************************************
//*      SETIPADDRS
//*      This function builds the IP header.
//************************************************************
void setipaddrs(void)
{
   //move IP source address to destination address
   packet[ip_destaddr]=packet[ip_srcaddr];
   packet[ip_destaddr+1]=packet[ip_srcaddr+1];
   //make ethernet module IP address source address
   packet[ip_srcaddr]=ipaddri[0];
   packet[ip_srcaddr+1]=ipaddri[1];
   //move hardware source address to destinatin address
   packet[enetpacketDest01]=packet[enetpacketSrc01];
   packet[enetpacketDest23]=packet[enetpacketSrc23];
   packet[enetpacketDest45]=packet[enetpacketSrc45];
   //make ethernet module mac address the source address
   packet[enetpacketSrc01]=macaddri[0];
   packet[enetpacketSrc23]=macaddri[1];
   packet[enetpacketSrc45]=macaddri[2];

   //CALCULATE THE IP HEADER CHECKSUM
   packet[ip_hdr_cksum]=0x00;

   hdr_chksum =0;
   hdrlen = (LOW_BYTE(packet[ip_vers_len]) & 0x0F) *4;
   addr = &packet[ip_vers_len];
   cksum();
   packet[ip_hdr_cksum]= ~(hdr_chksum + ((hdr_chksum & 0xFFFF0000) >> 16));
 }


//****************************

      setipaddrs();

      //SWAP THE TCP SOURCE AND DESTINATION PORTS
      temp = packet[TCP_srcport];
      packet[TCP_srcport] = packet[TCP_destport];
      packet[TCP_destport] = temp;


//************************************************************
```

The AirDrop 802.11b driver code under the *//ASSEMBLE THE ACKNOWLEDGMENT* comment in Sub Snippet 14.8 takes SNOOPER's incoming sequence number and increments it by one. The AirDrop-P module must acknowledge SNOOPER's TCP segment with a sequence number that is composed of SNOOPER's incoming sequence number plus the number of data bytes the AirDrop-P module saw in SNOOPER's incoming TCP segment. The newly modified sequence number is placed in the AirDrop-P packet buffer's acknowledgment field.

Sub Snippet 14.8
```
**************************************************************
        //ASSEMBLE THE ACKNOWLEDGMENT
        client_seqnum=make32i(packet[TCP_seqnum],packet[TCP_seqnum+1]);
        client_seqnum = client_seqnum + swapbytes(tcpdatalen_in);
        packet[TCP_acknum] = (client_seqnum & 0xFFFF0000) >> 16;
        packet[TCP_acknum+1] = client_seqnum & 0x0000FFFF;
**************************************************************
```

Before sending an acknowledgment during the first part of the TCP initialization process, the AirDrop-P module must establish an initial sequence number of its own. The ISN (Initial Sequence Number) is a 32-bit pseudo random number that is ultimately placed into the Air-Drop-P packet buffer's TCP sequence number header field.

In this portion of the 3-way handshake process, the AirDrop module must respond with the SYN and ACK flag bits set in the TCP header. The TCP header flags area is cleared as shown in Sub Snippet 14.9 before the TCP header flags are set.

Sub Snippet 14.9
```
**************************************************************
#define SYN_OUT     packet[TCP_hdrflags] |= 0x0200 //0000 0010 0000 0000
#define ACK_OUT     packet[TCP_hdrflags] |= 0x1000 //0001 0000 0000 0000
*****************************

        if(++ISN == 0x0000 || ++ISN == 0xFFFF)
           ISN = 0x1234;
        my_seqnum = 0xFFFF3412;

        packet[TCP_seqnum] = my_seqnum & 0x0000FFFF;
        packet[TCP_seqnum+1] = (my_seqnum & 0xFFFF0000) >> 16;
        //CLEAR TCP HEADER FLAGS
        packet[TCP_hdrflags] &= 0xC0FF;
        //SET THE SYN AND ACK FLAGS
        SYN_OUT;
        ACK_OUT;


**************************************************************
```

Nothing in the Ethernet and Internet worlds can live without a checksum. Fortunately, we won't have to learn yet another new checksum algorithm as the TCP checksum is calculated exactly like the UDP checksum with the code in Sub Snippet 14.10. However, unlike UDP, the TCP checksum is not an option.

Sub Snippet 14.10
****

```
//************************************************************
//*     CHECKSUM CALCULATION ROUTINE
//************************************************************
void cksum()
{
        //hdr_chksum = 0;
     while(hdrlen > 1)
     {
          hdr_chksum = hdr_chksum + *addr++;
        hdrlen -=2;
        }
     if(hdrlen > 0)
        {
        hdr_chksum = hdr_chksum + (*addr & 0x00FF);
     --hdrlen;
        }
}

unsigned int swapbytes(unsigned int val)
{
     char temphi,templo;

     temphi = make8(val,1);
     templo = make8(val,0);
     return(make16(templo,temphi));
}

//************************************************************
//*     Echo Packet Function
//*     This routine does not modify the incoming packet size and
//*     thus echoes the original packet structure.
//************************************************************
char echo_packet()
{
     char rc;
     unsigned int TxFID;

     TxFID = get_free_TxFID();
     if(TxFID == 0)
      rc = 1;
     else
      {
     rc = bap_write(TxFID,rxdatalength_offset,(char*)packet,packet[enetpacke
tLen11]+16);
        if(rc=send_command(TransmitReclaim_Cmd,TxFID))
             printf("\r\nTransmit failed");

     do_event_housekeeping();
```

```
        }
        return(rc);
}


* * * * * * * * * * * * * * * * * * * * * * * * * * *

        packet[TCP_cksum] = 0x00;

        hdr_chksum =0;
        hdrlen = 0x08;
        addr = &packet[ip_srcaddr];
        cksum();
        hdr_chksum = hdr_chksum + 256*HIGH_BYTE(packet[ip_ttlproto]);
        tcplen = swapbytes(packet[ip_pktlen]) - (LOW_BYTE(packet[ip_vers_len])
& 0x0F) *4;
        hdr_chksum = hdr_chksum + swapbytes(tcplen);
        hdrlen = tcplen;
        addr = &packet[TCP_srcport];
        cksum();
        packet[TCP_cksum]= ~(hdr_chksum + ((hdr_chksum & 0xFFFF0000) >> 16));
```

The 3-way handshake response image of the TCP segment we want to send to SNOOPER has been stuffed into the AirDrop-P packet buffer as shown in Array Table 14.2. All we have to do now is call the *echo_packet* function in Sub Snippet 14.11 to send the 802.11b frame, which contains our brand new TCP segment. Once SNOOPER receives and processes the AirDrop module's response, part 2 of the 3-way handshake will be complete.

Sub Snippet 14.11

```
        echo_packet();
    }
```

PACKET BUFFER LAYOUT			
Buffer Address	Array Slot	DATA	Description
0x06FA	[0]	0x0038	802.11b packet data length
0x06FC	[1]	0x0500	Destination MAC Address
0x06FE	[2]	0x033C	
0x0700	[3]	0x2FD9	
0x0702	[4]	0xE000	Source MAC Address
0x0704	[5]	0xBC98	
0x0706	[6]	0x6759	
0x0708	[7]	0x3800	802.3 packet data length
0x070A	[8]	0xAAAA	SNAP Header
0x070C	[9]	0x0003	UI/Protocol Identifier
0x070E	[10]	0x0000	Protocol Identifier

PACKET BUFFER LAYOUT			
Buffer Address	**Array Slot**	**DATA**	**Description**
0x0710	[11]	0x0008	IP Protocol
0x0712	[12]	0x0045	IP version/header length/TOS
0x0714	[13]	0x3000	IP packet length
0x0716	[14]	0x1FC3	IP datagram id
0x0718	[15]	0x0040	fragment offset
0x071A	[16]	0x0680	time to live/protocol = TCP
0x071C	[17]	0xB5B5	IP header checksum
0x071E	[18]	0xA8C0	Source IP Address
0x0720	[19]	0x9700	
0x0722	[20]	0xA8C0	Destination IP Address
0x0724	[21]	0x0B00	
0x0726	[22]	0x981F	TCP source port
0x0728	[23]	0x4307	TCP destination port
0x072A	[24]	0x3412	Initial Sequence Number
0x072C	[25]	0xFFFF	
0x072E	[26]	0x05C1	Acknowledgment Number
0x0730	[27]	0x2F70	
0x0732	[28]	0x1270	Data Offset/Reserved/Flags
0x0734	[29]	0x0040	Window
0x0736	[30]	0xD856	Checksum
0x0738	[31]	0x0000	Urgent Pointer
0x073A	[32]	0x0402	Options
0x073C	[33]	0xB405	
0x073E	[34]	0x0101	
0x0740	[35]	0x0204	

Array Table 14.2: The ACK flag bit is set and the Acknowledgment Number field in the AirDrop-P packet buffer is valid. Note that the value of the Acknowledgment Number field is one byte more than the Initial Sequence Number field in Array Table 14.1.

In Sniffer Text 14.2, the AirDrop module established its ISN as 305463295 and expects 305463296 acknowledged from SNOOPER. Note that both the SYN and ACK flag bits are set in the AirDrop module's reply TCP segment and the AirDrop module acknowledged with 3238359087 just as SNOOPER expected. The *Options* were left over in the AirDrop-P packet buffer and resent untouched. The same goes for other fields that the AirDrop 802.11b driver doesn't look at.

Sniffer Text 14.2

```
DLC: ----- DLC Header -----
     DLC:
     DLC: Frame 47 arrived at  10:15:03.1011; frame size is 80 (0050 hex)
bytes.
     DLC: Signal level              = 100%
```

```
        DLC: Channel                         = 1
        DLC: Data rate                       = 22 (11.0 Megabits per second)
        DLC:
        DLC: Frame Control Field #1 = 08
        DLC:                .... ..00 = 0x0 Protocol Version
        DLC:                .... 10.. = 0x2 Data Frame
        DLC:                0000 .... = 0x0 Data (Subtype)
        DLC: Frame Control Field #2 = 02
        DLC:                .... ...0 = Not to Distribution System
        DLC:                .... ..1. = From Distribution System
        DLC:                .... .0.. = Last fragment
        DLC:                .... 0... = Not retry
        DLC:                ...0 .... = Active Mode
        DLC:                ..0. .... = No more data
        DLC:                .0.. .... = Wired Equivalent Privacy is off
        DLC:                0... .... = Not ordered
        DLC: Duration                        = 149 (in microseconds)
        DLC: Destination Address             = Station Xircom03D92F
        DLC: Basic Service Set ID            = Station Netgea6FD3DA
        DLC: Source address                  = Station AbocomBC5967
        DLC: Sequence Control                = 0xCB70
        DLC: ...Sequence Number              = 0xCB7 (3255)
        DLC: ...Fragment Number              = 0x0   (0)
        DLC:
LLC:    ----- LLC Header -----
        LLC:
        LLC:  DSAP Address = AA, DSAP IG Bit = 00 (Individual Address)
        LLC:  SSAP Address = AA, SSAP CR Bit = 00 (Command)
        LLC:  Unnumbered frame: UI
        LLC:
SNAP:   ----- SNAP Header -----
        SNAP:
        SNAP: Type      = 0800 (IP)
        SNAP:
IP:  ----- IP Header -----
        IP:
        IP: Version = 4, header length = 20 bytes
        IP: Type of service = 00
        IP:      000. ....  = routine
        IP:      ...0 ....  = normal delay
        IP:      .... 0...  = normal throughput
        IP:      .... .0..  = normal reliability
        IP:      .... ..0.  = ECT bit - transport protocol will ignore the CE bit
        IP:      .... ...0  = CE bit - no congestion
        IP: Total length    = 48 bytes
        IP: Identification   = 49951
        IP: Flags           = 4X
        IP:      .1.. ....  = don't fragment
        IP:      ..0. ....  = last fragment
        IP: Fragment offset = 0 bytes
        IP: Time to live    = 128 seconds/hops
        IP: Protocol        = 6 (TCP)
```

```
       IP: Header checksum = B5B5 (correct)
       IP: Source address     = [192.168.0.151]
       IP: Destination address = [192.168.0.11]
       IP: No options
       IP:
TCP: ----- TCP header -----
       TCP:
       TCP: Source port          =  8088
       TCP: Destination port     =  1859
       TCP: Initial sequence number = 305463295
       TCP: Next expected Seq number= 305463296
       TCP: Acknowledgment number    = 3238359087
       TCP: Data offset             = 28 bytes (4 bits)
       TCP: Reserved Bits: Reserved for Future Use (6 bits)
       TCP: Flags                   = 12
       TCP:                 ..0. .... = (No urgent pointer)
       TCP:                 ...1 .... = Acknowledgment
       TCP:                 .... 0... = (No push)
       TCP:                 .... .0.. = (No reset)
       TCP:                 .... ..1. = SYN
       TCP:                 .... ...0 = (No FIN)
       TCP: Window                  = 16384
       TCP: Checksum                = 56D8 (correct)
       TCP: Urgent pointer          = 0
       TCP:
       TCP: Options follow
       TCP: Maximum segment size = 1460
       TCP: No-Operation
       TCP: No-Operation
       TCP: SACK-Permitted Option
       TCP:
ADDR   HEX                                              ASCII
0000: 08 02 95 00 00 05 3c 03 d9 2f 00 09 5b 6f d3 da | ..•...<.Ù/..[oÓÚ
0010: 00 e0 98 bc 59 67 70 cb aa aa 03 00 00 00 08 00 | .à~¼YgpË ª ª......
0020: 45 00 00 30 c3 1f 40 00 80 06 b5 b5 c0 a8 00 97 | E..0Ã.@.€.µµÀ¨.—
0030: c0 a8 00 0b 1f 98 07 43 12 34 ff ff c1 05 70 2f | À¨...~.C.4ÿÿÁ.p/
```

Sniffer Text 14.2: TCP/IP business cards get exchanged in this sequence.

Now that all of the sequence numbers have been exchanged, the only thing standing between the AirDrop module's and SNOOPER's exchange of real data is the final acknowledgment from SNOOPER.

In Array Table 14.3, which is SNOOPER's final acknowledgment from the AirDrop-P packet buffer point of view, we see that SNOOPER acknowledges with 305463296 and satisfies the next sequence number expected by the AirDrop module (see Sniffer Text 14.2). The *Sequence number* and *Next expected Seq number* fields contained within Sniffer Text 14.3 are identical, which means that there is no data in the SNOOPER's TCP acknowledgment segment.

PACKET BUFFER LAYOUT			
Buffer Address	Array Slot	DATA	Description
0x06FA	[0]	0x0030	802.11b packet data length
0x06FC	[1]	0xE000	Destination MAC Address
0x06FE	[2]	0xBC98	
0x0700	[3]	0x6759	
0x0702	[4]	0x0500	Source MAC Address
0x0704	[5]	0x033C	
0x0706	[6]	0x2FD9	
0x0708	[7]	0x3000	802.3 packet data length
0x070A	[8]	0xAAAA	SNAP Header
0x070C	[9]	0x0003	UI/Protocol Identifier
0x070E	[10]	0x0000	Protocol Identifier
0x0710	[11]	0x0008	IP Protocol
0x0712	[12]	0x0045	IP version/header length/TOS
0x0714	[13]	0x2800	IP packet length
0x0716	[14]	0x20C3	IP datagram id
0x0718	[15]	0x0040	fragment offset
0x071A	[16]	0x0680	time to live/protocol = TCP
0x071C	[17]	0xBCB5	IP header checksum
0x071E	[18]	0xA8C0	Source IP Address
0x0720	[19]	0x0B00	
0x0722	[20]	0xA8C0	Destination IP Address
0x0724	[21]	0x9700	
0x0726	[22]	0x4307	TCP source port
0x0728	[23]	0x981F	TCP destination port
0x072A	[24]	0x05C1	Sequence Number
0x072C	[25]	0x2F70	
0x072E	[26]	0x3512	Acknowledgment Number
0x0730	[27]	0x0000	
0x0732	[28]	0x1050	Data Offset/Reserved/Flags
0x0734	[29]	0x7044	Window
0x0736	[30]	0x2C7F	Checksum
0x0738	[31]	0x0000	Urgent Pointer

Array Table 14.3: This is the final installment of the 3-way handshake. Note that the Initial Sequence Number is now called the Sequence Number. Notice also that the Acknowledgment Number field is now open for business.

No data and no options appear in the final response from the SNOOPER laptop. However, all of the numbers line up correctly.

Sniffer Text 14.3

```
DLC: ----- DLC Header -----
      DLC:
      DLC: Frame 51 arrived at  10:15:03.1026; frame size is 72 (0048 hex)
bytes.
      DLC: Signal level             = 100%
      DLC: Channel                  = 1
      DLC: Data rate                = 22 (11.0 Megabits per second)
      DLC:
      DLC: Frame Control Field #1 = 08
      DLC:              .... ..00 = 0x0 Protocol Version
      DLC:              .... 10.. = 0x2 Data Frame
      DLC:            0000 .... = 0x0 Data (Subtype)
      DLC: Frame Control Field #2 = 02
      DLC:              .... ...0 = Not to Distribution System
      DLC:              .... ..1. = From Distribution System
      DLC:              .... .0.. = Last fragment
      DLC:              .... 0... = Not retry
      DLC:              ...0 .... = Active Mode
      DLC:              ..0. .... = No more data
      DLC:              .0.. .... = Wired Equivalent Privacy is off
      DLC:              0... .... = Not ordered
      DLC: Duration                = 149 (in microseconds)
      DLC: Destination Address     = Station AbocomBC5967
      DLC: Basic Service Set ID    = Station Netgea6FD3DA
      DLC: Source address          = Station Xircom03D92F
      DLC: Sequence Control        = 0xCB80
      DLC: ...Sequence Number      = 0xCB8 (3256)
      DLC: ...Fragment Number      = 0x0   (0)
      DLC:
LLC: ----- LLC Header -----
      LLC:
      LLC:  DSAP Address = AA, DSAP IG Bit = 00 (Individual Address)
      LLC:  SSAP Address = AA, SSAP CR Bit = 00 (Command)
      LLC:  Unnumbered frame: UI
      LLC:
SNAP: ----- SNAP Header -----
      SNAP:
      SNAP: Type     = 0800 (IP)
      SNAP:
IP: ----- IP Header -----
      IP:
      IP: Version = 4, header length = 20 bytes
      IP: Type of service = 00
      IP:       000. .... = routine
      IP:       ...0 .... = normal delay
      IP:       .... 0... = normal throughput
      IP:       .... .0.. = normal reliability
      IP:       .... ..0. = ECT bit - transport protocol will ignore the CE bit
      IP:       .... ...0 = CE bit - no congestion
      IP: Total length    = 40 bytes
```

```
        IP: Identification    = 49952
        IP: Flags             = 4X
        IP:        .1.. ....  = don't fragment
        IP:        ..0. ....  = last fragment
        IP: Fragment offset   = 0 bytes
        IP: Time to live      = 128 seconds/hops
        IP: Protocol          = 6 (TCP)
        IP: Header checksum    = B5BC (correct)
        IP: Source address     = [192.168.0.11]
        IP: Destination address = [192.168.0.151]
        IP: No options
        IP:
TCP: ----- TCP header -----
        TCP:
        TCP: Source port            =  1859
        TCP: Destination port       =  8088
        TCP: Sequence number        = 3238359087
        TCP: Next expected Seq number= 3238359087
        TCP: Acknowledgment number   = 305463296
        TCP: Data offset            = 20 bytes (4 bits)
        TCP: Reserved Bits: Reserved for Future Use (6 bits)
        TCP: Flags                  = 10
        TCP:            ..0. ....  = (No urgent pointer)
        TCP:            ...1 ....  = Acknowledgment
        TCP:            .... 0...  = (No push)
        TCP:            .... .0..  = (No reset)
        TCP:            .... ..0.  = (No SYN)
        TCP:            .... ...0  = (No FIN)
        TCP: Window                 = 17520
        TCP: Checksum               = 7F2C (correct)
        TCP: Urgent pointer         = 0
        TCP: No TCP options
        TCP:
ADDR   HEX                                                         ASCII
0000: 08 02 95 00 00 e0 98 bc 59 67 00 09 5b 6f d3 da | ..•..à˜¼Yg..[oÓÚ
0010: 00 05 3c 03 d9 2f 80 cb aa aa 03 00 00 00 08 00 | ..<.Ù/€Ëªª......
0020: 45 00 00 28 c3 20 40 00 80 06 b5 bc c0 a8 00 0b | E..(Ã @.€.µ¼À¨..
0030: c0 a8 00 97 07 43 1f 98 c1 05 70 2f 12 35 00 00 | À¨.—.C.˜Á.p/.5..
0040: 50 10 44 70 7f 2c 00 00                         | P.Dp□,..
```

Sniffer Text 14.3: Introductions have been made and business cards have been handed out. Now it's time to get down to doing business.

The ACK flag is set in Sniffer Text 14.3, the TCP destination port is that of the AirDrop module and there is no data in the TCP segment. That combination of events forces the AirDrop 802.11b driver's *tcp* function to branch to one of its internal ACK handlers. Sub Snippet 14.12 shows us how the AirDrop 802.11b driver handles that scenario.

Sub Snippet 14.12

```
*****************************************************************
    //IF AN ACK IS RECEIVED AND THE DESTINATION PORT ADDRESS IS VALID AND NO
DATA IS IN THE PACKET
    if(ACK_IN && portaddr == MY_PORT_ADDRESS && tcpdatalen_in == 0x00)
    {
       //ASSEMBLE THE ACKNOWLEDGMENT NUMBER FROM THE INCOMING PACKET
       incoming_ack =make32i(packet[TCP_acknum],packet[TCP_acknum+1]);

       //IF THE INCOMING PACKET IS A RESULT OF SESSION ESTABLISHMENT
       if(bsynflag)
       {
          //CLEAR THE SYN FLAG
          clr_synflag;

          //THE INCOMING ACKNOWLEDGMENT IS MY NEW SEQUENCE NUMBER
          my_seqnum = incoming_ack;

          //SEND THE TELNET SERVER BANNER
          //LIMIT THE CHARACTER COUNT TO 40 DECIMAL
          j = sizeof(telnet_banner);
          //LENGTH OF THE BANNER MESSAGE
          tcpdatalen_out = j;
          i = 0;
          k=0;
          do{
            packet[TCP_data+k] = (telnet_banner[i+1] << 8) | telnet_banner[i];
            i+=2;
            ++k;
          }while(j-- > 1);
          do{
            packet[TCP_data+k] = telnet_banner[i];
          }while(j-- > 0);

          //EXPECT TO GET AN ACKNOWLEDGMENT OF THE BANNER MESSAGE
          expected_ack = my_seqnum + swapbytes(tcpdatalen_out);

          //SEND THE TCP/IP PACKET
          send_tcp_packet();
       }
    }
*****************************************************************
```

Depending on future circumstances, we may have to perform some calculations with the acknowledgment number. So, in Sub Snippet 14.13, the incoming acknowledgment number is reassembled into a 32-bit value (*incoming_ack*) from the values found in the AirDrop-P packet buffer words that hold the incoming acknowledgment number.

Sub Snippet 14.13

```
************************************************************
    if(ACK_IN && portaddr == MY_PORT_ADDRESS && tcpdatalen_in == 0x00)
    {
      //ASSEMBLE THE ACKNOWLEDGMENT NUMBER FROM THE INCOMING PACKET
      incoming_ack =make32i(packet[TCP_acknum],packet[TCP_acknum+1]);
************************************************************
```

Earlier, the *synflag* bit was set to signal that we were participating in the 3-way hand-shake/session establishment process. The 3-way handshake process is now complete and SNOOPER and the AirDrop module are ready to exchange data. A set *synflag* bit signals the AirDrop 802.11b driver to send the AirDrop Telnet banner message. In Sub Snippet 14.14, the *synflag* bit is cleared to indicate that the Telnet banner has been transmitted.

Sub Snippet 14.14

```
************************************************************
char flags;
#define synflag 0x01
#define bsynflag       flags & synflag
#define clr_synflag    flags &= ~synflag
****************************
      //IF THE INCOMING PACKET IS A RESULT OF SESSION ESTABLISHMENT
      if(bsynflag)
      {
        //CLEAR THE SYN FLAG
        clr_synflag;
************************************************************
```

The incoming acknowledgment number we assembled and pushed into the *incoming_ack* variable is the AirDrop module's starting sequence number for data transmission in Sub Snippet 14.15.

Sub Snippet 14.15

```
************************************************************
        //THE INCOMING ACKNOWLEDGMENT IS MY NEW SEQUENCE NUMBER
        my_seqnum = incoming_ack;
************************************************************
```

The *telnet_banner* is hard-coded in AirDrop 802.11b driver. To include the Telnet banner in our transmitted TCP segment, we must move the *telnet_banner* characters into the TCP data area of the AirDrop-P packet buffer. The length of the *telnet_banner* message is calculated and the specified number of *telnet_banner* characters are then moved into the TCP data area of the AirDrop-P packet buffer with the code you see in Sub Snippet 14.16.

Sub Snippet 14.16

```
************************************************************
        //SEND THE TELNET SERVER BANNER
        //LIMIT THE CHARACTER COUNT TO 40 DECIMAL
        j = sizeof(telnet_banner);
        //LENGTH OF THE BANNER MESSAGE
        tcpdatalen_out = j;
```

```
    i = 0;
    k=0;
    do{
      packet[TCP_data+k] = (telnet_banner[i+1] << 8) | telnet_banner[i];
      i+=2;
      ++k;
    }while(j-- > 1);
    do{
      packet[TCP_data+k] = telnet_banner[i];
    }while(j-- > 0);
```
****

We know how many bytes we'll be sending in our Telnet banner message and we know our starting TCP sequence number. We can calculate the expected acknowledgment number that SNOOPER should return after the Telnet banner message is received and processed by SNOOPER with the line of code shown in Sub Snippet 14.17.

Sub Snippet 14.17
****
```
    //EXPECT TO GET AN ACKNOWLEDGMENT OF THE BANNER MESSAGE
    expected_ack = my_seqnum + swapbytes(tcpdatalen_out);
```
****

We're ready to send the Telnet banner message to SNOOPER. Array Table 14.4 shows us how the Telnet banner frame looks from inside the PIC18LF8621.

PACKET BUFFER LAYOUT			
Buffer Address	**Array Slot**	**DATA**	**Description**
0x06FA	[0]	0x003D	802.11b packet data length
0x06FC	[1]	0x0500	Destination MAC Address
0x06FE	[2]	0x033C	
0x0700	[3]	0x2FD9	
0x0702	[4]	0xE000	Source MAC Address
0x0704	[5]	0xBC98	
0x0706	[6]	0x6759	
0x0708	[7]	0x3D00	802.3 packet data length
0x070A	[8]	0xAAAA	SNAP Header
0x070C	[9]	0x0003	UI/Protocol Identifier
0x070E	[10]	0x0000	Protocol Identifier
0x0710	[11]	0x0008	IP Protocol
0x0712	[12]	0x0045	IP version/header length/TOS
0x0714	[13]	0x3500	IP packet length
0x0716	[14]	0x20C3	IP datagram id
0x0718	[15]	0x0040	fragment offset
0x071A	[16]	0x0680	time to live/protocol = TCP
0x071C	[17]	0xAFB5	IP header checksum
0x071E	[18]	0xA8C0	Source IP Address
0x0720	[19]	0x9700	

0x0722	[20]	0xA8C0	Destination IP Address
0x0724	[21]	0x0B00	
0x0726	[22]	0x981F	TCP source port
0x0728	[23]	0x4307	TCP destination port
0x072A	[24]	0x3512	Seqence Number
0x072C	[25]	0x0000	
0x072E	[26]	0x05C1	Acknowledgment Number
0x0730	[27]	0x2F70	
0x0732	[28]	0x1050	Data Offset/Reserved/Flags
0x0734	[29]	0x7044	Window
0x0736	[30]	0x8C8B	Checksum
0x0738	[31]	0x0000	Urgent Pointer
0x073A	[32]	0x0A0D	Data
0x073C	[33]	0x6941	
0x073E	[34]	0x4472	
0x0740	[35]	0x6F72	
0x0742	[36]	0x2D70	
0x0744	[37]	0x3E50	
0x0746	[38]	0x0D00	

Array Table 14.4: The last byte of data is actually 0x00 in slot [38], which is the string terminator byte (null or 0x00).

The packet buffer arrangement you see in Array Table 14.4 got some extra help from the *send_tcp_packet* function in Sub Snippet 14.18 before being transmitted by the AirDrop module's 802.11b CompactFlash NIC.

Sub Snippet 14.18
```
**********************************************************

//*********************************************************
//*    Send TCP Packet
//*    This routine assembles and sends a complete TCP/IP packet.
//*    40 bytes of IP and TCP header data is assumed.
//*********************************************************
void send_tcp_packet()
{
        unsigned int temp;

    //COUNT IP AND TCP HEADER BYTES.. TOTAL = 40 BYTES
    ip_packet_len = 40 + tcpdatalen_out;
    packet[ip_pktlen] = swapbytes(ip_packet_len);
    temp = packet[enetpacketLen11];
    temp += tcpdatalen_out;
    packet[enetpacketLen11] = temp;
    packet[enetpacketLen03] = swapbytes(temp);
```

```
    setipaddrs();

        //SWAP THE TCP SOURCE AND DESTINATION PORTS
        temp = packet[TCP_srcport];
        packet[TCP_srcport] = packet[TCP_destport];
        packet[TCP_destport] = temp;

    //ASSEMBLE THE ACKNOWLEDGMENT
    client_seqnum=make32i(packet[TCP_seqnum],packet[TCP_seqnum+1]);
    client_seqnum = client_seqnum + swapbytes(tcpdatalen_in);
    packet[TCP_acknum] = (client_seqnum & 0xFFFF0000) >> 16;
    packet[TCP_acknum+1] = client_seqnum & 0x0000FFFF;
    packet[TCP_seqnum+1] = my_seqnum & 0x0000FFFF;
    packet[TCP_seqnum] = (my_seqnum & 0xFFFF0000) >> 16;

    packet[TCP_hdrflags] &= 0xC0FF;
    ACK_OUT;
    if(bfinflag)
    {
        FIN_OUT;
        clr_finflag;
    }

    packet[TCP_cksum] = 0x00;

    hdr_chksum =0;
    hdrlen = 0x08;
    addr = &packet[ip_srcaddr];
    cksum();
    hdr_chksum = hdr_chksum + 256*HIGH_BYTE(packet[ip_ttlproto]);
    tcplen = swapbytes(packet[ip_pktlen]) - (LOW_BYTE(packet[ip_vers_len]) &
0x0F) *4;
    hdr_chksum = hdr_chksum + swapbytes(tcplen);
    hdrlen = tcplen;
    addr = &packet[TCP_srcport];
    cksum();
    packet[TCP_cksum]= ~(hdr_chksum + ((hdr_chksum & 0xFFFF0000) >> 16));

        echo_packet();
}

*****************************

        //SEND THE TCP/IP PACKET
        send_tcp_packet();
    }
    }

***********************************************************
```

Since this implementation of TCP/IP is not by any means a full-blown implementation, I've cut to the chase and assumed no options or frills would be included in the TCP or IP headers. That would set the TCP and IP headers at a constant 20 bytes each. That's where the number 40 comes from. The *tcpdatalen_out* value is equal to the length of our Telnet banner message. Counting the TCP data bytes in Array Table 14.4 reveals a 13-byte Telnet banner message. The AirDrop-P packet buffer's 802.11b packet length in slot [0] plots the 802.11b frame length at 0x003D bytes. Let's see if we can account for them all.

ip_packet_len = 40 + tcpdatalen_out = 40 + 13 = 53 (0x35 in slot [13])

Utilizing the services of the AirDrop 802.11b driver code in Sub Snippet 14.19, we simply count all of the bytes between the SNAP Header field at AirDrop-P packet buffer slot [8] and the end of the TCP data at AirDrop-P packet buffer slot [38], which totals to 61 bytes decimal, or equivalently, 0x003D bytes hexadecimal.

Sub Snippet 14.19

```
//************************************************************
//*     Send TCP Packet
//*     This routine assembles and sends a complete TCP/IP packet.
//*     40 bytes of IP and TCP header data is assumed.
//************************************************************
void send_tcp_packet()
{
        unsigned int temp;

    //COUNT IP AND TCP HEADER BYTES.. TOTAL = 40 BYTES
    ip_packet_len = 40 + tcpdatalen_out;
    packet[ip_pktlen] = swapbytes(ip_packet_len);
    temp = packet[enetpacketLen11];
    temp += tcpdatalen_out;
    packet[enetpacketLen11] = temp;
    packet[enetpacketLen03] = swapbytes(temp);
```

I'm rather sure you can tell me about what's going on in the *setipaddrs* function in Sub Snippet 14.20. I'm also positive that you know why we have to do a swap-do-dangle on the TCP ports.

Sub Snippet 14.20

```
    setipaddrs();

        //SWAP THE TCP SOURCE AND DESTINATION PORTS
        temp = packet[TCP_srcport];
        packet[TCP_srcport] = packet[TCP_destport];
        packet[TCP_destport] = temp;
```

There was no data in SNOOPER's last TCP segment. Thus, there is nothing for the AirDrop module to acknowledge. That's why the *Acknowledgment Number* field remains unchanged from Array Table 14.3 to Array Table 14.4.

The AirDrop module must add the total of incoming bytes to SNOOPER's last incoming sequence number to assemble and acknowledgment in Sub Snippet 14.21. In this case, no data was received and zero was added to the SNOOPER's sequence number. Therefore, the SNOOPER's sequence number was not changed. In a similar fashion, the AirDrop module sends its sequence number to the client expecting to get an acknowledgment of the 13 bytes of data in the Telnet banner message, which through the acknowledgment from SNOOPER will increase the AirDrop module's sequence number by 13 bytes. The TCP header Acknowledge flag bit will always be set after the client/server (SNOOPER/AirDrop module) connection is established.

Sub Snippet 14.21
****

```
//ASSEMBLE THE ACKNOWLEDGMENT
client_seqnum=make32i(packet[TCP_seqnum],packet[TCP_seqnum+1]);
client_seqnum = client_seqnum + swapbytes(tcpdatalen_in);
packet[TCP_acknum] = (client_seqnum & 0xFFFF0000) >> 16;
packet[TCP_acknum+1] = client_seqnum & 0x0000FFFF;
packet[TCP_seqnum+1] = my_seqnum & 0x0000FFFF;
packet[TCP_seqnum] = (my_seqnum & 0xFFFF0000) >> 16;
```
****

SNOOPER hasn't sent a FIN to end the connection yet. So, the AirDrop module won't set its FIN flag with the FIN_OUT code segment in Sub Snippet 14.22 this time around. The TCP checksum is calculated and the frame is sent along its way.

Sub Snippet 14.22
****

```
packet[TCP_hdrflags] &= 0xC0FF;
ACK_OUT;
if(bfinflag)
{
    FIN_OUT;
    clr_finflag;
}

packet[TCP_cksum] = 0x00;

hdr_chksum =0;
hdrlen = 0x08;
addr = &packet[ip_srcaddr];
cksum();
hdr_chksum = hdr_chksum + 256*HIGH_BYTE(packet[ip_ttlproto]);
tcplen = swapbytes(packet[ip_pktlen]) - (LOW_BYTE(packet[ip_vers_len]) &
0x0F) *4;
hdr_chksum = hdr_chksum + swapbytes(tcplen);
hdrlen = tcplen;
```

```
    addr = &packet[TCP_srcport];
    cksum();
    packet[TCP_cksum]= ~(hdr_chksum + ((hdr_chksum & 0xFFFF0000) >> 16));

        echo_packet();
}
```
****

The SNOOPER saw the frame as we see it in Sniffer Text 14.4. If you do some simple substraction with the *Sequence number* and the *Next expected Seq number* fields in the TCP Header area of Sniffer Text 14.4, you'll find that 13 bytes of data will be acknowledged by SNOOPER.

Sniffer Text 14.4

```
DLC: ----- DLC Header -----
     DLC:
     DLC: Frame 55 arrived at  10:15:03.1069; frame size is 85 (0055 hex)
bytes.
     DLC: Signal level              = 100%
     DLC: Channel                   = 1
     DLC: Data rate                 = 22 (11.0 Megabits per second)
     DLC:
     DLC: Frame Control Field #1 = 08
     DLC:                   .... ..00 = 0x0 Protocol Version
     DLC:                   .... 10.. = 0x2 Data Frame
     DLC:                   0000 .... = 0x0 Data (Subtype)
     DLC: Frame Control Field #2 = 02
     DLC:                   .... ...0 = Not to Distribution System
     DLC:                   .... ..1. = From Distribution System
     DLC:                   .... .0.. = Last fragment
     DLC:                   .... 0... = Not retry
     DLC:                   ...0 .... = Active Mode
     DLC:                   ..0. .... = No more data
     DLC:                   .0.. .... = Wired Equivalent Privacy is off
     DLC:                   0... .... = Not ordered
     DLC: Duration                   = 149 (in microseconds)
     DLC: Destination Address        = Station Xircom03D92F
     DLC: Basic Service Set ID       = Station Netgea6FD3DA
     DLC: Source address             = Station AbocomBC5967
     DLC: Sequence Control           = 0xCB90
     DLC: ...Sequence Number         = 0xCB9 (3257)
     DLC: ...Fragment Number         = 0x0   (0)
     DLC:
LLC: ----- LLC Header -----
     LLC:
     LLC: DSAP Address = AA, DSAP IG Bit = 00 (Individual Address)
     LLC: SSAP Address = AA, SSAP CR Bit = 00 (Command)
     LLC: Unnumbered frame: UI
     LLC:
SNAP: ----- SNAP Header -----
     SNAP:
     SNAP: Type      = 0800 (IP)
     SNAP:
IP: ----- IP Header -----
```

```
     IP:
     IP: Version = 4, header length = 20 bytes
     IP: Type of service = 00
     IP:       000. .... = routine
     IP:       ...0 .... = normal delay
     IP:       .... 0... = normal throughput
     IP:       .... .0.. = normal reliability
     IP:       .... ..0. = ECT bit - transport protocol will ignore the CE bit
     IP:       .... ...0 = CE bit - no congestion
     IP: Total length    = 53 bytes
     IP: Identification   = 49952
     IP: Flags            = 4X
     IP:       .1.. .... = don't fragment
     IP:       ..0. .... = last fragment
     IP: Fragment offset = 0 bytes
     IP: Time to live    = 128 seconds/hops
     IP: Protocol        = 6 (TCP)
     IP: Header checksum = B5AF (correct)
     IP: Source address      = [192.168.0.151]
     IP: Destination address = [192.168.0.11]
     IP: No options
     IP:
TCP: ----- TCP header -----
     TCP:
     TCP: Source port          = 8088
     TCP: Destination port     = 1859
     TCP: Sequence number      = 305463296
     TCP: Next expected Seq number= 305463309
     TCP: Acknowledgment number    = 3238359087
     TCP: Data offset          = 20 bytes (4 bits)
     TCP: Reserved Bits: Reserved for Future Use (6 bits)
     TCP: Flags                = 10
     TCP:             ..0. .... = (No urgent pointer)
     TCP:             ...1 .... = Acknowledgment
     TCP:             .... 0... = (No push)
     TCP:             .... .0.. = (No reset)
     TCP:             .... ..0. = (No SYN)
     TCP:             .... ...0 = (No FIN)
     TCP: Window               = 17520
     TCP: Checksum             = 8B8C (correct)
     TCP: Urgent pointer       = 0
     TCP: No TCP options
     TCP: [13 Bytes of data]
     TCP:
ADDR HEX                                                          ASCII
0000: 08 02 95 00 00 05 3c 03 d9 2f 00 09 5b 6f d3 da  | ...•...<.Ù/..[oÓÚ
0010: 00 e0 98 bc 59 67 90 cb aa aa 03 00 00 00 08 00  | .à˜¼Yg Ëªª......
0020: 45 00 00 35 c3 20 40 00 80 06 b5 af c0 a8 00 97  | E..5Ã @.€.µ¯À¨.—
0030: c0 a8 00 0b 1f 98 07 43 12 35 00 00 c1 05 70 2f  | À¨...˜.C.5..Á.p/
0040: 50 10 44 70 8b 8c 00 00 0d 0a 41 69 72 44 72 6f  | P.Dp‹Œ....AirDro
0050: 70 2d 50 3e 00                                   | p-P>.
```

Sniffer Text 14.4: I've highlighted the 13 bytes of Telnet banner message sent to SNOOPER's Telnet session.

Sending those bytes using TCP/IP was quite a bit more complicated than doing the same with UDP. In addition, we had to call upon a canned Telnet application on the SNOOPER side to complete the data transfer. The I-always-have-to-know-where-you-are nature of TCP/IP is why TCP is known as a connection-oriented protocol. UDP and IP's I-don't-care-where-you-are attitudes place them in the connectionless protocol category.

We've examined two of the four *tcp* function branches. The third *tcp* function branch deals with incoming data from a Telnet session. There's really nothing to the code. If after validating the TCP destination port address, the *tcp* function determines that there is data in the incoming TCP segment with the ACK flag bit set, the data from the incoming TCP segment is collected byte by byte into a small auxiliary data buffer (*aux_data*). Once all of the incoming TCP segment's data is accounted for, the TCP application code function shown in Sub Snippet 14.23 is called. The application code function does nothing but echo the characters in the *aux_data* buffer or on an incoming CRLF (carriage return line feed) sequence, resend the Telnet banner message. The application code is pure simplicity that is surrounded by great complexity.

Sub Snippet 14.23
****

```
//************************************************************
//*    Application Code
//*    Your application code goes here.
//************************************************************
void application_code()
{
    char i,j,k;

    i=0;
    if(aux_data[i] == 0x0D && aux_data[i+1] == 0x0A)
    {
        tcpdatalen_out = 0x00;
        j = sizeof(telnet_banner);
        tcpdatalen_out = j;            //length of the banner message
        i=0;
        k=0;
        while(j > 1)
        {
            packet[TCP_data+k] = (telnet_banner[i+1] << 8) | telnet_banner[i];
            i+=2;
            ++k;
            j-=2;
        }
        while(j > 0)
        {
            packet[TCP_data+k] = telnet_banner[i];
            --j;
        }
```

```
    }
    else
        tcpdatalen_out = tcpdatalen_in;
}


****************************

    //IF AN ACK IS RECEIVED AND THE PORT ADDRESS IS VALID AND THERE IS DATA
    IN THE INCOMING PACKET
    if(ACK_IN && portaddr == MY_PORT_ADDRESS && tcpdatalen_in)
    {
        //RECEIVE THE DATA AND PUT IT INTO THE INCOMING DATA BUFFER
            temp = tcpdatalen_in;
            i=0;
        while(temp > 1)
            {
                    aux_data[i] = LOW_BYTE(packet[TCP_data+i]);
                    aux_data[i+1] = HIGH_BYTE(packet[TCP_data+i]);
                    i+=2;
                    temp-=2;
            }
            while(temp > 0)
            {
                    aux_data[i] = LOW_BYTE(packet[TCP_data+i]);
                    --temp;
            }
            //RUN THE TCP APPLICATION
            application_code();

        //ASSEMBLE THE ACKNOWLEDGMENT NUMBER FROM THE INCOMING PACKET
        incoming_ack =make32i(packet[TCP_acknum],packet[TCP_acknum+1]);

        //CHECK FOR THE NUMBER OF BYTES ACKNOWLEDGED
        //DETERMINE HOW MANY BYTES ARE OUTSTANDING AND ADJUST THE OUTGOING
         SEQUENCE NUMBER ACCORDINGLY
        if(incoming_ack <= expected_ack)
            my_seqnum = expected_ack - (expected_ack - incoming_ack);

        //MY EXPECTED ACKNOWLEDGMENT NUMBER
        expected_ack = my_seqnum + swapbytes(tcpdatalen_out);


        send_tcp_packet();
    }

*************************************************************
```

It's rather fitting that the last of the four *tcp* function branches is the FIN branch. Upon receiving a FIN, the AirDrop 802.11b driver attempts to make sure that any data that may have arrived with the FIN segment is processed by the AirDrop 802.11b driver's TCP application. The *finflag* is set and a byte is added to the outgoing acknowledgment number (++tcpdatalen_in). When the code falls through to the *send_tcp_packet* function in Sub Snippet 14.24, the ACK and FIN flag bits are set in the outgoing TCP segment. See YA!

Sub Snippet 14.24
****

```
//this code segment processes a FIN from the Telnet client
//and acknowledges the FIN and any incoming data.
if(FIN_IN && portaddr == MY_PORT_ADDRESS)
{
    if(tcpdatalen_in)
    {
        for(i=0;i<tcpdatalen_in;++i)
        {
            aux_data[i] = packet[TCP_data+i];
            application_code();
        }
    }

    set_finflag;

    ++tcpdatalen_in;

    incoming_ack =make32i(packet[TCP_acknum],packet[TCP_acknum+1]);
    if(incoming_ack <= expected_ack)
        my_seqnum = expected_ack - (expected_ack - incoming_ack);

    expected_ack = my_seqnum + swapbytes(tcpdatalen_out);
    send_tcp_packet();

}
}
```
****

You've Done It!

You have mastered the AirDrop and the 802.11b driver code that propels it. All of the major internet protocols are at your feet. You are now protocol-fluent in wireless ARP, ICMP, UDP and, the feared, TCP/IP. You're flying really high and fast and there's only one way to go. UP! So, let's take on the next chapter and look at how to secure the data we send with our AirDrop module across an 802.11b link.

WEP and the AirDrop

I should have said, "Attempt to secure the data we send with our AirDrop module across an 802.11b link" in the last sentence of the previous chapter. WEP (Wired Equivalent Privacy) is an attempt to give wireless LANs the security that wired LANs enjoy by default.

It's relatively simple to add WEP capability to the AirDrop 802.11b driver. So, this chapter will be short and sweet. Let's do it.

Incorporating WEP into the AirDrop 802.11b Driver

Sub Snippet 15.1 is the code segment that makes up the *main* function of the AirDrop 802.11b driver that is shipped with the assembled and tested EDTP AirDrop variants. As you can see in the Sub Snippet, WEP capability is very easy to add.

Sub Snippet 15.1
```
*************************************************************
void main(void)
{
        unsigned int i,temp,evstat_data;
        char rc;

        init_USART1();
        init_cf_card();
        airdrop_cfg(BSS);
        airdrop_cfg(SSID);
        airdrop_cfg(MAX_DATALEN);
        airdrop_cfg(NIC_RATE);
        airdrop_cfg(WEP_KEY_128);
        airdrop_cfg(WEP_KEYID);
        airdrop_cfg(WEP_AUTH_SK);
        airdrop_cfg(WEP_ON);
        temp = rid_read(RID_cfgOwnMACAddress);
        for(i=0;i<6;i++)
        {
         if(i%2)
          macaddrc[i] = fidrid_buffer[2+i/2]>>8;
         else
          macaddrc[i] = fidrid_buffer[2+i/2]& 0xFF;
        }
        for(i=0;i<3;++i)
         macaddri[i] = fidrid_buffer[i+2];
```

```
                ipaddri[0] = make16(ipaddrc[1],ipaddrc[0]);
                ipaddri[1] = make16(ipaddrc[3],ipaddrc[2]);

            allocate_xmit_buffers();
            wr_cf_io16(0xFFFF, EvAck_Register);
            if(rc=send_command(EnableMAC_Cmd,0))
             printf("MAC Enable Error\r\n");
            do{
                  temp=rid_read(RID_PortStatus);
                }while(fidrid_buffer[2]!=4);
            temp = rid_read(RID_CurrentBSSID);
            for(i=0;i<6;i++)
            {
             if(i%2)
              bssidc[i] = fidrid_buffer[2+i/2]>>8;
             else
              bssidc[i] = fidrid_buffer[2+i/2]& 0xFF;
                }
            for(i=0;i<3;++i)
             bssidi[i] = fidrid_buffer[i+2];

        while(1){
            do{
                  evstat_data = rd_cf_io16(EvStat_Register);
                }while(!(evstat_data & EvStat_Rx_Bit_Mask));
            get_frame();
          }
        }
```

****

Before we do anything to the AirDrop module, we must setup the AP to be WEP capable. Screen Capture 15.1 is all it takes for the Netgear MR814.

Code Snippet 15.1 is a compilation of all of the supporting code needed to implement WEP on an AirDrop module. The WEP key that was loaded into the AIRDROP_NETWORK AP is hard-coded as a string in the AirDrop module's PIC18LF8621 program flash.

Code Snippet 15.1

****

```
#define   RID_cfgWEPDefaultKeyID          0xFC23 //which key will do encryption
#define   RID_cfgWEPDefaultKeyID_Length   2
#define   RID_cfgWEPDefaultKeyID_Value0   0x0000 //encryption key0
#define   RID_cfgWEPDefaultKeyID_Value1   0x0001 //encryption key1
#define   RID_cfgWEPDefaultKeyID_Value2   0x0010 //encryption key2
#define   RID_cfgWEPDefaultKeyID_Value3   0x0011 //encryption key3


#define   RID_Default128Key_Length        8      //128-bit encryption with a
                                                 13-byte key
#define   RID_Default64Key_Length         4      //64-bit encryption with a 5-
                                                 byte key and a 24-bit IV
```

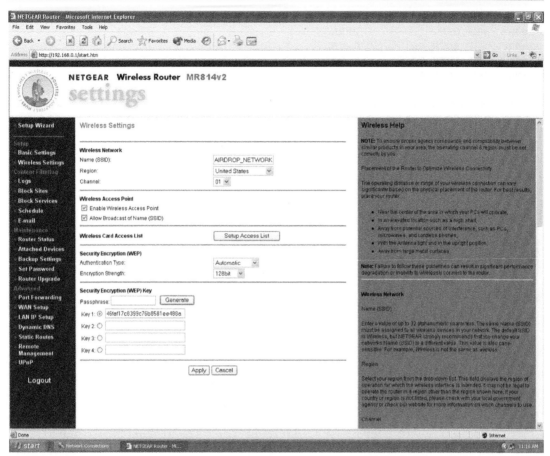

Screen Capture 15.1: I've chosen to use 128-bit encryption, which requires a 13-byte key versus 64-bit encryption and its 5-byte key. The AirDrop 802.11b driver will do both 128-bit and 64-bit WEP encryption.

```
#define   RID_cfgDefaultKey0              0xFC24 //key bytes for key0
#define   RID_cfgDefaultKey1              0xFC25 //key bytes for key1
#define   RID_cfgDefaultKey2              0xFC26 //key bytes for key2
#define   RID_cfgDefaultKey3              0xFC27 //key bytes for key3
#define   RID_cfgWEPFlags                 0xFC28 //turn WEP ON or OFF and
                                                 decide how much to use IV
#define   RID_cfgWEPFlags_Length          2
#define   RID_cfgWEPFlags_Value_ON        0x0001 //encryption/decryption en-
abled and IV reuse every frame
#define   RID_cfgWEPFlags_Value_OFF       0x0000 //encryption/decryption
                                                 disabled
#define   WEP_KEYID         0x06          //use WEB Key 0-3
#define   WEP_KEY_64        0x07
#define   WEP_KEY_128       0x08          //WEP key value
```

```
#define    WEP_ON              0x09
#define    WEP_OFF             0x0A
#define    WEP_AUTH_SK         0x0B
#define    WEP_AUTH_OS         0x0C
#define    WEP_AUTH_ALL        0x0D

const char wepkey[] = {0x46,0xFA,0xF1,0x7C,0x83,0x99,0xC7,0x6B,0x85,0x81,0xE
E,0x48,0x8A,0x00};

//**********************************************************************
//* AIRDROP CONFIGURATION
//**********************************************************************
char airdrop_cfg(unsigned int cmd)
{
 char rc;
 unsigned int i,j,k;

 switch(cmd)
 {
  case BSS:
   fidrid_buffer[0] = RID_cfgPortType_Length;
   fidrid_buffer[1] = RID_cfgPortType;
   fidrid_buffer[2] = RID_cfgPortType_Infrastructure;
   rc = rid_write(RID_cfgPortType, fidrid_buffer, RID_cfgPortType_Length);
   break;
  case IBSS:
   fidrid_buffer[0] = RID_cfgPortType_Length;
   fidrid_buffer[1] = RID_cfgPortType;
   fidrid_buffer[2] = RID_cfgPortType_IBSS;
   rc = rid_write(RID_cfgPortType, fidrid_buffer, RID_cfgPortType_Length);
   break;
  case SSID:
   fidrid_buffer[0] = RID_DesiredSSID_Length;
   fidrid_buffer[1] = RID_cfgDesiredSSID;
   fidrid_buffer[2] = strlen(ssid);
   j = strlen(ssid);
   i=0;
   k=3;
   while(j > 1)
  {
    fidrid_buffer[k] = (ssid[i+1] << 8) | ssid[i];
    i+=2;
    ++k;
    j-=2;
  }
 while(j > 0)
 {
   fidrid_buffer[k] = ssid[i];
   --j;
 }
```

```
 rc = rid_write(RID_cfgDesiredSSID,fidrid_buffer,RID_DesiredSSID_Length);
 break;

case MAX_DATALEN:
 fidrid_buffer[0] = RID_cfgMaxDataLength_Length;
 fidrid_buffer[1] = RID_cfgMaxDataLength;
 fidrid_buffer[2] = RID_cfgMaxDataLength_Value;   //0x05DC - 1500 decimal
 rc = rid_write(RID_cfgMaxDataLength, fidrid_buffer, RID_cfgMaxDataLength_
Length);
 break;

 case NIC_RATE:
  fidrid_buffer[0] = RID_TxRateControl_Length;
  fidrid_buffer[1] = RID_TxRateControl;
  fidrid_buffer[2] = RID_TxRateControl_Value_ALL;
  rc = rid_write(RID_TxRateControl, fidrid_buffer, RID_TxRateControl_Length);
  break;

case WEP_KEYID:
  fidrid_buffer[0] = RID_cfgWEPDefaultKeyID_Length;
  fidrid_buffer[1] = RID_cfgWEPDefaultKeyID;
  fidrid_buffer[2] = RID_cfgWEPDefaultKeyID_Value0;
  rc = rid_write(RID_cfgWEPDefaultKeyID, fidrid_buffer, RID_cfgWEPDefaultKey-
ID_Length);
  break;

case WEP_KEY_64:
  fidrid_buffer[0] = RID_Default64Key_Length;
  fidrid_buffer[1] = RID_cfgDefaultKey0;
  j = strlen(wepkey);
  i=0;
  k=2;
  while(j > 1)
  {
   fidrid_buffer[k] = (wepkey[i+1] << 8) | wepkey[i];
   i+=2;
  ++k;
   j-=2;
 }
 while(j > 0)
 {
  fidrid_buffer[k] = wepkey[i];
  --j;
 }
 rc = rid_write(RID_cfgDefaultKey0,fidrid_buffer,RID_Default128Key_Length);
 break;

case WEP_KEY_128:
 fidrid_buffer[0] = RID_Default128Key_Length;
 fidrid_buffer[1] = RID_cfgDefaultKey0;
 j = strlen(wepkey);
 i=0;
```

```
 k=2;
 while(j > 1)
 {
  fidrid_buffer[k] = (wepkey[i+1] << 8) | wepkey[i];
  i+=2;
  ++k;
  j-=2;
 }
 while(j > 0)
 {
  fidrid_buffer[k] = wepkey[i];
  --j;
 }
 rc = rid_write(RID_cfgDefaultKey0,fidrid_buffer,RID_Default128Key_Length);
 break;

case WEP_ON:
 fidrid_buffer[0] = RID_cfgWEPFlags_Length;
 fidrid_buffer[1] = RID_cfgWEPFlags;
 fidrid_buffer[2] = RID_cfgWEPFlags_Value_ON;
 rc = rid_write(RID_cfgWEPFlags, fidrid_buffer, RID_cfgWEPFlags_Length);
 break;

case WEP_OFF:
 fidrid_buffer[0] = RID_cfgWEPFlags_Length;
 fidrid_buffer[1] = RID_cfgWEPFlags;
 fidrid_buffer[2] = RID_cfgWEPFlags_Value_OFF;
 rc = rid_write(RID_cfgWEPFlags, fidrid_buffer, RID_cfgWEPFlags_Length);
 break;

case WEP_AUTH_SK:
 fidrid_buffer[0] = RID_cfgAuthentication_Length;
 fidrid_buffer[1] = RID_cfgAuthentication;
 fidrid_buffer[2] = RID_cfgAuthentication_Value_SK;
 rc = rid_write(RID_cfgAuthentication, fidrid_buffer, RID_cfgAuthentication_
Length);
 break;

case WEP_AUTH_OS:
 fidrid_buffer[0] = RID_cfgAuthentication_Length;
 fidrid_buffer[1] = RID_cfgAuthentication;
 fidrid_buffer[2] = RID_cfgAuthentication_Value_OS;
 rc = rid_write(RID_cfgAuthentication, fidrid_buffer, RID_cfgAuthentication_
Length);
 break;

case WEP_AUTH_ALL:
 fidrid_buffer[0] = RID_cfgAuthentication_Length;
 fidrid_buffer[1] = RID_cfgAuthentication;
 fidrid_buffer[2] = RID_cfgAuthentication_Value_OK;
 rc = rid_write(RID_cfgAuthentication, fidrid_buffer, RID_cfgAuthentication_
Length);
```

```
  break;

default:
 rc = 1;
 break;
 }
  return(rc);
}
```

★★★★★★★★★★★★★★★★★★★★★★★★★★★★

Code Snippet 15.1: Everything in the AirDrop 802.11b driver that has to do with generating AirDrop module WEP capability is found here and in Sub Snippet 15.1.

After the AP is setup to do WEP, we can begin our WEP enabling process on the AirDrop module. The first order of WEP business is to establish a matching WEP key.

The source lines in Sub Snippet 15.2 call the WEP unit of the *airdrop_cfg* function. We'll take them on one by one beginning with the loading of the 128-bit WEP key value.

Sub Snippet 15.2
★★
```
void main(void)
{
        unsigned int i,temp,evstat_data;
        char rc;

        init_USART1();
        init_cf_card();
        airdrop_cfg(BSS);
        airdrop_cfg(SSID);
        airdrop_cfg(MAX_DATALEN);
        airdrop_cfg(NIC_RATE);
        airdrop_cfg(WEP_KEY_128);
        airdrop_cfg(WEP_KEYID);
        airdrop_cfg(WEP_AUTH_SK);
        airdrop_cfg(WEP_ON);
```
★★

The actual WEP key value is 13-bytes long. The 0x00 at the end of the *wepkey* array is the string terminator byte. If you've truly read and understood the earlier chapters, there should be nothing new about the RID write process you see in Sub Snippet 15.3.

Sub Snippet 15.3
★★
```
const char wepkey[] = {0x46,0xFA,0xF1,0x7C,0x83,0x99,0xC7,0x6B,0x85,0x81,0xE
E,0x48,0x8A,0x00};
```
★★★★★★★★★★★★★★★★★★★★★★★★★★★★
```
case WEP_KEY_128:
 fidrid_buffer[0] = RID_Default128Key_Length;
 fidrid_buffer[1] = RID_cfgDefaultKey0;
 j = strlen(wepkey);
```

```
i=0;
k=2;
while(j > 1)
{
 fidrid_buffer[k] = (wepkey[i+1] << 8) | wepkey[i];
 i+=2;
 ++k;
 j-=2;
}
while(j > 0)
{
 fidrid_buffer[k] = wepkey[i];
 --j;
}
rc = rid_write(RID_cfgDefaultKey0,fidrid_buffer,RID_Default128Key_Length);
break;
*************************************************************
```

Up to four WEP key values can be defined and stored. However, you can only use one encryption key at a time. To keep it as simple as possible, I've stored the AIRDROP_NET-WORK WEP key value in the 0x00 position in Sub Snippet 15.3 and I select key value position 0x00 in Sub Snippet 15.4.

Sub Snippet 15.4
```
*************************************************************
case WEP_KEYID:
  fidrid_buffer[0] = RID_cfgWEPDefaultKeyID_Length;
  fidrid_buffer[1] = RID_cfgWEPDefaultKeyID;
  fidrid_buffer[2] = RID_cfgWEPDefaultKeyID_Value0;
  rc = rid_write(RID_cfgWEPDefaultKeyID, fidrid_buffer, RID_cfgWEPDefaultKey-
ID_Length);
  break;
*************************************************************
```

Shared Key and/or Open System authentication are supported by the AirDrop 802.11b driver. In this case, it would make sense to use the Shared Key algorithm over the Open System algorithm and that's exactly what I did in Sub Snippet 15.5.

Sub Snippet 15.5
```
*************************************************************
case WEP_AUTH_SK:
 fidrid_buffer[0] = RID_cfgAuthentication_Length;
 fidrid_buffer[1] = RID_cfgAuthentication;
 fidrid_buffer[2] = RID_cfgAuthentication_Value_SK;
 rc = rid_write(RID_cfgAuthentication, fidrid_buffer, RID_cfgAuthentication_
Length);
 break;
*************************************************************
```

Let there be WEP.

Sub Snippet 15.5

```
************************************************************
case WEP_ON:
 fidrid_buffer[0] = RID_cfgWEPFlags_Length;
 fidrid_buffer[1] = RID_cfgWEPFlags;
 fidrid_buffer[2] = RID_cfgWEPFlags_Value_ON;
 rc = rid_write(RID_cfgWEPFlags, fidrid_buffer, RID_cfgWEPFlags_Length);
 break;
************************************************************
```

Ladies and Gentlemen, that's all there is to WEP on the AirDrop.

An Experimental AVR AirDrop Variant

There are times when the old brain works well with the hands. I've been reading nonstop since January a year ago now in preparation to write the text you've been reading in this book. For the past few months I've been concentrating on some additional 802.11b hardware ideas that I could turn into yet another working variant of the AirDrop series. All of that thinking and reading resulted in a finished piece of 802.11b hardware coupled with a slightly different set of 802.11b firmware routines. As always, my goal along with getting on an 802.11b wireless LAN is to keep it all as simple as possible. So, here we go..

The New Experimental AirDrop Hardware

As you can see in Photo 16.1, I'm sticking with the original TRENDnet TEW-222CF-compatible CompactFlash 802.11b interface and matching up an Xterasys CWB1K NIC with an Atmel ATmega128L. I'll stray from the convention of posting the AirDrop 802.11b driver source lines in the text here as the entire AirDrop 802.11b driver source code library for the experimental AirDrop module is included for your review on the CDROM that accompanies this book. You have already seen 99% of the experimental AirDrop 802.11b driver code in the previous chapters of this book. There are very few modifications and additions to the production AirDrop code in the experimental version.

Photo 16.1: The design in Photo 16.1 is a variant of the production AirDrop-A. The board was lengthened a bit to accommodate the 128K SRAM and its latch.

The ATmega128L has a very straight forward I/O interface and in this version of the hardware I simply use whatever I/O pins I need for interfacing to the CompactFlash NIC and leave the rest for use by the outside world. I also decided to add a fast 128K × 8 SRAM to the original AirDrop-A design. With the exception of the additional SRAM there are no other frills on the development board other than the original RS-232 serial port and the Atmel ISP programming portal. I decided to keep the AVR JTAG debugging/programming port used on the production AirDrop-A to make it easier to debug the new AirDrop-A 802.11b driver. As with the production AirDrop-A, the Atmel AVR JTAG port is wired to support the Atmel JTAGICE and the newer Atmel JTAGICE mkII. For those of you that don't breath and sleep with Atmel microcontrollers, the new JTAGICE mkII is identical in operation to the legacy JTAGICE but includes support for the new single-wire debugWIRE interface found on the smaller, low pin count AVR microcontrollers. The JTAGICE mkII also supports the use of USB communication between the JTAGICE mkII and the AVR Studio IDE, which made the AirDrop 802.11b driver debugging cycles move along quite a bit faster.

For those of you that are familiar with the EDTP Electronics, Inc. product line, you'll recall that EDTP marketed an ATA Hard Drive Controller development board that was based on the ATmega128. I took quite a bit of flack in the AVR forums for implementing the 64K x 16 SRAM IC on the ATA Hard Drive Controller "my way". Instead of utilizing the ATA Hard Drive Controller ATmega128's built-in external memory management scheme, I tied the ATA Hard Drive Controller's 64K × 16 SRAM IC directly to the 16-bit bus of the IDE hard drive. The idea was to be able to capture the outgoing IDE data in the fast SRAM with little to no intervention required by the ATA Hard Drive Controller's ATmega128 microcontroller. The downside to doing it "my way" was that I destroyed the AVR programmer's ability to use the AVR's external memory management system as it was designed to be used. As it turned out, the AVR programmer's that were interested in the ATA Hard Drive Controller worked their way around all of the obstacles I put in their way.

I didn't employ the 128 × 8 SRAM device "my way" on the new AVR-based 802.11b Air-Drop module. Instead, I wired the 128K of SRAM in as two 64K banks of external SRAM. SRAM bank selection is controlled by the SRAM's A16 address line, which is connected to I/O pin PD6 of the ATmega128L. Because the ATmega128L contains an internal 4K block of SRAM, normal addressing methods will only allow 60K of the 64K bank of SRAM to be accessed. If you take a close look at the ATmega128L data sheet in the *AVR ATmega128 Memories* section, you'll see that there are masking methods that can be used to "fake out" the SRAM's address lines to get the full use of the 64K bank of SRAM if your project requires it. Using the A16 banking I/O line is easier and safer than twiddling mask bits and using the banking I/O pin gives you 120K of extra SRAM.

The inclusion of the external 128 × 8 SRAM meant that I had to comb through the data sheets and select a suitable SRAM IC and address latch. I finally settled on the Cypress CY7C1019CV33 SRAM IC. The Cypress CY7C1019CV33 is plenty fast even with a 15 nS access time. I chose the 12 nS part, which has an access time that is well below my calculated AVR t_{RLDV} (Read Low to Data Valid) value of 76 nS. I came up with 76 nS using the following algorithm, which I gleaned from the ATmega128L data sheet:

$t_{RLDV} = 1.0t_{CLCL} - 60 \text{ nS} = 76 \text{ nS}$

Where:

AirDrop module AVR oscillator frequency = 7.3728 MHz

$t_{CLCL} = 1/7.3728 \text{ MHz} = 136 \text{ nS}$

With the CY7C1019CV33's 12 nS access time, it's obvious that no AVR wait states will be needed when accessing the external SRAM.

The ATmega128L external memory interface was designed to be used with a 74AHC series latch. My latch selection decision was based on +3.3 VDC operation versus the address latch guidelines put forth in the ATmega128L data sheet. I'm partial to 573 latch layouts and eliminated any latch with a 373 designation. I ended up selecting the Texas Instruments 3.3VDC SN74LVTH573PWR in the 20-TSSOP package. Compared to the address latch, the CY7C1019CV33 is a relatively large part as it is packaged as a 32-lead 400-Mil Molded SOJ.

The ATmega128L guarantees a minimum address hold time of 5 nS once the ALE signal falls low and the ATmega128L data sheet warns the hardware designer to choose the address latch accordingly. As it turns out, the SN74LVTH573PWR only needs a 1.5 nS window to capture the address data.

Another critical factor concerning the selection of the address latch is the address valid setup time (t_{AVLLC}) before the ALE signal drops low. The minimum t_{AVLLC} value with a clock of 7.3728 MHz and a 3.3 VDC operating voltage is:

$t_{AVLLC} = 0.5 \, t_{CLCL} - 10 = 58 \text{ nS}$

Things are good as the SN74LVTH573PWR only needs 0.7 nS of setup time. We're also OK with the ALE pulse width. The minimum ALE pulse width of our ATmega128L configuration with a 7.3728 MHz clock comes in at 121 nS. The SN74LVTH573PWR only needs 3 nS of ALE pulse width. All of these timings assume the capacitance on the bus lines is not excessive. So, I checked the capacitance loading figures the best I could. I couldn't get a handle on the 802.11b CompactFlash NIC as the CompactFlash card data sheet I have doesn't have any indepth electrical specifications. However, the SRAM data sheet says the CY7C1019CV33 comes in at only 8 pF. There really aren't enough devices on the AirDrop module's address bus or data bus to throw the capacitance loading into high gear. So, I feel pretty safe with the designated address latch/SRAM combination.

The SRAM system design is flexible as the programmer can choose to use the ATmega128L's external SRAM control scheme or turn off the AVR automatic SRAM control subsystem and bit bang data into the SRAM by manually manipulating the SRAM control lines and the address/data bus. The CompactFlash card and the SRAM cannot operate concurrently when the ATmega128L's external SRAM is under the control of the AVR. However, it is possible to talk to both the CompactFlash card and the SRAM when the ATmega128L's external SRAM control bits are in the disabled state. As you can see Schematic 16.1, the rest of the hardware is very similar to the production AirDrop-A hardware. So, let's move on and talk about the firmware behind the new hardware.

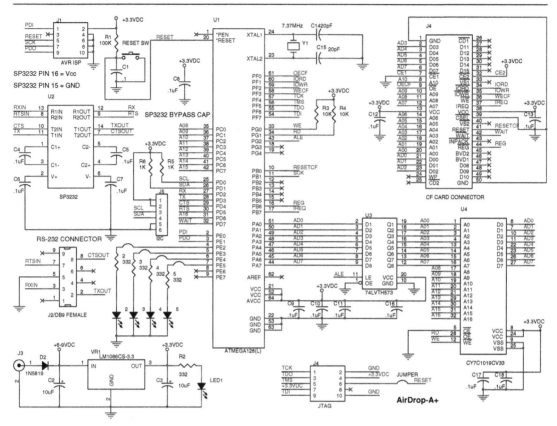

Schematic 16.1: A quartet of LEDs, the CY7C1019CV33 and a 74LVTH573 are the only additions made to the base AirDrop-A circuitry.

The Experimental AirDrop Firmware

Now that you've explored the AirDrop module hardware and the AirDrop 802.11b driver, you know that the same types of Ethernet packets that flow across wire are identical to the Ethernet packets that float across thin air. That relationship makes it pretty easy for me to reuse some time-tested code from my Easy Ethernet series of hard-wired Ethernet devices and the AirDrop 802.11b driver code we've discussed in this text. The Easy Ethernet code is basically a minimal TCP/IP stack with just enough oomph to get a message from one point to another using standard internet protocols. Easy Ethernet devices come in two microcontroller flavors, PIC and AVR. Since I have chosen to use an ATmega128L in this project, the AirDrop module base code will be ported from the Easy Ethernet AVR device firmware, which is the basis for the AirDrop 802.11b driver. The original Easy Ethernet AVR code was written using ImageCraft's ICCAVR. To keep things like I like them (simple), I'll stick with the ImageCraft C compiler for the experimental AirDrop module as well. I'll pull heavily from the AirDrop-A firmware, which will have to be modified to accommodate the additional SRAM circuitry.

The first obstacle I fell upon was that I really needed to rethink and rewrite all of the basic AirDrop 802.11b driver I/O routines to support the additional SRAM and address latch. I then set out to rework all of my I/O port definitions to match the new SRAM-equipped hardware. Honestly, that's all I figured I really had to do to get the AirDrop web server on the air as I'm using the port of the working AirDrop-A firmware. However, after glaring at a dead AirDrop web server module, I did some quick single-step debugging and determined that the basic CompactFlash card reads and writes from the original AirDrop 802.11b driver simply did not work.

Sub Snippet 16.1 is a before and after compilation of the code needed to read and write the CompactFlash card's attribute registers.

Sub Snippet 16.1
```
**********************************************************
```

```c
***** BEFORE *****
void wr_cf_addr(unsigned int addr)
{
        addr_hi = (make8(addr,1)) & 0x07;
        addr_lo = addr & 0x00FF;
}
void wr_cf_reg(unsigned int reg_data,unsigned int reg_addr)
{
        TO_NIC;
        wr_cf_addr(reg_addr);
        data_out = reg_data;
        clr_WE;
        delay_ms(2);
        set_WE;
        FROM_NIC;
}
char rd_cf_reg(unsigned int reg_addr)
{
        char data,i;
        wr_cf_addr(reg_addr);
        clr_OE;
        i=1;
        while(--i);
        data = data_in;
        set_OE;
        NOP();
        return(data);
}
***** AFTER *****
void wr_cf_addr(unsigned int addr)
{
    TO_NIC;
    set_ALE;
    addr_hi = make8(addr,1);
    addr_lo = make8(addr,0);
```

```
    clr_ALE;
}
void wr_cf_reg(unsigned int reg_data,unsigned int reg_addr)
{
        wr_cf_addr(reg_addr);
        data_out = reg_data;
        clr_WECF;
        delay_ms(2);
        set_WECF;
        FROM_NIC;
}
char rd_cf_reg(unsigned int reg_addr)
{
        char data,i;

        wr_cf_addr(reg_addr);
        FROM_NIC;
        clr_OECF;
        i=1;
        while(--i);
        data = data_in;
        set_OECF;
        NOP();
        return(data);
}
```

****

Note that in the "before" code snippet the CompactFlash card I/O control lines have no rivals. Adding the SRAM duplicated some of the I/O signal names like *OE and *WE. Thus, all of the I/O control labels that include the letters CF belong to the CompactFlash card. Also, the "before" code always left a function with the data bus, which was separate from the address bus in the AirDrop-A scheme, in input mode. Now that we've added the 74LVTH573 latch and combined what used to be the data bus into a combination address/data bus the old data bus state logic no longer applies. After some more debugging runs, what I determined was that the combined address/data bus I/O pins at the inputs of the address latch were actually in input mode when the ATmega128L was attempting to write and latch the lower 8-bits of the target address. In the "after" code you can see that I still leave the combined address/data bus in input mode when I leave a function but the *TO_NIC* macro, which puts the combined address/data bus into output mode, is only issued in the write address (wr_cf_addr) function.

After squashing a few more minor bugs I found in my port I/O pin direction assignments, which are opposite that of the PIC I/O pin direction bits, I finally got the tuple hashing routines to run successfully. Although the tuples can tell us important things about the CompactFlash card, right now I'm only interested in making sure I get the correct I/O portal address so I can issue commands to the 802.11b CompactFlash NIC. I've stopped the AirDrop module firmware just short of enabling the 802.11b CompactFlash NIC in Screen Capture 16.1. The *cor_addr* variable holds the valid CompactFlash card I/O portal address, which in this case is 0x03E0.

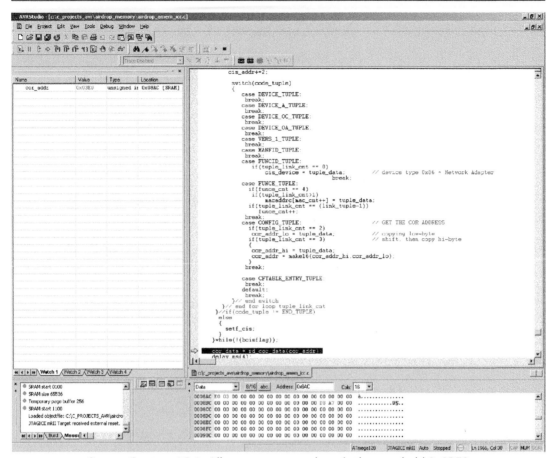

Screen Capture 16.1: All we want to see here is the good old 0x03E0.

At this point our troubles aren't over. The 802.11b CompactFlash NIC enabled just fine but that's where it ended. Using the Netasyst Sniffer, I captured what I thought would be the 802.11b CompactFlash NIC's probe session with the AP. However, I didn't see any radio activity from the Xterasys CWB1K. That could only mean that my 802.11b CompactFlash NIC register read and write routines were still in the ditch. After some further debugging I found that the Xterasys CWB1K NIC was never receiving an initialize command. The *rd_cf_io16* and *wr_cf_io16* functions in Sub Snippet 16.2 were not working.

Sub Snippet 16.2
```
*************************************************************
char   send_command(unsigned int cmd, unsigned int parm0)
{
       char cmd_code,rc;
       unsigned int cmd_status,evstat_data;

       do{
```

```
        cmd_data = rd_cf_io16(Command_Register);
      }while(cmd_data &  CmdBusy_Bit_Mask);

  wr_cf_io16(parm0, Param0_Register);
  wr_cf_io16(cmd, Command_Register);

  do{
      cmd_data = rd_cf_io16(Command_Register);
      }while(cmd_data &  CmdBusy_Bit_Mask);

  do{
      evstat_data = rd_cf_io16(EvStat_Register);
      }while(!(evstat_data & EvStat_Cmd_Bit_Mask));

  cmd_status = rd_cf_io16( Status_Register);
  cmd_code = cmd_status &  Status_CmdCode_Mask;
  rc = cmd_status &  Status_Result_Mask;
  wr_cf_io16( EvStat_Cmd_Bit_Mask, EvAck_Register);

  switch(rc)
  {
        case 0x00:
                rc = 0;
                break;
        case 0x01:
                rc = 1;
                break;
        case 0x05:
                rc = 2;
                break;
        case 0x7F:
                rc = 3;
                break;
        default:
                //return_code = result_code;
                break;
  }
  return(rc);
```

****

Sub Snippet 16.3 is another "before and after" code snippet featuring the 16-bit 802.11b CompactFlash NIC register read and write routines. Now that I'm only issuing the *TO_NIC* macro in the *wr_cf_addr* function, the "before" *rd_cf_io16* function will never work as the combination address/data bus is always in the output mode following the address write and latch operation. The cure here is made by the same medicine that got us through the tuple hash. In the Sub Snippet 16.3 "after" code snippet I only issue the *TO_NIC* macro in the address write/latch function and leave the 16-bit read routines to call out the *FROM_NIC* input macro as they need it.

Sub Snippet 16.3

```
*************************************************************
```

```
***** BEFORE *****
void wr_cf_io16(unsigned int data16,unsigned int addr)
{
    char i;

        wr_cf_addr(addr);
        data_out = LOW_BYTE(data16);
        TO_NIC;
        clr_IOWR;
        i=1;
        while(--i);
        set_IOWR;
        NOP();

        wr_cf_addr(addr+1);
        data_out = HIGH_BYTE(data16);
        clr_IOWR;
        i=1;
        while(--i);
        set_IOWR;
        NOP();
        FROM_NIC;
}
unsigned int rd_cf_io16(unsigned int addr)
{
        char data_lo,data_hi,i;
        unsigned int data16;

        wr_cf_addr(addr);
        clr_IORD;
        i=2;
        while(--i);
        data_lo=data_in;
        set_IORD;

        wr_cf_addr(addr+1);
        clr_IORD;
        i=2;
        while(--i);
        data_hi=data_in;
        set_IORD;
        NOP();
        data16 = make16(data_hi,data_lo);
        return(data16);
}
***** AFTER *****
void wr_cf_io16(unsigned int data16,unsigned int addr)
{
```

```
    char i;

        wr_cf_addr(addr);
        data_out = LOW_BYTE(data16);
        clr_IOWR;
        i=1;
        while(--i);
        set_IOWR;
        NOP();

        wr_cf_addr(addr+1);
        data_out = HIGH_BYTE(data16);
        clr_IOWR;
        i=1;
        while(--i);
        set_IOWR;
        NOP();
        FROM_NIC;
}
unsigned int rd_cf_io16(unsigned int addr)
{
        char data_lo,data_hi,i;
        unsigned int data16;

        wr_cf_addr(addr);
        FROM_NIC;
        clr_IORD;
        i=2;
        while(--i);
        data_lo=data_in;
        set_IORD;

        wr_cf_addr(addr+1);
        FROM_NIC;
        clr_IORD;
        i=2;
        while(--i);
        data_hi=data_in;
        set_IORD;
        NOP();
        data16 = make16(data_hi,data_lo);
        return(data16);
```

****

After making the modifications you see in the Sub Snippet 16.3 "after" code, the newly ported AirDrop 802.11b driver fired up and successfully joined the AIRDROP_NETWORK BSS.

Coding a Simple 802.11b Web Server

The first step in getting the 802.11b web server off the ground is to make sure that I'm receiving the HTTP request correctly from the client. My Easy Ethernet/AirDrop-A TCP routines are good enough for simple Telnet transactions but just weren't quite right for processing HTTP requests. Basically, all I need to do is receive the HTTP request, parse the incoming HTTP request data, send an HTML page and disconnect the TCP/IP session.

I stopped the code in Screen Capture 16.2 just before testing for the "GET /" string in the HTTP request I initiated from one of the Florida room personal computers. I pulled the first 32 bytes from the incoming HTTP request looking for only the first five characters. I sized the *http_temp* array to allow for parsing things following the initial "GET /" string. For instance, following the "GET /" string one could specify a filename to retrieve or include some codes to control or monitor I/O pins on the server. Right now, all I really want to do is set up very simple HTTP server framework and send some HTML to the client browser. Once that is accomplished, then I can add the code to serve a real web page from the AirDrop.

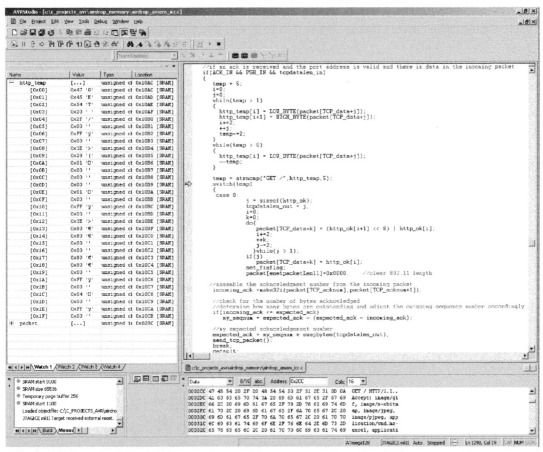

Screen Capture 16.2: This AVR C source code looks just like the PIC source code you've seen throughout this book. In fact, most of it is the same. That's a statement in C portability.

Let's start at the beginning and fast forward up to the breakpoint in Screen Capture 16.2. The AirDrop web server module has initialized and joined the BSS (Basic Service Set) controlled by the AIRDROP_NETWORK AP with an IP address of 192.168.0.151. A simple HTTP request (http://192.168.0.151) was issued from the SNOOPER laptop, which is 802.11b enabled. A TCP/IP session was established between SNOOPER (the client) and the AirDrop web server via the AIRDROP_NETWORK AP. The establishment of the TCP/IP session went just as it was described to you in Chapter 14. Once the TCP/IP session is established, the client, SNOOPER, can exchange data with the AirDrop web server. SNOOPER issues the HTTP request via its web browser application. The AirDrop web server received the HTTP request data, determined that it was addressed to 192.168.0.151 port 80 (the well-known HTTP port) and parsed the incoming HTTP data looking for the "GET /" string. If all goes as planned, the AirDrop web server should find the "GET /" string, push some HTML into its transmit buffer, tack on a FIN in the TCP header and transmit the HTTP reply back to SNOOPER.

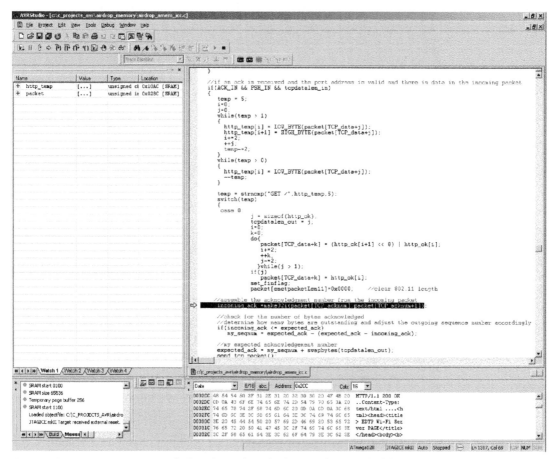

*Screen Capture 16.3: It's not obvious here, but the **http_ok** string holds the entire web page, which in this scenario is super simple and just consists of a text message.*

The HTTP response has been loaded into the AirDrop web server's transmit buffer in Screen Capture 16.3. Where did that come from? Take a look at Sub Snippet 16.4. That's our "web page" complete with the HTTP high sign. This whole thing started and ended with SNOOPER's web browser. The results of SNOOPER's HTML request are shown in Screen Capture 16.4. When you look over and follow the web server thread in the AirDrop web server source code, you'll see that it really didn't take much coding effort to put that one-liner in the client's web browser window.

Sub Snippet 16.4

```
*********************************************************
const char http_ok[]="HTTP/1.1 200 OK \r\nContent-Type: text/html \r\n\r\
n<html><head><title> EDTP AirDrop Web Server PAGE</title></head><body><b><u>
BUILDING AND CODING 802.11b DEVICES IS EASY.</b></u></body></html>";
*********************************************************
```

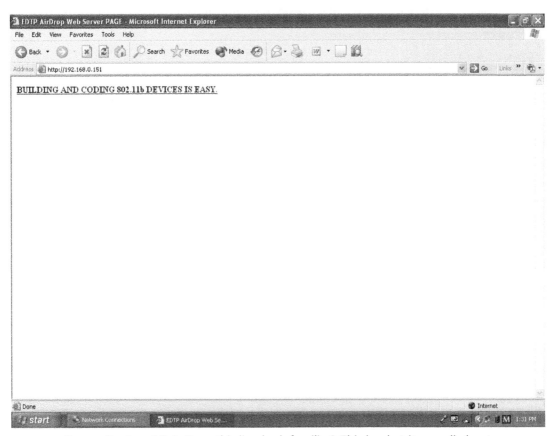

Screen Capture 16.4: Does this line look familiar? This is what it was all about.

The AirDrop SRAM

Now that I've shown you how easy it is to create a simple wireless web server you can add your personality to the code and generate some pages yourself. You are sure to want to store stuff along the way but if you look closely at the data memory layout in the experimental AirDrop 802.11b driver source code, you'll see that the AirDrop web server's little TCP/IP stack uses most of the ATmega128L's on-chip SRAM. I included the 128K × 8 SRAM on the AirDrop web server board to avoid you having to limit your web server application due to a lack of available data memory.

The way you access the external SRAM depends on how your assembler or compiler handles the extra memory area. Screen Capture 16.5 is a code snippet and debugger view of ImageCraft's external SRAM handler mechanism. The ImageCraft scheme allows the programmer to allocate chunks of external SRAM space. In my example I've carved out a 256 byte block of external SRAM called *xmem* beginning at data memory address 0x1100,

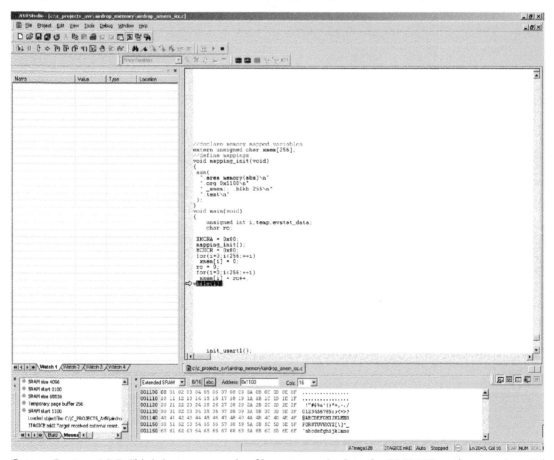

Screen Capture 16.5: This is just an example of how to manipulate the SRAM IC on the experimental AirDrop module. I'm sure you'll find something to do with all of that extra memory.

which is the beginning of the external SRAM memory area and immediately follows the ATmega128L's 4K of on-chip SRAM. The real work is done with the inline assembler routine, which was generated by the ICCAVR Application Builder. Using the ICCAVR Application Builder application, all I had to do was designate the external SRAM address, name the memory block and enter the number of bytes to reserve. My hand-written code simply clears the SRAM block to zero and then writes the block with ascending ASCII characters beginning at ASCII 0x00. Nothing to it.

Although there's plenty of SRAM to go around, you'll probably still want to store your large web pages in program Flash. I noticed that when I told the ImageCraft C compiler that I had 64K of external SRAM, it took it upon itself to start storing variables there on its own. There's a warning that comes up when you tell the compiler you have the extra data memory. It basically says that if you want to control the content of the external SRAM don't tell the compiler about it.

This chapter was all about the basics of serving web pages from a wireless 802.11b station. Add another line to your 802.11b resume.

CHAPTER **17**

A New Kid in Town
Who Calls Himself ZigBee

Yeah, I know. This is supposed to be a book based totally on 802.11b technology. While I was traveling the Internet world in search of the Holy Grail of 802.11b, I came across this rabbit. Nope, that's another story. I came across this ZigBee stuff. Everybody seems to be making a big deal about it. So, I decided to get my hands on a ZigBee development kit and see what all of the hoopla was about.

Zig What???

I love my wife. So, ladies don't take this in the wrong way. When I first put ZigBee and IEEE 802.15.4 in the same sentence, I immediately thought of my wife's closet. It's full of shoes. I thought to myself, here's ZigBee, based on the IEEE 802.15.4 standard, another pair of 802 dot something shoes in the IEEE 2.4 GHz closet. All I know is what I have read about ZigBee. So, instead of listening to ZigBee hearsay, I decided to take a hard look at ZigBee myself. You and I are about to open a box full of ZigBee goodies from Microchip.

What's in the ZigBee Box

Packed neatly inside the standard Microchip development kit two-piece carton were four printed circuit boards; two 9V batteries, two radio cards (they had the word *Antenna* silked on them) and a pair of larger printed circuit boards that looked like what you would expect a Microchip development board to look like. I mated one of the radio printed circuit boards to the ZigBee development board labeled COORD and had them smile for the camera in Photo 17.1. A CDROM was also in the box and I pulled it out of its sleeve and inserted it into one of the EDTP Florida room personal computers to have a look at it.

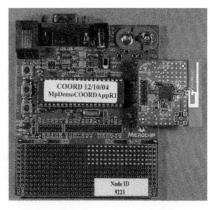

Photo 17.1: Pretty cool. I didn't have to tear the radio apart to see what was inside. And, there seem to be no "secrets" to this ZigBee stuff.

One of the first things I learned was what to call all of this. Officially, the gaggle of boards is called the PICDEM Z Demonstration Kit. I'm going to refer to it as the PICDEM Z for the remainder of this text. The two pairs of printed circuit boards make up a two-node ZigBee network. The printed circuit board marked COORD is the coordinator node and the printed circuit board marked RFD is the Reduced Function Device node. The ZigBee radios that were included in the PICDEM Z are based on Chipcon CC2420 devices. In addition to a ton of source code files, there's a big ZigBee application note on the CDROM.

Before I wandered off into a corner of the EDTP Florida room with the ZigBee application note AN965 in tow, I wanted to know a bit more about the hardware. Both of the 40-pin PIC microcontroller identification markings on each of the motherboard microcontrollers are covered with opaque labels. A quick look at the PICDEM Z schematic revealed that both ZigBee motherboards are supported by a PIC18LF4620, which is one of the new Microchip nanoWatt microcontrollers comprised of 64K of program flash and 3.9K of on-chip SRAM. These speedy little microcontrollers are only clocked at 4 MHz and I figure that's for power conservation as both motherboards have 9V battery power terminals in addition to wall wart power receptacles. The other reason why a 4 MHz oscillator would be used is to allow the PIC's internal PLL circuitry to produce a 4x-oscillator frequency, 16 MHz instruction clock. A 28-pin socket is nestled inside the wings of the 40-pin PIC18LF4620 socket to allow the human ZigBee programmer to evaluate the PICDEM Z using the smaller footprint PIC18LF2620. Two user-defined pushbutton switches and two user-defined LEDs attached to the PIC18LF4620's I/O subsystem are provided for the ZigBee programmer's convenience. These pushbutton switches and LEDs are also used by the ZigBee demo that comes preprogrammed into the PICDEM Z motherboards' PIC18LF4620 microcontrollers. There's also a SPI-attached TC77 temperature sensor, a standard RS-232 serial port and the Microchip standard 6-pin ICSP programming/debugging port. The PICDEM Z's 6-pin interface was designed to utilize the services of the Microchip MPLAB ICD 2 or Microchip PM3. As will all future PICDEM Z ZigBee radios, the Chipcon-based ZigBee radio mates with a Samtec 12-pin male/female connector on the PICDEM Z motherboard, which supplies +3.3VDC, a SPI bus, ground and a couple of necessary radio I/O control lines that are terminated at the PIC18LF4620. If you're interested in power consumption, there's even a jumper you can cut to wire in your ammeter. Need to measure the PICDEM Z's power draw with an analog-to-digital converter? No problem, there's a pad set for a voltage drop resistor in the same current measurement area of the ZigBee motherboard. All of the aforementioned circuitry is pure Microchip and you can download the PICDEM Z schematics from the Microchip website.

My interest in the radio technology behind 802.11b rolled right over to the ZigBee radios. So, I downloaded a Chipcon CC2420 data sheet and added it to my ZigBee reading list. I guess deep down I really want my own pointy hat with moons and stars on it. Without going into detail, the CC2420 is designed for ZigBee. It is IEEE 802.15.4 compliant operates at 2.4 GHz and can modulate magnetic fields at rates up to 250 kbps. Like most of the 802.11b radio chip sets, the CC2420 does lots of neat things in hardware. Without microcontroller intervention the CC2420 buffers data, encrypts data, authenticates data and supervises the link. All of the configuration and data transfer between the C2420 and the host, which is most likely to be a microcontroller, is done via a simple SPI interface. From what I could gather from the rest of the CC2420 data sheet, there are also lots of things about the way the CC2420 does those things that RF things do that will make an RF guru very happy as well. A nose-to-ZigBee-radio view of one of the Chipcon ZigBee radios that came with my PICDEM Z can be seen in Photo 17.2.

Photo 17.2: What do you know, a PIFA is etched into this ZigBee radio printed circuit board. That stands to reason giving the operational band of the ZigBee radio.

Making ZigBee Talk

The ZigBee radio/motherboard combinations that come with the PICDEM Z package are supported by the Microchip ZigBee Stack, which you can get free from the Microchip website. The Microchip ZigBee Stack isn't ZigBee-protocol compliant but it does the job. I have the cutting-edge version of the Microchip ZigBee Stack and there are a few things that are not yet supported. The Microchip ZigBee Stack documentation points all of those shortcomings out. The Microchip ZigBee Stack documentation also states that the Microchip ZigBee Stack is evolving and as time passes more features and more radio types will be supported.

The PICDEM Z motherboard comes preprogrammed with Coordinator and Reduced Function Device (RFD) applications that utilize Microchip ZigBee Stack API calls. The PICDEM Z designers and programmers didn't expect you to be a ZigBee expert. So, they splashed generous amounts of debug text all through the demo application code. The PIC-DEM Z programmers at Microchip also created a nice menu system that encapsulates some of the Microchip ZigBee Stack API action into a simple menu driven interface that you can interact with using a personal computer and a terminal emulator program.

I kicked off two instances of Tera Term Pro in Screen Capture 17.1. The COM1 Tera Term Pro instance is attached to the PICDEM Z coordinator node and the COM2 Tera Term Pro instance in Screen Capture 17.1 is attached to the PICDEM Z's RFD node.

At this point we're talking to the application code only and the ZigBee radios are silent. To enter this NO-RF zone, I powered up both of the PICDEM Z's motherboards with the motherboard's configuration switch depressed. To give you an idea of how these menus affect the PICDEM Z motherboards, I started the coordinator node (Run application menu item) and after the network was successfully started, I performed a network join task at the RFD node. Screen Capture 17.2 shows what happened in the terminal emulator windows when the RFD left and rejoined the little ZigBee network that was created.

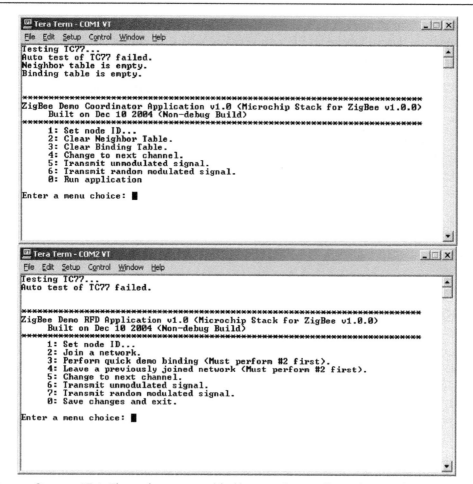

Screen Capture 17.1: These shots are enabled by entering configuration mode at power up. Neither the Coordinator nor the RFD ZigBee radio is active at this point. If you have all of the proper RF analysis equipment, you can choose Options 5 and Option 6 on the menu.

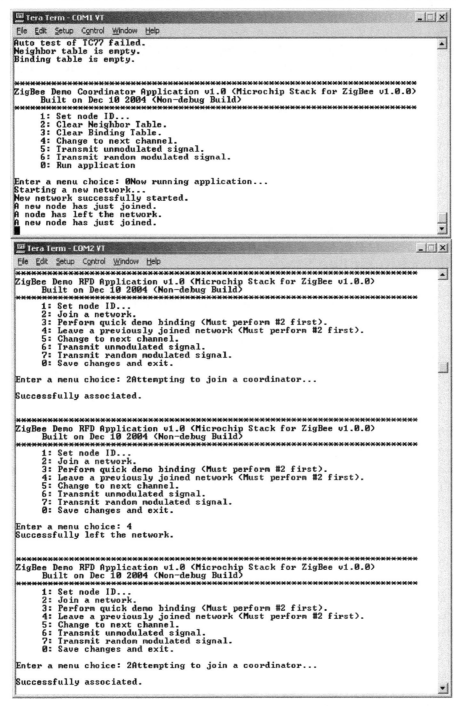

*Screen Capture 17.2: Running the ZigBee Coordinator application fired up the Coordinator's ZigBee radio. I turned on the RFD ZigBee radio by selecting the **Join a network** menu item.*

The PICDEM Z demo consists of a coordinator application and an RFD application that allows either PICDEM Z ZigBee device to toggle a LED on the other ZigBee device via one of the pushbuttons on each PICDEM Z motherboard. The PICDEM Z ZigBee network is predefined so that the coordinator node is associated with the RFD node. The pushbutton-to-LED bindings are also predefined. The quick demo binding menu option also calls all of the associated Microchip ZigBee Stack code to perform the pushbutton-to-LED binding process I just described.

The PICDEM Z coordinator node is brought up first and establishes a new network. Then the PICDEM Z RDF node is enabled and rejoins the coordinator's newly hatched ZigBee network. When the PICDEM Z ZigBee network is started in the manner I just described, the coordinator Tera Term Pro window will issue the message "New network successfully started" followed by the message "A familiar node has just rejoined." The RFD Tera Term Pro window posts the message "Rejoin successful."

The Microchip ZigBee Stack

The current Microchip ZigBee Stack supports Microchip's PIC18 line of microcontrollers. Microchip's own C18 C compiler is supported as well as HI-TECH's PICC-18 C compiler. The Microchip ZigBee Stack source code is "self-adjusting" as far as the C compilers are concerned. The individualities of the C18 and PICC-18 compilers are sorted out in the source and brought into play depending on which C compiler you use. A modular top-down approach makes the Microchip ZigBee Stack source easier to understand and apply.

From top to bottom, the Microchip ZigBee Stack consists of five distinct layers. All of the layers communicate directly with the layer below and above them. The user application resides at the penthouse level of the stack. If necessary, the user application can choose to call any function from any of the lower layers as well. Each layer interacts with the layer below it using APIs. This top-down, layer-to-layer communication is usually done with C macros, which call functions from the layer below. Each Microchip ZigBee Stack layer is coded as a separate C source file.

The Application Layer, or APL, is subordinate to the user application. An Application Support Sublayer (APS) rides A6 Intruder style (that's side-by-side for you civilians) in the Application Layer and provides lower-level support for the Application Layer much like the A6's Bombardier/Navigator, or BN, does for the A6 pilot. The APL is a high-level stack layer used to manage the Microchip ZigBee Stack functionality. As I mentioned earlier, a C source file (zAPL.c) is the basis of the physical APL. The APIs supported by the APL are defined in an associated header file (zAPL.h).

A typical APL function is APLInit. APLInit initializes all of the Microchip ZigBee Stack modules by calling each module's initialization routine. Typically, the next call will be to APLEnable, which enables the ZigBee MAC. A ZigBee network is established by calling APLNetworkInit. The APLNetworkInit call forces the ZigBee coordinator node to scan the medium and establish itself as an identifiable ZigBee network in its domain. I think you get the idea behind the APL. I've pulled the sequence I've just described into Code Snippet 17.1.

Code Snippet 17.1

```
************************************************************
//** START APL **
// Read MAC long address from nonvolatile memory.
 NVMRead(&macInfo.longAddr,
    (ROM void*)&macLongAddr,
       sizeof(macInfo.longAddr));

// This is main Zigbee stack initialization call.
// MAC must be enabled when application is ready.
APLInit();

// We have to enable MAC separately. This way you can disable and enable as
per your
// requirement.
APLEnable();

// First of all establish a network.
APLNetworkInit();

//** END APL **
//** START APS **
// Private copy of handle to lamp endpoint.
EP_HANDLE hD1;
void D1Init(void)
{
    // We will assume that main application would already have initialized
D1 as dig output.

    // Open endpoint for lamp.
    hD1 = APSOpenEP(EP_D1,                       // Lamp EP number
                   PICDEMZ_DEMO_CLUSTER_ID,      // Lamp cluster number
                   0,     // Dest EP not relevant as this is an indirect EP
                   FALSE);  // This is an indirect (i.e. via Coordinator)

}
//** END APS **
//** START zMAC.c TX_INDIRECT_BUFFER **

#if defined(MAC_USE_TX_INDIRECT_BUFFER)
// MAC_USE_TX_INDIRECT_BUFFER is used by coordinator only to hold outgoing
(indirect) frames until
// remote end device request it.

    #define TX_INDIRECT_BUFFERS          (6)
    #define INVALID_INDIRECT_BUFFER      (0xFF)

    typedef struct
    {
        unsigned char *pBuffer; // Pointer to dynamically allocated TX buffer
```

```
    struct                      // Flags about current frame.
    {
        unsigned int hFrame:7;
        unsigned int bIsSent:1;
        unsigned int bIsInUse:1;
    } Flags;

    BYTE frameDSN;              // DSN assigned to this frame.
    SHORT_ADDR destShortAddr;  // Destination that it is addressed to.
    TICK lastTick;             // Tick to calculate timeout condition.
    } TX_INDIRECT_BUFFER;
//** END zMAC.c TX_INDIRECT_BUFFER **
```

****

Code Snippet 17.1: The Microchip ZigBee Stack comes arranged as a set of individual C source files. You can program at the upper levels of the stack or dip into the low-level details of supporting Microchip ZigBee Stack files. The really good news is that the ZigBee stack support HI-TECH PICC-18 C compiler natively.

ZigBee endpoint interface support is provided by the APS. The user application uses the APS to open and close ZigBee endpoints as well as send and receive data associated with the endpoints. The ZigBee binding table is maintained by the APS. With the help of flash programming routines provided by the Microchip ZigBee Stack's zNVM.c file, the APS stores binding information in tables located in the PIC18LF4620's nonvolatile flash memory area. The zNVM.c file is unique to the Microchip ZigBee Stack and can be replaced with a user-defined set of flash binding table storage utilities.

I attached my MPLAB ICD 2 to the COORD ZigBee module in an attempt to show you a binding table entry. Due to the use of a custom binding routine, the normal binding table entries never got populated. So, instead I pulled an LED endpoint initialization routine that is issuing the APSOpenEP call into Code Snippet 17.1 for you.

Right now, the Microchip ZigBee Stack only supports a non-slotted star network. That means that all ZigBee traffic must flow through a coordinator node. Since the typical ZigBee RFD node will be sleeping most of its life, the coordinator node must buffer ZigBee packets destined to the sleepy subordinate ZigBee nodes. This is done by the APS, which maintains what is called an indirect transmit buffer. If the buffered frame is addressed to one of the nodes in the coordinator's binding entry table, the frame is held in the indirect transmit buffer until the recipient node retrieves it or a predefined timeout period occurs. I did manage to catch an entry in the PICDEM Z coordinator's indirect transmit buffer in Screen Capture 17.3. I've included the code associated with the Watch Window and File Registers display inside of Code Snippet 17.1.

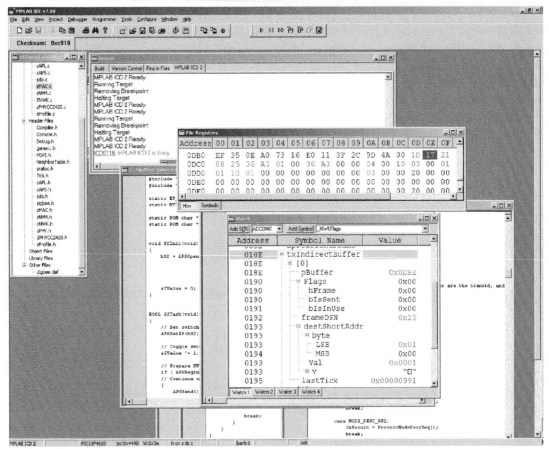

Screen Capture 17.3: This Screen Capture gives you an idea of what can be done by simply connecting an MPLAB ICD 2 to one of the PICDEM Z's ZigBee nodes. Using well placed breakpoints, you can turn the MPLAB ICD 2 into a rudimentary ZigBee packet sniffer.

The Network Layer lies below the Application Layer/Application Support Sublayer slot and does what every other network layer does. In the ZigBee world, the Network Layer handles the establishment and maintenance of the network connection. As a ZigBee application programmer using the Microchip ZigBee Stack, you won't have to worry about interfacing to the Network Layer. The same is true for the MAC and Physical layers of the Microchip ZigBee Stack. The Microchip ZigBee Stack MAC layer supports all of the IEEE 802.15.4 functions. Since the PICDEM Z will be able to use differing radios, the MAC and PHY can change with the type of radio. To help keep a handle on the MAC and PHY layers, the Microchip ZigBee Stack will include a set of MAC/PHY files that are unique to the radio being used. For instance, the indirect transmit buffer code snippet in Code Snippet 17.1 was taken from the PICDEM Z's zMAC.c file.

While I had my MPLAB ICD 2 attached to the PICDEM Z's coordinator node, I went to the ZigBee.def file and uncommented *#define ENABLE_DEBUG*. While I was looking around inside the ZigBee.def file, my assumption about using the 4 MHz crystal and the internal PLL was confirmed. The *CLOCK_FREQ* definition is set for 16 MHz with the comment stating that this clock speed was a result of using the HS-PLL clocking option. As you can see for yourself in Screen Capture 17.4, following the recompilation and reprogramming of the PIC18LF4620, the real time ZigBee information in my Tera Term Pro emulator windows became more detailed.

Screen Capture 17.4: ZigBee gibberish. I guess we'll get the hang of it by and by. Maybe ZigBee will get big enough to write a book about someday.

In a nutshell, ZigBee is a slower form of wireless data communications with a limited range and lower bandwidth than 802.11b. I'll keep a watchful eye on this wireless newcomer.

CHAPTER **18**

Parting Frames

I hope you have found this to be an enjoyable and enlightening journey. This book is only the beginning of your 802.11b experience. I've set up a Yahoo AirDrop user's group. There you will find some pretty savvy folks doing all sorts of things with both variants of the AirDrop. I highly recommend that you join the AirDrop Yahoo forum. Here's the information you need to join the AirDrop Yahoo group:

Group name: airdrop_user

Group home page: http://groups.yahoo.com/group/airdrop_user

Group email address: airdrop_user@yahoogroups.com

All of the AirDrop hardware, including the experimental AirDrop variant you saw, can be obtained from EDTP Electronics, Inc. The EDTP website (http://www.edtp.com) is also a repository for the AirDrop 802.11b drivers for both the AirDrop-A and the AirDrop-P. You can also find the experimental AirDrop 802.11b driver on the EDTP website.

I've had a blast researching and writing this book. I'm always there for my readers and if you have questions of comments, just fire off a note to: fred@edtp.com.

Conventions

This book's pages are very heavily laden with AirDrop 802.11b driver source code. To help you successfully navigate and understand the various AirDrop 802.11b driver source code segments, I've attempted to put a predictable pattern into place that follows through all of the source code segments you will encounter in the chapters of this book.

Numeric Notation

You will not see any octal representations of numbers in this text. However, you will see hexadecimal, binary and decimal numeric notation.

Decimal notation will always be explicit. "The result is 37 decimal."

Hexadecimal notation will always prefix the numeric value with 0x. "The variable's value is 0x55."

Binary notation will always prefix the numeric value with 0b. "The binary equivalent of 0x55 is 0b01010101."

Source Code Presentation

Major code segments of the AirDrop 802.11b driver will be presented as Code Snippets. A Code Snippet will always be captioned and delineated with a short star bar. All AirDrop module Code Snippets in this text are HI-TECH PICC-18 C source code segments unless otherwise noted at the beginning of the Code Snippet. For example:

Code Snippet 14.1
★★★★★★★★★★★★★★★★★★★★★★★★★★★★★

```
//*************************************************************
//*    IP Protocol Types
//*************************************************************
#define PROT_ICMP          0x01
#define PROT_TCP           0x06
#define PROT_UDP           0x11

    if(HIGH_BYTE(packet[ip_ttlproto]) == PROT_ICMP)
     icmp();
    else if(HIGH_BYTE(packet[ip_ttlproto]) == PROT_UDP)
     udp();
```

```
    else if(HIGH_BYTE(packet[ip_ttlproto]) == PROT_TCP)
    tcp();
```

★★★★★★★★★★★★★★★★★★★★★★★★★★★★

Code Snippet 14.1: These definitions are used by the AirDrop 802.11b driver to determine how to process an incoming 802.11b frame.

Sub Snippets

Sub Snippets are pieces of larger Code Snippets that are brought to the front for detailed analysis. Here's an example of a Sub Snippet taken from Code Snippet 14.1:

Sub Snippet 14.9

★★

```
    if(HIGH_BYTE(packet[ip_ttlproto]) == PROT_ICMP)
    icmp();
    else if(HIGH_BYTE(packet[ip_ttlproto]) == PROT_UDP)
    udp();
    else if(HIGH_BYTE(packet[ip_ttlproto]) == PROT_TCP)
    tcp();
```

★★

Sub Snippets may contain supporting code segments from related functions. A short star bar will be used to separate the main body of the Sub Snippet from the supporting code.

Sub Snippet 17.2

★★

```
       (SUPPORTING CODE)
char flags;
#define synflag 0x01
#define bsynflag        flags & synflag
#define clr_synflag    flags &= ~synflag
```
★★★★★★★★★★★★★★★★★★★★★★★★★★★★
```
       (MAIN BODY)
    //IF THE INCOMING PACKET IS A RESULT OF SESSION ESTABLISHMENT
    if(bsynflag)
    {
       //CLEAR THE SYN FLAG
       clr_synflag;
```

★★

Sub Snippets will always be surrounded by long star bars and will never be augmented with a caption. Sub Snippets will always be referred to in detail in the text that immediately surrounds them.

Netasyst Sniffer Capture Text Presentation

In addition to graphic screen captures, Netasyst Sniffer captures will be presented in a text format. The use of both graphics and text will enhance your view of the Netasyst Sniffer capture. All text Netasyst Sniffer captures will follow this format:

Sniffer Text 19.5

```
DLC: ----- DLC Header -----
      DLC:
      DLC: Frame 56 arrived at  10:15:03.1070; frame size is 10 (000A hex)
bytes.
      DLC: Signal level                    = 100%
      DLC: Channel                         = 1
      DLC: Data rate                       = 22 (11.0 Megabits per second)
      DLC:
      DLC: Frame Control Field #1 = D4
      DLC:                 .... ..00 = 0x0 Protocol Version
      DLC:                 .... 01.. = 0x1 Control Frame
      DLC:                 1101 .... = 0xD Acknowledgment (ACK) (Subtype)
      DLC: Frame Control Field #2 = 00
      DLC:                 .... ...0 = Not to Distribution System
      DLC:                 .... ..0. = Not from Distribution System
      DLC:                 .... .0.. = Last fragment
      DLC:                 .... 0... = Not retry
      DLC:                 ...0 .... = Active Mode
      DLC:                 ..0. .... = No more data
      DLC:                 .0.. .... = Wired Equivalent Privacy is off
      DLC:                 0... .... = Not ordered
      DLC: Duration                       = 0 (in microseconds)
      DLC: Receiver Address               = Station Netgea6FD3DA
      DLC: Implied Transmitter Address    = Station Xircom03D92F
ADDR  HEX                                         ASCII
0000: d4 00 00 00 00 09 5b 6f d3 da      | Ô.....[oóÚ
```

Sniffer Text 19.5: This is an 802.11b ACK frame.

Sniffer Text will always be accompanied by a caption.

Mini Sniffs

Mini Sniffs are smaller pieces of Sniffer Text modules. Mini Sniffs will always contain associated highlighted fields and will never be supported by a caption. Like Sub Snippets, Mini Sniffs are referenced in detail by their surrounding text.

Mini Sniff 12.3

```
TCP: ----- TCP header -----
     TCP:
     TCP: Source port                = 1859
     TCP: Destination port           = 8088
     TCP: Initial sequence number = 3238359086
     TCP: Next expected Seq number= 3238359087
     TCP: Data offset                = 28 bytes (4 bits)
     TCP: Reserved Bits: Reserved for Future Use (6 bits)
     TCP: Flags                      = 02
     TCP:                   ..0. .... = (No urgent pointer)
     TCP:                   ...0 .... = (No acknowledgment)
     TCP:                   .... 0... = (No push)
     TCP:                   .... .0.. = (No reset)
     TCP:                   .... ..1. = SYN
     TCP:                   .... ...0 = (No FIN)
     TCP: Window                     = 16384
     TCP: Checksum                   = 691D (correct)
     TCP: Urgent pointer             = 0
     TCP:
     TCP: Options follow
     TCP: Maximum segment size = 1460
     TCP: No-Operation
     TCP: No-Operation
     TCP: SACK-Permitted Option
     TCP:
ADDR  HEX                                             ASCII
0000: 08 02 95 00 00 e0 98 bc 59 67 00 09 5b 6f d3 da | ..•..à~¼Yg..[oÓÚ
0010: 00 05 3c 03 d9 2f 60 cb aa aa 03 00 00 00 08 00 | ..<.Ù/`Ëªª......
0020: 45 00 00 30 c3 1f 40 00 80 06 b5 b5 c0 a8 00 0b | E..0Ã.@.€.µµÀ..
0030: c0 a8 00 97 07 43 1f 98 c1 05 70 2e 00 00 00 00 | À".−.C.~Á.p....
0040: 70 02 40 00 69 1d 00 00 02 04 05 b4 01 01 04 02 | p.@.i......´....
```

Index

A

Access command, 117, 123, 377
Access point, 4, 10-14, 21, 23, 153-154, 156, 162-163, 173, 80-82, 97, 102-103, 377, 410-411, 484-490
Ad hoc system, 81-82, 377
Airdrop_cfg, 113-114, 116, 123, 126, 128-130, 134, 138, 147, 164, 167, 335-338, 341, 377-378, 507-511
allocate_buffer function, 144, 381-382
allocate_xmit_buffers function, 138, 381-382
allocate_xmit_buffers(), 137-138, 140, 147, 164, 167, 335-336, 381-382
AP, 4, 80, 93-96, 99-104, 114, 137, 142, 148, 150-153, 155-160, 162-164, 172, 174-175, 209-210, 229, 250-251, 282-284, 296, 336, 341, 350-351, 355-356, 377-379, 381-390, 404-412, 422-424, 427, 456-457, 495-500
ARP frame, 168-169, 171, 174, 184, 381-385
Association, 137, 160-163, 81-82, 381-386, 478-479
Atmel JTAGICE mkII, 6, 346, 381-387, 456-458
Authentication, 137, 158-160, 342, 81-82, 108, 377-379, 387, 439-443, 456-461, 478-479, 484-486
AVR JTAG, 17, 346, 381-388, 433-437

B

BAP, 79, 105-107, 109, 112-117, 121, 123-125, 128, 140-141, 145, 189, 191-192, 198-200, 203-204, 216-217, 222, 228, 255, 261, 275-279, 289, 315, 387-389, 416-429, 433-437, 478-479, 484-486
bap_write, 112-116, 124, 140-141, 145, 198-200, 204, 222, 228, 261, 289, 315, 387-389, 422-424
bap_write function, 114-116, 124, 387-389
Beacon, 80-82, 94-103, 153-154, 156, 160, 377-379, 381-385, 387-390, 422-427, 450-452, 456-459, 473-476, 478-484
big endian, 191-192, 387-391
BSS, 80-82, 93-96, 99-104, 108-109, 113-114, 117-118, 123, 125-127, 129, 134, 137-138, 142, 147, 152, 156, 158, 163-164, 166-167, 171, 335-338, 341, 353-356, 377-379, 381-384, 387-396, 422-425, 444-448, 456-457, 462-464, 490-494
BSSID, 99-100, 149, 151-152, 155, 164-166, 172, 174-175, 184, 381-393, 478-480, 484-490

C

Channel, 80-82, 93-99, 103, 108, 149, 153-159, 161-163, 169, 171, 206, 209, 229-230, 250, 269, 272, 294, 296, 317-318, 320-321, 330, 375, 393-397, 456-461
CIS, 40-43, 46-48, 50-54, 56-62, 64-66, 68-75, 77-78, 88, 387, 393-399, 410-412, 422-426, 433-435, 444-447, 456-457, 473-475, 490, 501, 507-515
Command_Register, 83-85, 87-90, 109-111, 118-120, 138-139, 195-197, 351-352, 399-401
CompactFlash, 2-7, 9-15, 19-21, 23, 25-29, 31-43, 45-46, 50-54, 56-57, 62-72, 74-

79, 81-83, 86-91, 93-96, 100, 105, 108-109, 114, 117, 125-127, 130, 132-135, 137, 143, 146, 148, 158, 163, 178, 184, 191-192, 202-204, 212, 218-219, 223, 228, 243, 249, 326, 345-352, 377-390, 393-403, 416-417, 422-429, 433-436, 444-452, 456-464, 467-481, 490-491, 495-502, 507-514

COR, 41-43, 51, 56-57, 62-63, 66, 68-78, 350, 393-395, 404-405, 427-429, 444-445, 450-455, 478-483, 495-498

create_comframestructure function, 132, 141, 191, 404-406

D

data function, 115, 117, 404-408, 467-470, 507-513

DCF-660W, 26-32, 35, 37, 404-408

Direct Sequence Spread, 103, 410-413

Distribution System, 142-143, 149-150, 153, 155, 158-162, 169-172, 175, 206, 209, 229-230, 250, 269, 272, 294-297, 317-318, 320-321, 330, 81-82, 96, 99, 375, 410-413, 427, 495-500

do_event_housekeeping function, 204-205, 228, 410-414

DSAP, 169-170, 173, 177, 206-207, 209, 211, 229-230, 250, 257, 269, 272, 294-295, 297, 317-318, 320-321, 330, 410-415

E

echo_packet function, 222, 228, 261, 267, 289, 315-316, 410-416

EDTP, 2-5, 17, 21, 94, 244-246, 249, 275, 294, 335, 346, 357, 361-362, 371, 404-405, 416, 484-489, 507-512

EnableMAC command, 147-148, 416-418

ESS, 102-103, 416-419

Event Status Register, 90, 168, 204, 416-420

Exponent, 62-70, 404-408, 422

F

FID, 79, 106-107, 116, 121-125, 127,

140-141, 145-146, 189-193, 204-206, 217-218, 255-256, 275-280, 422-426, 433-439, 444-446, 473-475, 484-488, 507-513

fid_read, 121-125, 127, 422-424

G

get_frame function, 168, 219, 228, 251, 257, 275, 427-429

get_free_bap function, 114-116, 427-429

H

HI-TECH PICC-a8 C compile, 427-432

I

I/O mode, 42, 54, 74-78, 433-434

IBSS, 80-81, 102-103, 108-109, 118, 164, 172, 336-338, 393-396, 433-434, 478-481, 507

ICMP, 185-186, 190-191, 193-194, 209-214, 217-226, 228-241, 244, 252, 256-257, 264, 275-276, 281, 284-286, 334, 373-374, 377-381, 416-419, 422-425, 433-435, 439-446, 456-458, 462-466, 473-476, 478-487, 490-494

ICMP_cksum, 186, 214, 221-223, 225-226, 377-381, 433-435

ImageCraft's ICCAVR, 6, 347-348, 433-436

Internet Test Panel, 244-249, 262-264, 268-271, 273, 404-409, 416, 439-440, 484-495, 507-509

IP address, 94-95, 133-134, 137, 146, 168, 174, 178-186, 193, 195, 201-202, 209, 213, 219-221, 224, 226, 232, 239, 243-244, 246-248, 252, 257, 260, 265-269, 275-276, 281-282, 284-288, 313, 355-356, 410-411, 439-441, 490-495, 507-508

L

Linksys WCF12, 33-35, 65-71, 410-415, 444-447

Linux, 1-3, 25, 61-64, 71, 79, 105-107, 116,

404-408, 444-447, 456-461
little endian, 191-193, 444-448

M
MAC, 4, 27-29, 31-35, 51, 56-57, 62-63, 66-
70, 73-74, 79-82, 95, 99-100, 102, 108-
109, 118, 130-133, 137, 142-143, 145-
148, 152, 164-168, 172-174, 176-183,
185, 192-195, 199, 202-203, 208-209,
212-213, 219-221, 223-226, 228, 252,
260, 265-267, 275-276, 282-288, 312-
313, 335-336, 366-367, 369, 377-384,
387-401, 404-406, 410-411, 416-420,
422-424, 427-430, 433-435, 439-453,
456-458, 473-475, 484-495, 501-515
main function, 48, 126, 129-130, 134, 137,
164, 167, 335, 444-450
Mantissa, 62-70, 404-408, 450-451
MPLAB ICD2, 4-5, 10-14, 450-455

N
Netasyst Sniffer, 94-100, 102, 374-375, 381-
385, 142, 146-147, 151, 155, 157-158,
168-169, 174-175, 183, 207, 209, 233,
240, 265, 270, 294, 296, 298, 303-304,
308, 312, 350-351, 422-423, 456-457,
478-480, 484-487, 495-504, 507-509
Netgear MA701, 27-30, 68, 410-415, 456-
457

P
PCMCIA, 1-3, 25, 27-29, 37-38, 42, 47, 61,
93, 95, 433-435, 444, 462-465, 490-494,
501-503
PIC18LF8621, 4-6, 9-14, 19, 21, 50, 59,
174, 179, 191-192, 248-249, 265-266,
271, 273, 294, 309, 325, 336, 450-453,
462-467
PING command, 168, 174, 467-468
PRISM, 2-4, 25, 27-37, 45, 52, 70-71, 76,
79-81, 87, 105-106, 116, 125, 128, 130,
133-134, 141, 144, 177, 191-192, 200,
204, 387-389, 393-395, 404-408, 416-

420, 422-424, 427-430, 439, 444-447,
456-464, 467-472, 478-480, 495-497,
501-506
Probe Requests, 137, 148, 439-442, 467-472

R
ransmitReclaim_Cmd, 473-475
rd_cf_io16, 84-85, 87-88, 90, 109-113,
115-116, 118-122, 124, 138-140, 145,
167-168, 187-190, 192-193, 195-199,
204-205, 208, 215-218, 253-256, 275-
280, 335-336, 351-354, 473-475
rd_cf_io16 function, 87, 115, 124, 352, 473-
475
RID, 79, 106-109, 113-118, 122-134, 138,
146-147, 163-165, 167-168, 335-343,
377, 381-382, 393-397, 404-411, 416-
418, 422-424, 433-448, 456-461, 467-
470, 473-475, 478-481, 484-486, 490-
491, 501-503, 507-513
rid_read, 118, 122-125, 130, 132, 134, 138,
147, 163-165, 167-168, 335-336, 422-
424, 478-481
rid_read function, 118, 123, 478-481
rid_write function, 114, 117, 128, 478-481
RS-232, 4-6, 10-13, 20-22, 48, 50, 55, 71,
91, 243, 346, 362, 377-379, 444-447,
478-482, 484-488, 507-508

S
SanDisk, 37-38, 76, 444-445, 478-483
Send_command function, 86, 123, 144, 484-
485
send_tcp_packet function, 326, 334, 484-485
setipaddrs function, 223, 266, 312, 328, 484-
486
SNAP, 169-170, 173-174, 177, 184, 202,
206-209, 211, 219-220, 229-230, 250-
251, 257, 265-266, 269, 272-273, 294-
295, 297, 317-318, 320-321, 328, 330,
484-488
SSAP, 169-170, 173, 177, 206-207, 209, 211,
229-230, 250, 257, 269, 272, 294-295,

297, 317-318, 320-321, 330, 490-491

SSID, 81-82, 94-95, 103, 108, 125-129, 134, 137-138, 147, 152, 156, 164, 167, 335-338, 341, 381-384, 393-396, 410-411, 444-448, 450-451, 456-457, 462-464, 490-491

Status Register, 84-85, 90, 109-111, 118-120, 138-140, 168, 195-197, 204, 351-352, 416-420, 490-492

swapbytes function, 309, 490-494

T

tcp function, 275, 290, 308-309, 311-312, 322, 332-333, 495-496

TEW-222CF, 4-5, 10-15, 21, 23, 25-32, 46-48, 50, 59, 61-66, 77, 393-396, 490-491, 495-497, 501

TO_NIC, 51, 56, 75-76, 84-85, 88-89, 109-111, 118-120, 187-188, 195-197, 214-216, 253-254, 275-279, 349-350, 352-353, 495-500

Tuples, 45-47, 52, 54, 57, 59, 61, 65, 68-71, 77, 350, 393-399, 410-415, 478-483, 501-502

U

UDP, 185-187, 190-191, 193-194, 212-214, 217-218, 220, 224, 232, 237, 241, 243-252, 257-259, 261-276, 281-286, 314,

332, 334, 373-374, 377-381, 387-390, 393-395, 399-407, 410-411, 416-420, 422-425, 433-437, 439-445, 450-454, 456-461, 473-474, 478-479, 484-488, 495-512

udp function, 258-259, 261, 263, 266-267, 501-505

USART, 48-50, 54-55, 125, 137, 377-382, 387-388, 410-416, 433-439, 450-454, 456-460, 462-467, 495-498, 501-508

W

WEP, 80, 99, 102-103, 108, 335-343, 377-379, 393-396, 404-409, 422-427, 433-437, 450-451, 456-457, 507-511

wr_cf_data function, 115, 117, 467-470, 507-513

wr_cf_io16 function, 88-89, 507-513

wr_cf_io16 I/O, 192, 507-513

Z

Zonet ZCF1100, 31-35, 69-71, 507-515

ELSEVIER SCIENCE CD-ROM LICENSE AGREEMENT

PLEASE READ THE FOLLOWING AGREEMENT CAREFULLY BEFORE USING THIS CD-ROM PRODUCT. THIS CD-ROM PRODUCT IS LICENSED UNDER THE TERMS CONTAINED IN THIS CD-ROM LICENSE AGREEMENT ("Agreement"). BY USING THIS CD-ROM PRODUCT, YOU, AN INDIVIDUAL OR ENTITY INCLUDING EMPLOYEES, AGENTS AND REPRESENTATIVES ("You" or "Your"), ACKNOWLEDGE THAT YOU HAVE READ THIS AGREEMENT, THAT YOU UNDERSTAND IT, AND THAT YOU AGREE TO BE BOUND BY THE TERMS AND CONDITIONS OF THIS AGREEMENT. ELSEVIER SCIENCE INC. ("Elsevier Science") EXPRESSLY DOES NOT AGREE TO LICENSE THIS CD-ROM PRODUCT TO YOU UNLESS YOU ASSENT TO THIS AGREEMENT. IF YOU DO NOT AGREE WITH ANY OF THE FOLLOWING TERMS, YOU MAY, WITHIN THIRTY (30) DAYS AFTER YOUR RECEIPT OF THIS CD-ROM PRODUCT RETURN THE UNUSED CD-ROM PRODUCT AND ALL ACCOMPANYING DOCUMENTATION TO ELSEVIER SCIENCE FOR A FULL REFUND.

DEFINITIONS

As used in this Agreement, these terms shall have the following meanings:

"Proprietary Material" means the valuable and proprietary information content of this CD-ROM Product including all indexes and graphic materials and software used to access, index, search and retrieve the information content from this CD-ROM Product developed or licensed by Elsevier Science and/or its affiliates, suppliers and licensors.

"CD-ROM Product" means the copy of the Proprietary Material and any other material delivered on CD-ROM and any other human-readable or machine-readable materials enclosed with this Agreement, including without limitation documentation relating to the same.

OWNERSHIP

This CD-ROM Product has been supplied by and is proprietary to Elsevier Science and/or its affiliates, suppliers and licensors. The copyright in the CD-ROM Product belongs to Elsevier Science and/or its affiliates, suppliers and licensors and is protected by the national and state copyright, trademark, trade secret and other intellectual property laws of the United States and international treaty provisions, including without limitation the Universal Copyright Convention and the Berne Copyright Convention. You have no ownership rights in this CD-ROM Product. Except as expressly set forth herein, no part of this CD-ROM Product, including without limitation the Proprietary Material, may be modified, copied or distributed in hardcopy or machine-readable form without prior written consent from Elsevier Science. All rights not expressly granted to You herein are expressly reserved. Any other use of this CD-ROM Product by any person or entity is strictly prohibited and a violation of this Agreement.

SCOPE OF RIGHTS LICENSED (PERMITTED USES)

Elsevier Science is granting to You a limited, non-exclusive, non-transferable license to use this CD-ROM Product in accordance with the terms of this Agreement. You may use or provide access to this CD-ROM Product on a single computer or terminal physically located at Your premises and in a secure network or move this CD-ROM Product to and use it on another single computer or terminal at the same location for personal use only, but under no circumstances may You use or provide access to any part or parts of this CD-ROM Product on more than one computer or terminal simultaneously.

You shall not (a) copy, download, or otherwise reproduce the CD-ROM Product in any medium, including, without limitation, online transmissions, local area networks, wide area networks, intranets, extranets and the Internet, or in any way, in whole or in part, except that You may print or download limited portions of the Proprietary Material that are the results of discrete searches; (b) alter, modify, or adapt the CD-ROM Product, including but not limited to decompiling, disassembling, reverse engineering, or creating derivative works, without the prior written approval of Elsevier Science; (c) sell, license or otherwise distribute to third parties the CD-ROM Product or any part or parts thereof; or (d) alter, remove, obscure or obstruct the display of any copyright, trademark or other proprietary notice on or in the CD-ROM Product or on any printout or download of portions of the Proprietary Materials.

RESTRICTIONS ON TRANSFER

This License is personal to You, and neither Your rights hereunder nor the tangible embodiments of this CD-ROM Product, including without limitation the Proprietary Material, may be sold, assigned, transferred or sub-licensed to any other person, including without limitation by operation of law, without the prior written consent of Elsevier Science. Any purported sale, assignment, transfer or sublicense without the prior written consent of Elsevier Science will be void and will automatically terminate the License granted hereunder.

TERM

This Agreement will remain in effect until terminated pursuant to the terms of this Agreement. You may terminate this Agreement at any time by removing from Your system and destroying the CD-ROM Product. Unauthorized copying of the CD-ROM Product, including without limitation, the Proprietary Material and documentation, or otherwise failing to comply with the terms and conditions of this Agreement shall result in automatic termination of this license and will make available to Elsevier Science legal remedies. Upon termination of this Agreement, the license granted herein will terminate and You must immediately destroy the CD-ROM Product and accompanying documentation. All provisions relating to proprietary rights shall survive termination of this Agreement.

LIMITED WARRANTY AND LIMITATION OF LIABILITY

NEITHER ELSEVIER SCIENCE NOR ITS LICENSORS REPRESENT OR WARRANT THAT THE INFORMATION CONTAINED IN THE PROPRIETARY MATERIALS IS COMPLETE OR FREE FROM ERROR, AND NEITHER AS-SUMES, AND BOTH EXPRESSLY DISCLAIM, ANY LIABILITY TO ANY PERSON FOR ANY LOSS OR DAMAGE CAUSED BY ERRORS OR OMISSIONS IN THE PROPRIETARY MATERIAL, WHETHER SUCH ERRORS OR OMIS-SIONS RESULT FROM NEGLIGENCE, ACCIDENT, OR ANY OTHER CAUSE. IN ADDITION, NEITHER ELSEVIER SCIENCE NOR ITS LICENSORS MAKE ANY REPRESENTATIONS OR WARRANTIES, EITHER EXPRESS OR IMPLIED, REGARDING THE PERFORMANCE OF YOUR NETWORK OR COMPUTER SYSTEM WHEN USED IN CONJUNCTION WITH THE CD-ROM PRODUCT.

If this CD-ROM Product is defective, Elsevier Science will replace it at no charge if the defective CD-ROM Product is returned to Elsevier Science within sixty (60) days (or the greatest period allowable by applicable law) from the date of shipment.

Elsevier Science warrants that the software embodied in this CD-ROM Product will perform in substantial compliance with the documentation supplied in this CD-ROM Product. If You report significant defect in performance in writing to Elsevier Science, and Elsevier Science is not able to correct same within sixty (60) days after its receipt of Your notification, You may return this CD-ROM Product, including all copies and documentation, to Elsevier Science and Elsevier Science will refund Your money.

YOU UNDERSTAND THAT, EXCEPT FOR THE 60-DAY LIMITED WARRANTY RECITED ABOVE, ELSEVIER SCIENCE, ITS AFFILIATES, LICENSORS, SUPPLIERS AND AGENTS, MAKE NO WARRANTIES, EXPRESSED OR IMPLIED, WITH RESPECT TO THE CD-ROM PRODUCT, INCLUDING, WITHOUT LIMITATION THE PROPRIETARY MATERIAL, AN SPECIFICALLY DISCLAIM ANY WARRANTY OF MERCHANTABILITY OR FITNESS FOR A PARTICULAR PURPOSE.

If the information provided on this CD-ROM contains medical or health sciences information, it is intended for professional use within the medical field. Information about medical treatment or drug dosages is intended strictly for professional use, and because of rapid advances in the medical sciences, independent verification f diagnosis and drug dosages should be made.

IN NO EVENT WILL ELSEVIER SCIENCE, ITS AFFILIATES, LICENSORS, SUPPLIERS OR AGENTS, BE LIABLE TO YOU FOR ANY DAMAGES, INCLUDING, WITHOUT LIMITATION, ANY LOST PROFITS, LOST SAVINGS OR OTHER INCIDENTAL OR CONSEQUENTIAL DAMAGES, ARISING OUT OF YOUR USE OR INABILITY TO USE THE CD-ROM PRODUCT REGARDLESS OF WHETHER SUCH DAMAGES ARE FORESEEABLE OR WHETHER SUCH DAMAGES ARE DEEMED TO RESULT FROM THE FAILURE OR INADEQUACY OF ANY EXCLUSIVE OR OTHER REMEDY.

U.S. GOVERNMENT RESTRICTED RIGHTS

The CD-ROM Product and documentation are provided with restricted rights. Use, duplication or disclosure by the U.S. Government is subject to restrictions as set forth in subparagraphs (a) through (d) of the Commercial Computer Restricted Rights clause at FAR 52.22719 or in subparagraph (c)(1)(ii) of the Rights in Technical Data and Computer Software clause at DFARS 252.2277013, or at 252.2117015, as applicable. Contractor/Manufacturer is Elsevier Science Inc., 655 Avenue of the Americas, New York, NY 10010-5107 USA.

GOVERNING LAW

This Agreement shall be governed by the laws of the State of New York, USA. In any dispute arising out of this Agreement, you and Elsevier Science each consent to the exclusive personal jurisdiction and venue in the state and federal courts within New York County, New York, USA.